Balancing on a Planet

Balancing on a Planet

The Future of Food and Agriculture

DAVID A. CLEVELAND

UNIVERSITY OF CALIFORNIA PRESS

Berkeley   Los Angeles   London

University of California Press, one of the most distinguished university presses in the United States, enriches lives around the world by advancing scholarship in the humanities, social sciences, and natural sciences. Its activities are supported by the UC Press Foundation and by philanthropic contributions from individuals and institutions. For more information, visit www.ucpress.edu.

University of California Press
Berkeley and Los Angeles, California

University of California Press, Ltd.
London, England

Library of Congress Cataloging-in-Publication Data

Cleveland, David Arthur.
    Balancing on a planet : the future of food and agriculture / David A. Cleveland.
        pages cm — (California studies in food and culture ; 46)
    Includes bibliographical references and index.
    ISBN 978-0-520-27741-0 (cloth : alk. paper)
    ISBN 978-0-520-27742-7 (pbk. : alk. paper)
    1. Food industry and trade.   2. Agricultural industries—Environmental aspects.   3. Sustainable agriculture—Economic aspects.   4. Food supply.   I. Title.   II. Series: California studies in food and culture ; 46.
    HD9000.5C585 2014
    338.1—dc23   2013035775

Manufactured in the United States of America

23 22 21 20 19 18 17 16 15 14
10 9 8 7 6 5 4 3 2 1

The paper used in this publication meets the minimum requirements of ANSI/NISO Z39.48–1992 (R 2002) (Permanence of Paper).

I dedicate this book to my colleague and partner, Daniela Soleri, whose intellectual insights and inspiration have been essential to this book; to our daughter, Xina C.C. Soleri, who has helped in many ways—with fieldwork, with ideas, and always with a fresh outlook on everything; and to the small-scale farmers of the world whose knowledge and aspirations are critical resources for creating more sustainable agrifood systems.

CONTENTS

ILLUSTRATIONS

TABLES

A Personal History

My grandparents planted the hill behind their farmhouse in upstate New York, all the way to where the woods of maple, oak, and beech began, with dozens of apple varieties. Every fall the apples harvested from that orchard filled many barrels, and they were transformed into cider and apple pies all through the bitter winter. That was the story I was told. But my grandfather died in the woods above the orchard when a tree he and the hired men were cutting fell on him. My mother was a just a young girl, and my grandmother and her children struggled to keep up the farm, but as the children grew they were not much interested in farming, and most moved away to nearby towns. In my earliest boyhood memories of the orchard it was overgrown, thick with a jumble of native vegetation—blackberry brambles, elderberry trees, thistles, and grasses. By then the farm was becoming more of a memory than a real operation; there was not much left of its four barns, smokehouse, icehouse, dairy cattle, sheep, chickens, and fields of grain, hay, and tobacco.

In the summers I hacked paths through the dense vegetation under the wild, old unpruned apple trees, their branches broken by storms, some rotting and dying, others already dead, their dark limbs stark against the lush green of the hillside and the blue sky. I climbed the living apple trees too. My favorite was a Pound Sweet tree that produced huge golden-green apples full of sweet juice, a mid-nineteenth-century New England variety that I have never eaten or seen since.

I continued to visit the farm often after my family moved to the suburbs. Almost every drive back to visit brought new scenes of abandonment and decay of the agricultural landscape—barns and fields melting away to be replaced by housing developments. The stories my grandmother and great-aunts and great-uncles told of their own youth, growing up on farms, seemed more and more distant. I was witnessing the local reality of a massive national change in the structure of the U.S. agrifood system, a change that was happening in industrial countries around the world. From the first U.S. census of 1840 until 1935, the number of farms had been growing steadily, but with the rise of modern industrial agriculture, with its increased inputs and greater yields, small family farms were withering up and blowing away, and people were moving to cities. During the forty-three years between 1935 and 1978, U.S. agriculture changed dramatically—the number of farms declined 64 percent from its all-time peak of 6,812,350 to 2,478,642, while the average farm size increased by 168 percent, from 63 to 168 ha (Hoppe and Banker 2010:4, USDA NASS 2012b).

My family was a part of that change; my grandparents were the last really rural generation. But food and how it is grown remained strong interests for me, and those interests, together with my time teaching in a secondary school in Zambia, central Africa, led me to graduate work at the University of Arizona, and to dissertation fieldwork in northeast Ghana. That fieldwork was part of a project investigating causes and responses to the Sahelian drought and the food shortages and famines that followed in its path (Cleveland 1980), funded by the U.S. Agency for International Development (USAID) and conducted by the Ghanaian Council for Scientific and Industrial Research, based in Accra.

My fieldwork investigated the relationship between population dynamics and agriculture in the Kusasi village of Zorse, a farming community of about twenty-two hundred people spread over an area of about 18 km² (fig. 0.1). My year and a half of research included many interviews, observations, and measurements in 145 individual households (a 50 percent random sample) dispersed across the savanna, and it involved a lot of time moving between them. To expedite fieldwork I consolidated and repaired a few Polish bicycles from a dozen or so left in unusable condition by a previous project in the area, and I hired several research assistants, most from Zorse, who were bilingual in English and Kusaal, the language of the Kusasi.

On breaks from data collection I got to participate in the life of the village, helping in the fields of sorghum, millet, Bambara groundnut, sesame, okra, and other crops. Planting began with the rainy season, when the village turned from dusty brown to a sea of bright-green fields. As soon as the shoots emerged, the race with the weeds was on. The weeds were mostly grasses that to me looked exactly like the millet and sorghum seedlings, because these crops are also grasses. I felt clumsy and ignorant when weeding in the fields alongside Kusasi farmers, and my muscles ached afterward. I felt awe and admiration for the farmers' skill in growing almost all the food they ate in the rocky soil using only fire, a short–handled hoe, and, in the case of a very few households, an ox-drawn plow.

During my time in Zorse, the importance of social relations in that farming community became more and more evident. They governed the formation of work parties, the allocation of land, and the regulation of livestock. One incident in particular impressed on me the social cohesion of the village. It was the dry season; the harvest was over and the fields had completed another cycle, shifting back from green to brown. We were pushing our bicycles along the top of a treeless ridge, studded with sharp chunks of granite, difficult terrain for a bicycle and likely to give you a flat tire if you dared to try riding. We trudged along under a hot, cloudless sky, colored beige with dust from the vastness of the Sahara, blown by the powerful Harmattan winds out of the northeast. From our vantage point I looked out over the landscape, dotted with leafless dawadawa, baobab, and other native trees, sentinels in the midst of the now-empty fields, trees left by farmers for their useful products. Dawadawa seedpods have a sweet, bright yellow pulp with a delicious flavor, and the seeds are proc-

FIGURE 0.1. Zorse, Bawku, Ghana, in the dry season. The village is spread over the savanna landscape and consists of house compounds of round mud and stick structures with conical grass roofs and a few newer buildings with metal roofs. Compounds are surrounded by their most intensely managed fields, which are bare in the dry season. Also visible are bundles of 3–4 m sorghum stalks, economically important trees scattered in the fields, and walled dry season gardens following the course of a stream bed. Photo © David A. Cleveland.

essed into a dense, protein-rich substance that reminds me very much of miso made from soybeans and is used in soups and other dishes. Baobabs provide fresh green leaves for soup in April, when they leaf out before the rains begin, and when food supplies are running very low, with the first harvest of early millet still three months away. When the baobab fruits mature, kids puncture holes in the pods, each 15–20 cm long, add water, shake, and suck out the sweet-sour drink made by the dry, white matrix dissolving in the water.[1]

Resting with our bikes on that ridge, we looked out along the narrow, shallow brown valleys punctuated with splotches of green gardens filled with evergreen mangoes and citrus trees, sugarcane and vegetables, irrigated with water from shallow hand-dug wells. Next to the house compounds were *soks,* or shelters from the dry-season dust and wind, made by placing 4-meter-tall bundles of sorghum and millet stalks around the outside of a pole frame and across its top. The rainy season would bring a shift back to green, with the millet and sorghum growing everywhere, the native trees leafing out, and the compounds disappearing as the stalks of sorghum and millet towered over them.

Thinking about flat tires, about when the rains would start, and about needing to get as quickly as possible to the next household to do another set of interviews, I suddenly became aware of several scattered lines of people in the flat area below, about a half kilometer away. They were converging on the small cluster of mud-and-thatch compounds that included the one where I lived with the family of my chief field assistant, John Nbod. I stopped and asked John what was going on. He of course had already noticed the lines of people and had known immediately that they must be mourners coming to celebrate the death of an old woman who had been our neighbor, but whom I had seen only on one occasion because she was blind from onchocerciasis, a parasitic disease, and had leprosy, so she seldom left the confines of her room.

Later that night the celebration began with drinking, eating, dancing, and a trio of musicians playing traditional instruments. There was lots of *dam*, the earthy tasting, mildly alcoholic beer made from sprouted red sorghum grain, served in calabashes, rusty red with flecks of white, living yeast floating on it. Someone came up with an impromptu song about me—"There's a white man in our village"—and goaded me into joining the dance on the graves of those who had gone before the old woman with leprosy, lying in shallow holes beneath the giant baobab trees, graves shaped like wide-bottomed vases with a narrow neck open to the sky for the body to enter and sealed with an inverted clay pot.

Everyone sang and danced until the early hours of the morning. For the first time after nine months in the village, I felt more than superficially connected to the people, the soils, the crops, the food, the place, but most of all the community that was all of these together, this unique place in space and time. A small village in the savanna where generations of farmers had gained intimate knowledge of soils, plants, and weather patterns and had learned how to feed themselves. A village where increased population growth, spurred by the social changes brought by incorporation into the global economy, vaccinations, and roads seemed to be bringing a much more tenuous and uncertain future.

During two seasons of drought in this small village in the great east-west sweep of the savanna of West Africa I kept up my rigorous schedule of collecting quantitative data on farming, nutritional status, human fertility, mortality and migration, and labor allocation. Guilt and confusion flooded over me when I was greeted on the path with smiles and the stock phrase *Com sabit* ("Hunger is biting me") or when I doled out pieces of *kulikuli* (fried peanut paste) to the little children in the houses I visited to conduct interviews. I sought refuge in transforming the information from the interviews into the precise numbers in my data forms. Would the data I was gathering somehow help to increase understanding of the crisis in the human-nature relationship, help to make things better for these people whose survival here in the West African savanna sometimes seemed so precarious?

What my research was showing was that before European colonialists invaded and dominated Kusasi and other Ghanaian groups militarily, those groups had a system that was locally self-sufficient and provided some balance between population growth and food supply. The British colonialists' goal was to replace local self-sufficiency with integration into the global economy so that they could capture the benefits of local resources and labor. Kusasi were forced to migrate and work in the cacao farms, harbors, and mines in the southern part of Ghana (then under the colonial name of the Gold Coast). (I discuss the results of my research in Zorse further in chapter 1.)

As I lay awake at night in my mud-and-wattle hut, I felt there must be something drastically wrong with an agrifood system that rewarded hard work and insightful management not only with good harvests and joyful celebrations, in some years, but also with periodic famines, high infant death rates, and extensive degradation of soil, vegetation, and water resources. I tried to make sense of what I was learning from the people of Zorse. It didn't fit with the information I got from my infrequent trips to visit other researchers at the University of Science and Technology in Kumasi and from government and international development professionals in the capital, Accra (Cleveland 1990). I was frustrated by my ignorance about how to reconcile the conflicting views of the situation held by Kusasi farmers and by these experts.

From the perspective of most of these experts, the problem was that Kusasi and other small-scale African farmers were "backward" and needed to modernize, to catch up with the rest of the world by using more inputs such as manufactured fertilizer and seeds of modern crop varieties. And if they could not do that, they needed to leave their traditional farming and move to work as laborers on larger farms or in nonfarming jobs elsewhere. It seemed in some ways like a newer version of the colonial agenda. Indeed, the agricultural initiative in Africa today that receives the most funding is the Alliance for a Green Revolution in Africa (AGRA) and its associated programs, funded by industrial world governments, foundations, and agribusiness corporations, whose agenda seems in many ways like a high-speed version of those of the colonial era and of my time in Ghana (AGRA 2012).

The perspective of most Kusasi farmers was quite different; they were proud of their skills and accomplishments, and they were carrying on the farming, religious, and social life of their ancestors. Still, the hard work of farming and the uncertainty of their harvests in some years made them ready for changes that would improve their lives yet not require that they relinquish control of them. But what possibilities were there in a world changed irrevocably in the last several generations? As the situation became more and more difficult, many people were leaving Zorse. These included young people seeking education, some of whom ended up working as veterinarians or researchers or administrators, not only in Ghana but in countries around the world.

After returning to the United States from Ghana and finishing my dissertation, I had the opportunity to collaborate with ethnoecologist Daniela Soleri on work with Hopi and Zuni farmers in the southwestern United States and with Zapotec farmers in Oaxaca, Mexico. Their situation was much like that of the Kusasi. Native Americans had been farming throughout most of the North and South American continents for generations. In the United States, almost all native farming and food had been exterminated by Europeans, along with many of the tribes themselves. A very small number of tribes, such as the Hopi and Zuni, were able to survive, along with many of their crop varieties, foods, and farming traditions (Cleveland, Bowannie, et al. 1995, Soleri and Cleveland 1993), only because most Europeans considered their arid high desert land unfit for farming compared with the fertile valleys they stole from the less fortunate tribes. In Mexico, traditional small-scale farmers faced more and more difficult economic conditions under increasingly unfriendly neoliberal economic policies, such as the North American Free Trade Agreement (NAFTA) (Fox and Haight 2010).

It is a long drive north from Tucson, where we were based, before the turnoff onto Arizona State Route 264 running east-west through the high mesa country of Hopi land. It is easy to look past the sandy, scrub-covered terrain that lies beyond the side of the road. Every so often a pickup truck has pulled off the road, no one around, nothing that would seem to be a reason for stopping. But if you do stop, and take the time to walk away from the road until you can get a view down the slope beyond, chances are you will see a Hopi farmer, or farm family, checking on their sandy field—a field filled with scattered, dense bundles of long, bright-green leaves. This is a Hopi maize field.[2] Hopis, and their neighbors, the Zuni and Acoma, have been growing maize in fields like this for at least two thousand years.

When Daniela and I worked at Hopi, farmers invited us to join them in their fields—it was a lot easier for all of us than asking them to answer our questions while sitting in their houses. We saw firsthand that Hopi maize fields are not scattered at random over the landscape. Hopi, Zuni, and other farmers in the area learned through many generations to plant where the scrub-covered slopes indicate water drainage that brings the snowmelt and later the runoff from summer rains to their maize plants, along with nutrients in organic matter and minerals (Norton et al. 2007). Water is held in the fields by a deep layer of heavy, clayey soil, and the thick sandy layer above acts as a mulch, slowing down loss of that water by evaporation. Seeds are planted with a digging stick, traditionally made of greasewood but now often made of other materials, such as recycled crankshafts (fig. 0.2). Farmers use the sticks to scoop out narrow holes 30 cm and more below the surface where the clayey layers begin, layers that have trapped the runoff of the melting snow. Farmers place the seeds there, and then replace the soil, in the same order as it was removed—the moister, cooler, clayey soil next to the seeds, the warm, dry, sandier soil on the surface.

For generations Hopi and Zuni farmers succeeded in feeding their families and communities in some of the most arid agricultural conditions in the world. Among other things, they developed farming methods, selected crop populations, organized their labor and other resources, and devised storage techniques that made this success possible. Today these farmers struggle to find ways of keeping their agriculture, food, language, and culture alive, even while they are surrounded by and often participate successfully in the larger society around them, including the huge cattle ranches and the cotton, alfalfa, and vegetable fields of the mainstream agrifood system.

This highly industrial, mainstream agriculture contrasts with traditional Hopi and Zuni agriculture in the inputs used, crops grown, and methods deployed, although both are labor intensive (Khan, Martin, et al. 2004). In the orchards, vineyards, and fields of vegetables of the U.S. south and west it is common to see farm workers stooped over, weeding, thinning, or picking, seeming out of place in the large-scale landscape—factories in the field (McWilliams 2000 [1939]). They are there instead of in their own fields and pueblos in Mexico and Central America because of the money they can earn. For some it is an adventure, and for

FIGURE 0.2. Hopi farmer Jerry Honawa planting bean seed. The order of layers of soil removed from the planting hole has been preserved; soil layers will be returned to the hole in that order. Photo © Daniela Soleri. Used with permission of photographer and subject.

others it is a sacrifice they make reluctantly, leaving their families and communities for a foreign land.

The home of many of the people who work in the fields of Arizona and California is southern Mexico, including the state of Oaxaca. In the middle of a summer with good rains, the Valley of Oaxaca, at 1,500 meters above sea level, is a huge bowl of green maize plants. Maize was first domesticated about 350 km from here nearly nine thousand years ago, and some of the oldest known cobs of early maize were found mere kilometers from where we worked. Today maize is still revered throughout Mexico, and there is rapidly growing interest in Mexico City and other urban areas in traditional foods made from traditional varieties of maize (Soleri, Cleveland, and Aragón Cuevas 2008), of which there are hundreds in Mexico (Aragón-Cuevas et al. 2006, Wellhausen et al. 1952). The farming families we became friends with in Oaxaca take special pride in the foods they make with the maize they grow on their farms and sometimes grind by hand on a *metate*—including *tejate,* a refreshing drink made from maize, cacao, and other ingredients, a version of which has likely been drunk in Oaxaca for millennia (Soleri, Cleveland, and Aragón Cuevas 2008, Sotelo et al. 2012).

Yet after hundreds of generations, a way of life seems to be vanishing. Many Zapotec farmers have a deep sadness from their sense of having no control of the change, no good choices. They are caught between trying to grow enough food to stay alive in their homes or abandoning the villages of their ancestors—the soil, the maize, the mountains they love—and migrating to Mexico City or El Norte, the United States, some of them to work in the industrial agriculture fields. There are Oaxacan villages where most of the working-age women and men are migrants, gone for much of the year, or for several years at time, or even permanently, breaking the chain of knowledge about how to plant maize, save seeds, make *tejate*. Recently this trend is changing due to factors in both the United States and Mexico; it remains to be seen whether this will also result in slowing the shift away from farming by small-scale farmers in Mexico, which has been going on for many decades.

It is not only large-scale industrialized agribusiness that relies on a noncompetitive, mostly migrant labor supply. Small-scale farmers in California, who are part of a national movement in the United States to localize the food system, are also often dependent on migrant workers from villages in Oaxaca and other places in southern Mexico and Central America. Alternative ways of growing food, such as "organic," are even more labor intensive, for example, because they have to control weeds mechanically, including by hand, instead of using herbicides (Shreck et al. 2006). Shoppers at farmers' markets buying "fresh and local" are for the most part unaware of the long, arduous hours of work that produce the bounty on offer, or of the effect on the home communities of the migrants who do that work.

Many in the grassroots local food movement believe that localization is the best solution to the problems caused by the mainstream global agrifood system. In Santa Barbara County, where I live, local food is booming, as it is in much of the rest of the United States. Yet the system is still largely a centralized export-import business—Santa Barbara County produces more than ten times the fruits and vegetables it consumes, yet 99 percent of this production is exported, while over 95 percent of the produce consumed in the county is imported (Cleveland et al. 2011b).

A major challenge to localization is that most produce and other food is sold through national supermarket and hyper store chains. One of the most heated arguments about local food is whether the large corporations that control the current system can contribute to localization, or whether their involvement in the movement will co-opt and destroy it. What does it mean when global retailers such as Walmart—the largest food retailer globally and with a bad reputation among local food advocates, community groups, and labor unions—promotes sustainable agriculture, local food, or better nutrition (Huber 2011)? When Will Allen's organization Growing Power, a star in the U.S. local food movement, accepted a large donation from Walmart it generated much controversy and soul searching (e.g., Fisher 2011).

Debates about our agrifood system and what direction it should be moving in are often dominated by a contrast between supporters of the dominant mainstream, global industrial system and supporters of a variety of alternatives, which are often smaller in scale and focused more on environmental and social issues and less on economic issues. The current world food crisis that began in 2007–2008 has intensified these debates, highlighting the contrast between the 15 percent of the world's poor who are chronically hungry and the excessive consumption of the rich. So has the biggest environmental crisis of our time, perhaps of all human history—global climate warming. Agriculture is not only one of the largest causes of global warming and other environmental problems but will also suffer the negative consequences of this change.

My experiences and research—from the family farm in upstate New York to the villages in West Africa; from Hopi, Zuni, and Oaxaca to Santa Barbara—have led me to believe that learning to identify and analyze these perspectives, and the data and values they are based on, can empower us to participate in creating a better future. For example, to discuss localization we first have to agree on whose definition to use. Even if we agree on a definition, we will need to figure out how local an agrifood system is by measuring some sort of indicator, such as food miles. Food miles have become a popular localization indicator, yet often they may not accurately reflect the goals of localization—which include greater profits, improved nutrition, environmental stewardship, and food justice—as my students and I found in our research in Santa Barbara, discussed in more detail in chapter 9 (Cleveland et al. 2011b).

In these debates, sorting out the data used and the assumptions made from different perspectives is essential, but not easy. How can we do this? In this book I present one way, using key concepts to make sense of the range of data and opinions about our agrifood systems and the dramatically different ideas about how they should change.

ACKNOWLEDGMENTS

I am especially indebted to Daniela Soleri, who not only commented on the entire manuscript, parts of it many times, but who has also provided intellectual stimulation and emotional support as a collaborator on research since the early 1980s. Particularly inspiring has been her remarkable ability to establish rapport with farmers and to formulate questions and scenarios based on scientific concepts that elucidate farmers' deep knowledge of and values in farming and food. I also happily acknowledge the assistance of Xina Soleri, our daughter, who accompanied us during many research trips and whose keen sense of observation helped open my eyes to things I would not have otherwise noticed.

During and after the time he served as chair of my doctoral committee at the University of Arizona, Bob Netting introduced me to the academic study of agriculture and population issues, and we had many lively discussions we had on this topic. Bob's professional career was focused on elucidating the workings of small-scale agriculture characterized by a deep respect for farmers' intelligence and creativity, and he encouraged my own dedication to this topic. The title for this book was inspired by his seminal work on the demography of a small farming community in Switzerland, *Balancing on an Alp*.

For comments on parts of the manuscript I thank Daniela Soleri, Quentin Gee, Salvatore Ceccarelli, Josh Schimel, Jennifer King, Mel Manalis, Kyle Meisterling, and Fred Estes.

Research reported in this book in which I have been involved has been carried out in collaboration with colleagues, farmers, and others. Among those many I especially want to thank are the following.

Farmers and others in the agrifood system: In Zorse, Ghana, the Zorse Naba (chief) and the subchiefs, Nbod Awindago and his family, especially the late John Anyagre Nbod, my chief assistant, and Asambo Abimbila; at Hopi, Jerry Honawa, Norman Honie, Sr., and Joyce and Morgan Saufkie; at Zuni, Fred Bowannie, Andy Laate, Lygatie Laate, and Donald Eracho; in Oaxaca, Lucilia Martínez Martínez, Marcial Gomez, Pulciano Gomez Martínez, Lorenza Castellanos, Delfina Castellanos, Maxima Castellanos, Mario Vasquez Olvera, Delfino Jesus Llandez Lopez, and Bernardino Castellanos Coseme; and in Santa Barbara County,

Bonnie Crouse, Terry Thomas, Shu Takikawa, Noey Turk, Anna Breaux, Wesley Sleight, Melissa Cohen, Shawn McMahon, Chris Thompson, B.D. Dautch, Tom Shepherd, and Ben Faber.

Colleagues who have helped my thinking on the topics of this book: Steve Smith, Tom Orum, Nancy Ferguson, Sadiq Awan, Flavio Aragón-Cuevas, Humberto Ríos Labrada, Salvatore Ceccarelli, Eva Weltzien, Paul Richards, Paul Gepts, Norman Ellstrand, Scott Lacy Carla D'Antonio, and Oliver Chadwick.

TAs for classes where material in this book has been presented: Robyn Clark, Quentin Gee, and Grayson Maas; undergraduate and graduate student researchers Corie Radka, Nora Müller, Niki Mazaroli, Noelle Phares, Robyn Weatherby, and Krista Nightingale.

Major funding for research mentioned has been from UCSB, the Wallace Genetic Foundation, and the National Science Foundation (SES-99779960, DEB-0409984)

At UC Press, Kate Marshall, Rachel Berchten, and Emily Park have been incredibly encouraging, supportive, and patient.

While indebted to all who have contributed in so many ways, I alone am responsible for any shortcomings. Any comments or suggestions for improvement would be welcome at cleveland@es.ucsb.edu.

Introduction

1. THE CURRENT WORLD FOOD CRISIS

The current world food crisis that began in 2007–2008 is in many ways similar to the hundreds or thousands of local and regional crises that have transpired since the beginning of agriculture. As we have seen, there are sharply contrasting perspectives on the causes and solutions of food crises past and present—and on how to prevent them in the future. The "mainstream" and "alternative" perspectives can sound superficially similar, yet they differ fundamentally in terms of their theories and assumptions, problem definitions and solutions.

The mainstream emphasizes the direction that brought us the most dramatic and significant successes in terms of increased food production—supporting the modern, large-scale, industrially based system that is most developed in the United States, Canada, Australia, and western Europe. The alternative emphasizes that the successes of the mainstream have also created many environmental and social problems, and that the conditions that allowed the mainstream approach to flourish no longer exist—the population is much larger, and new resources for production are more limited by scarcity and degradation. Therefore, many taking an alternative approach argue that we need to build on more traditional, less resource-intensive small-scale agriculture, while incorporating the best elements of modern science and technology, seeking solutions that nurture people and communities psychologically and socially as well as physically.[1]

The dichotomy between mainstream and alternative as I've just described it is a simple model, a sort of caricature, that nonetheless can be useful for understanding. It is important to remember that all our knowledge about the world is comprised of models and that all models are simplifications—the useful models are the ones that help us advance toward the goals we have chosen. I believe that this model can help us toward a deeper understanding of our agrifood system, which is the starting point for moving it in the direction of our goals. In chapters 3 and 4 I will elaborate on this model, and in part 2 I will explore its complexities and contradictions as I apply it to making sense of agrifood systems past, present, and future.

2. FEEDING THE FUTURE

Part of what makes finding solutions for agrifood system problems so controversial are the predictions of large increases in future demand for food, driven by increasing population—now past seven billion and expected to add at least two billion more before it stops growing—and increasing per capita consumption, driven by changing diets that include more processed and animal foods. For example, an analysis by the United Nations Food and Agriculture Organization (FAO) for the interval from 2005–2007 to 2050 projects an increase in per capita demand for meat and oil crops by 28 percent and 39 percent, respectively. Based on an estimated increase in population of 39 percent, this means an increase in production of 76 percent and 86 percent, respectively (table 0.1), and an overall increase in food production of approximately 60 percent (Alexandratos and Bruinsma 2012:21, 99). Even for cereals, which have a projected per capita increase in demand of only 5 percent, required production would have to increase by 30 percent because of population growth; this means that yields would have to increase by 30 percent, because arable land will likely increase very minimally (4 percent).

So how will this expected demand be met? The *mainstream* industrial agrifood system has been remarkably successful over the long run in increasing food production at a rate faster than population growth, with the promise of continuing to do so in the immediate future. Supporters of this system believe that a globally integrated agrifood system and technological breakthroughs, for example in genetic engineering of crop plants or precision agriculture, are key to providing enough food for the future (Evans 1998, Fedoroff et al. 2010). Advocates of *alternative* agrifood systems have a different perspective—they argue that the demand can be lowered via better diets (Eshel 2010) and reduction of waste, and that supply can be increased in more sustainable ways, with ecological agriculture based on traditional methods and more local control (IAASTD 2009). But the issue is far from settled, and it hinges on disagreements over values as well as facts. A major problem from an alternative perspective is that the mainstream agrifood system monopolizes the bulk of research and development resources, leaving little opportunity for developing the kinds of solutions needed to save the planet, nurture communities, and increase human happiness.

Yet, regardless of one's perspective, there is also shockingly bad news about every element of our agrifood systems—from the contamination of drinking water with agricultural chemicals to the deteriorating nutritional quality of the food supply and of child nutritional status, from the loss of crop genetic resources to loss of prime farmland. It seems that our

TABLE 0.1.   FAO Projections of Future Food Demand

	2005/2007	2050	Percent change, 2005/2007–2050
Population (million), UN 2008 revision	6,592	9,150	38.8
Cereals, food (kg/capita)	158	160	1.3
Cereals, all uses (kg/capita)	314	330	5.1
Meat, food (kg/capita)	38.7	49.4	27.6
Oilcrops (oil. equiv.), food (kg/cap)	12.1	16.2	33.9
Oilcrops (oil. equiv.), all uses (kg/cap)	21.9	30.5	39.3
Cereals, production (million tonnes)	2,068	3,009	45.5
Meat, production (million tonnes)	258	455	76.4
Oilcrops (oil. equiv.), Food (million tonnes)	80	148	85.8
Cereal yields (tonnes/ha; rice paddy)	3.3	4.3	29.5
Arable land area (million ha)	1,592	1,661	4.3

Source: Data from (Alexandratos and Bruinsma 2012:21), and calculations based on those data.

agrifood system has been going in a direction that is producing at least as many problems as solutions. While those in power have demanded more food and higher yields to maintain and expand their power for millennia, pushing farmers into practices that were environmentally and socially destructive (Diamond 2005), their effects were mostly localized. Today, however, we have a global system, highly degraded environments, and more than seven billion humans to feed, with one billion of those chronically hungry.

In order to move toward a more desirable future, we need to understand the successes and failures of our past and current agrifood systems and how they are linked in time and space. We also need to agree on how to define that future and on how we need to change our current system to get there. The goals of this book are meant to contribute to this process.

3. GOALS OF THIS BOOK

I have two main goals for this book. The *first* is to encourage critical thinking by explaining the concepts that I think are key to understanding the problems and potential solutions for the challenges facing our agrifood systems. This includes demonstrating how these concepts can be applied to specific situations so that readers can use them to analyze new situations and discuss their findings with others. I hope this results in better understanding of the challenges we face, where they come from, and the options for responding to them—empowering readers to participate in a critical and constructive way in the discussions and decisions that will determine the future of food and agriculture. My *second* main goal is to demonstrate how I have applied these concepts in my own thinking about agrifood systems; I share what I have concluded about the problems and solutions based on my own research and values. These two goals are synergistic in that if I achieve the first, it means that readers will be able to independently critique my application of the concepts and my conclusions.

In other words, this book is a guide to the concepts I have found useful in analyzing the agrifood system *and* to the conclusions that using these concepts has led me to. This means that *Balancing on a Planet* (hereafter, *BOP*) is different from many other books about agriculture and food in that it does *not* attempt a review of what we know about the history of food and agriculture or its current state around the world, nor is it simply a polemic in favor of a particular agenda for change. It provides a framework for analysis of empirical data and for explicit discussion of subjective goals, illustrated by case studies from around the world.

3.1. Critical Thinking

The key to achieving the first goal of understanding problems and potential solutions is critical thinking. This includes the ability to distinguish between how the world is and how we would like it to be. As I described in the preface, when I was living in Zorse I would often lie awake at night thinking about why people in the village were hungry—and I could name a number of proximal causes, such as drought, eroded soils, and loss of labor to migration, as well as intermediate causes, such as the undermining and brutalization of indigenous communities by European colonists, corrupt and ineffective foreign and national development workers, and changing climate patterns. But no matter what causal path I traced in my mind, all paths ended at lack of agreement among individuals and groups about how the world *does* work based on empirical data (empirically based assumptions) and lack of agreement about how the world *should* work based on values about what is good (value-based assumptions).[2] The disagreements are difficult to overcome in part because the analytical part of our brains tends to be lazy, so we often don't bother to make the effort to disentangle these two very different ways of thinking.[3]

The result is that we are usually unaware of how our value assumptions about how the world should work influence our empirical assumptions about how the world actually does work. For example, if we assume that the knowledge and culture of small-scale family farmers *should* be valued, and that they *should* have access to production resources, we may be more likely to assume that these farmers' loss of knowledge and resources *is* the cause of the food crisis (LVC 2010). Our empirical assumptions also influence our assumptions about the way the world should be. For example, if we assume that the food crisis *is* primarily due to a lack of food production and that yields on small-scale family farms *are* much lower than those on corporate, industrial farms, we may be more likely to assume that to solve the food crisis the former *should* be replaced by the latter (e.g., Collier 2008). While this kind of interaction between empirical and value-based assumptions exists among farmers, consumers, scientists, and all of us, to a greater or lesser extent, it is more serious in people and organizations operating at higher structural and geographical levels—both the amount of information they have to process and the consequences of their decisions are much greater.

Therefore, an important way of achieving the first goal of *BOP* is to analyze the assumptions underlying different perspectives about how the world is and should be, including our own assumptions. Throughout *BOP* I try to present as openly as possible my own conclusions and assumptions while also standing back and viewing them critically—that is, not becoming too attached to them and remaining open to new data, to alternative interpretations of data, and to appreciating different values. For example, my values include the assumptions that equity of resource access and use for all people is good and that interacting with the biophysical world in ways that maintain high biological and cultural diversity and ecosystem functioning is good, and my analysis of the data leads me to empirical assumptions that anthropogenic climate change is a real and immense threat and that small-scale, resource-poor farmers' behaviors are often based on insightful and efficacious understandings of their environments and crops.

3.2. The Results of My Critical Thinking

So, what have I concluded about the problems with our agrifood system and the best way to solve these problems? Explaining my conclusions and the process by which I reached them is the second goal of *BOP*. It was my fascination with the many different ways that humans grow and eat food that first led me to farm communities around the world. I have worked with farmers, gardeners, and scientists on research and development projects in northeast Ghana; in the Swabi valley in Pakistan; in the Central Valleys of Oaxaca, Mexico; and in the United States, on the Zuni and Hopi reservations and in Santa Barbara County, California. In addition, I have spent shorter periods of time researching agrifood systems in other places, including Burkina Faso, Egypt, India, Syria, Mali, and China. I have interviewed people and collected observational data, in addition to studying the research of others. Since climbing in those old apple trees as a boy, I have also become avid about food gardening and cooking, experiences that give me a personal connection to the process and experience of growing and preparing food. Finally, I have thought a great deal about the successes and problems of different ways of growing and eating food.

One of my central conclusions is that small-scale, traditional, locally oriented, low-external-input agrifood systems are an important resource for the future. Much of the Earth's remaining cultural and biological diversity is in the care of small-scale farmers. Many of the farmers I have worked with use knowledge and methods passed on through generations to grow locally adapted crop varieties, evaluating and incorporating new ideas from other farming traditions, from extension agents, and from scientists. I have celebrated with them their successful harvests and eaten special foods made from those harvests, rich with history, meaning, and flavor.

These farmers are often proud of what they do and know, and while they seek improvements in their farming and their lives in general, most do not want to abandon those things they value about their way of life. For example, in Oaxaca, Mexico, when farmers were asked as part of our research on crop diversity if they wanted their children to be maize and bean farmers like themselves, 91 percent said "yes" (Soleri, Cleveland, Castro García et al. n.d.). However, these same farmers see the world changing rapidly from the traditions of the many generations that preceded them—only 47 percent thought their children would actually grow up to be maize and bean farmers.

I have also seen farmers struggle to feed themselves and to understand the forces seemingly beyond their control that make the survival of their agrifood system almost impossible—population growth; environmental degradation; climate change; market fluctuations; privatization of water, land, and other resources; inappropriate development projects; and corrupt and incompetent governments and development organizations at home and abroad. I have also seen many young women and men, including many of my students, choose to work as small-scale farmers, food processors, chefs, and distributors instead of in careers that are less risky and more remunerative. In the midst of the most productive industrial agrifood system in the world, and with college degrees in hand, most of these students who choose to work in the agrifood system are moving away from the vision of mainstream agronomists and economists, choosing to create and participate in alternative ways of doing things.

While I see much potential in small-scale agriculture for solving the world food crises, I am also aware that small-scale farming is often physically and mentally grueling, and that most farmers are not well rewarded for their work. According to one estimate, the more than two billion people living on almost five hundred million small-scale (less than 2-ha) farms in the Third World include half of the world's undernourished people and the majority living in absolute poverty (IFAD 2011:1).[4] In short, I am not a nostalgic romantic. There is no going back to the small-scale agriculture of the past—doing so would be neither possible nor desirable. It was often a very hard life, and the world is a different place now, with more than seven billion humans to support. But simply continuing to promote the mainstream agrifood system is not the answer either.

I believe that an important aspect of creating alternatives for the future will be to combine small-scale, traditional agriculture with select aspects of modern, scientific agriculture in ways that provide solutions to the current food crisis—long-term solutions to balancing our biological need for food with our environmental impact in ways that also fulfill our cultural, social, and psychological needs. This means searching for basic principles that underlie both modern and traditional agriculture, both modern and traditional demographic behavior, and both modern and traditional values and social organizations. This is not a quick fix, but it may be one of the best ways to solve the present food crisis and to avoid future ones. As I will discuss in more detail in chapters 3 and 4, there are usually trade-offs between what is possible and our goals for the future, and also between the different goals we have for the future. We need to minimize these trade-offs, to look for ways to make the system work better for everyone. We need to think critically, holistically, systemically, and compassionately. And we need to get to work right away.

4. THE LAYOUT OF THIS BOOK

To address my goals, I have organized *BOP* into two parts. In part 1, *Agrifood Systems History and Future*, I focus on the food demand-supply problem, the underlying factors that drive our past and current agrifood crises and successes, and how basic concepts such as sustainability can be useful tools for understanding those factors and moving toward a better future. In part 2, Moving toward Sustainable Agrifood Systems: *A Balancing Act,* I give more detailed examples of how these concepts can be applied and how I have applied them in my own thinking.

Chapter 1 is about the relationship between the demand and supply sides of the agrifood system: the demand for food created by an increasing number of people and increasing per capita food consumption, and the supply of food based on the ability of technology and the environment to produce it. The fundamental concept was laid out by Malthus: population tends to increase geometrically and food supply arithmetically. Like other organisms, humans are selected for their success in reproducing, leading to growth in numbers, but the ability of the environment to feed the growing numbers is limited. There are four basic ways humans can respond to avoid the collision between demand and supply. Humans have been very good at avoiding this collision by increasing the production of food, but ultimately it can be avoided only by conscious personal and social planning. An example of calculating the Earth's human carrying capacity (HCC), based on demand—total human energy requirements—and supply of water for growing rice to meet that demand, illustrates the critical roles of efficiencies and assumptions in determining supply and demand.

In chapter 2 I explore in more depth the supply side of the equation: What determines how much food we can produce? How has agriculture evolved in ways that increase HCC? I begin with the first agricultural revolution, the Neolithic, and the way in which it dramatically changed the relationship of humans with other species, with the environment, and with other humans. We will see how changes in these three fundamental relationships have continued through time with the spread of agriculture from its centers of origin, the scientific-industrial revolution, the Green Revolution, and the biotech revolution. Because the focus of these revolutions has been overwhelmingly on increasing short-term production, the social and environmental costs have often been ignored, yet they have undermined HCC over the longer term in many places. Ensuring a future for our species will require balancing short-term strategies with long-term or sustainable strategies, the subject of chapters 3 and 4.

Chapter 3 describes how the sustainability revolution is a response to the problems caused by the supply-side approach and a discussion of how sustainability can shift the emphasis to dealing with the demand side—how to reduce growth in population, consumption, and inefficient technologies. I show that sustainability is a subjective concept about what we want the future to be and therefore requires discussion of values to reach agreement. It also requires objective analysis of the current situation and the effectiveness of different solutions. Important concepts for this analysis are how knowledge is generated, and how we can understand the similarities and differences in knowledge among farmers, among scientists, and between farmers and scientists.

Chapter 4 concludes part 1 by describing how the three main emphases in agrifood system sustainability—economic, environmental, and social—can have very different goals, theories, and solutions. These emphases also often have very different assumptions about the key concepts of the agrifood system, including markets, natural resources, human nature, discount rates, internalization of externalities, and risk management. In general, the mainstream perspective has an economic emphasis, and alternative perspectives have environmental and social emphases. The simplified characterization of the three emphases provides a framework for more nuanced understanding and analysis, illustrated by further examination of the current world food crisis.

Part 2 moves from the general discussion of problems and solutions to specific aspects of the long-term food crisis, and it provides examples of how to apply the concepts introduced in part 1. Chapters 5, 6, and 7 take up the three fundamental changes of the Neolithic—increased management of other species, ecosystems, and people—in more detail, showing how supply-side solutions worked through subsequent revolutions and how they can be combined with demand-side solutions to create a more sustainable alternative agrifood system, one that contrasts with the mainstream vision. In chapters 8 and 9 I address two of the biggest challenges to creating a more sustainable agrifood system—global climate disruption and economic globalization—and discuss the potential for diet change, food waste reduction, and localization to meet these challenges.

Chapter 5 is about the management of other species, focusing on the basics of plant breeding in a broad perspective that includes environmental and social as well as biological variables. Building on the introduction to farmer and scientist knowledge in chapter 3, I show how fundamental biological variables are understood and used in different ways with different results by small-scale farmer and professional scientist plant breeders, in many ways reflecting the contrasting alternative and mainstream perspectives. I illustrate this for three topics important for crop improvement: yield and yield stability and narrow versus wide adaptation, collaboration between farmers and scientists, and genetically engineered crop varieties. Similarities and differences among farmers, among plant breeders, and between farmers and breeders can often be accounted for by similarities and differences in their experiences of biophysical reality—that is, the germplasm and the growing environments they have worked with; their experiences of social reality, including social and institutional settings; and the way they create new knowledge as influenced by preexisting knowledge, technology, and practice.

Chapter 6 describes the development of ecosystems management, and how this is different in traditional and industrial agrifood systems. The move to sustainability can be thought of as a search for those unique "places" where the stability and diversity of traditional systems can be combined with the high yields of industrial systems. I describe how polyculture can produce greater yields than monoculture, illustrated with a case study from Yunnan, China, where growing traditional and modern rice varieties together eliminated the need for fungicides and increased yields and farmer income.

The ways in which humans manage themselves in order to manage agrifood system resources is the focus of chapter 7. It describes how resources can be categorized as private, public, or common pool, and how common-pool resources can be managed by private individuals or corporations, governments, or communities—or not managed at all. Common property management has the potential to internalize negative externalities in ways that optimize the equal distribution of benefits, including to future generations, yet it has been largely ignored or dismissed by the mainstream agrifood system. I show how game theory can help us understand success and failure of different management types, and I provide examples of the potential of common property management for irrigation water and crop genetic resources.

In chapter 8 I discuss anthropogenic climate change, to which the agrifood system is one of the largest—perhaps *the* largest—single contributor and which in turn has profound effects on the agrifood system. I look at the relationship of climate change to the evolution of biogeochemical cycles in the history of the Earth and to recent changes in the agrifood system. I give examples of two key cycles, carbon and nitrogen, and show how our agrifood system has affected these cycles in ways that make big contributions to climate change and therefore offers opportunities for mitigating that change. Solutions that receive the bulk of attention typically require a lot of additional research, technology development, and resources and entail a lot of uncertainty and risk, with benefits slow to materialize. Some of these approaches will need to be part of the longer-term solution—for example, increasing soil carbon sequestration, increasing the efficiency of nitrogen fertilizer use, and reducing food packaging and transport. My focus, however, is on strategies available to us right now: reducing the high level of waste from field to fork and adding more healthy plant foods to our diets while reducing processed and animal foods. These behavioral changes receive relatively little attention, yet they require few resources and can have dramatic and rapid benefits. Their biggest challenges are cultural, social, and economic.

In chapter 9 I examine what has become the most popular alternative to the problems caused by the mainstream globalized and industrial agrifood system—grassroots localization. In the industrial world, the push for localization seeks to reshape the economic, social, and physical infrastructure of agrifood systems; in the Third World, it seeks to conserve and improve what remains of local agrifood systems. Localization is a critical case study of how different values and goals for the future can lead to very different interpretations and actions. The battle over localization is a microcosm of the battle over who gets to set the goals of our agrifood system and select the paths to reach them, and it is embedded in the larger economic, environmental, and cultural struggle for the future of the planet. This is why we must all keep asking the key questions, carefully examining our empirical and value assumptions, and use indicators for sustainability that most accurately reflect our goals for the agrifood system.

PART ONE

Agrifood Systems History and Future

In the four chapters of part 1 I focus on the underlying factors that drive our current and past food crises and successes, the food demand-supply problem, and how basic concepts like sustainability can be useful tools for understanding those factors and moving toward a better future. Chapter 1 looks at the deep history of our planet and species, and the relationship between them based on food. Chapter 2 focuses on the last twelve thousand years, beginning with the Neolithic revolution and the changes it created, which are still ongoing. Chapters 3 and 4 introduce the sustainability revolution in our agrifood systems and three different emphases in pursuing it, based on very different assumptions. In part 2 I will apply these concepts to examples of how we can create a more sustainable agrifood system.

ONE

Eating Stardust

Population, Food, and Agriculture on Planet Earth

1. INTRODUCTION

Imagine that you are zooming outward from the chair you are sitting on while reading this book.[1] You see the place where you were sitting recede into its continent, and then the curvature of our planet Earth appears, growing smaller and smaller, joined by the other planets of our star, the Sun. Our solar system, too, grows smaller and seems to hover in emptiness against the background of our galaxy, the Milky Way. And then, as you continue zooming through the Milky Way it too becomes a speck and disappears, lost in the vastness of the universe.

Our solar system is about 12×10^9 km in diameter, but the next closest star is just over four light years away (Murphy 2006)—a distance of over three thousand times that diameter.[2] The Milky Way consists of about two hundred billion stars and is about one hundred thousand light years (~9.5×10^{17} km) in diameter. The farthest point of the universe we can see is about 13.5 billion light years away and contains a vast number of stars—and perhaps a vast number of planets? Our planet, and our Sun that provides the energy for life here, are truly insignificant at the scale of the universe. But how unique are humans as a form of intelligent life capable of thinking about our place in the universe?

Humans appear to have been fascinated for a long time by the possibility of extraterrestrial intelligent life—from our earliest ancestors' animations of star formations and planets to present-day astrobiologists quantifying the possibilities of life-supporting planets in the universe, such as NASA's Kepler mission, "a search for habitable planets" (Kepler 2013). Kepler will explore about 156,000 nearby stars, and as of the end of 2012 it had confirmed the existence over sixty planets.

It would be easy to jump to the conclusion that there must be intelligent life out there, given the vastness of the universe. However, Howard Smith, an astronomer at Harvard University, argues that available data suggest that the number of "even vaguely suitable stars with possible habitable planets" is very small, and the probability of intelligent life evolving on them is "much slimmer" (Smith 2011). He concludes that even if we waited for one hundred human generations, any signals reaching us would not be from further than 1,250 light years away, which is a tiny portion of the universe. So, even if the universe were infinitely large and lasted infinitely long, making the probability of other intelligent life evolving very high, the chances of us Earthlings detecting it, much less interacting with it, are pretty darn slim. We humans are most likely not going to get help from another planet—we are on our own in figuring out how to get out of the food and agriculture problems we have gotten ourselves into.

Nor will we likely get help from contemplating the meaning of it all because there is no evidence that the universe itself has any meaning or purpose that we can discover. As Steven Weinberg, the Nobel Prize–winning physicist, famously remarked, "The more the universe seems comprehensible, the more it seems pointless" (Weinberg 1994). Yet I think that seeing our human problems in deep time and space—in the context of the evolution of the universe—can help us to put things in perspective. It suggests to me that it is up to us—as individuals, as members of local communities, and ultimately as a global community—to define our purpose in the universe, including how we treat our planet and one another, and how we feed ourselves. We can't expect that the universe will provide a purpose, or that intelligent beings from another planet will come to our aid in helping us discern one—or in solving our food crisis.

Whether we do solve the world food crisis depends on our ability to understand the problems and conceive of solutions. It also depends on consciously deciding, individually and collectively, that coming up with a solution is our goal, and then committing the resources needed to achieve that goal. As a step toward understanding, this chapter takes a broad view of the relationship between the demand and supply sides of our agrifood systems, including basic principles of life in the context of our planet and universe, how human populations and their impacts on the Earth grow, the Malthusian conflict of potentially limitless growth of population and consumption and the limited human carrying capacity of the Earth, and finally, the options for human response when we enter the zone where human impact exceeds human carrying capacity.

2. THE ORIGINS OF THE UNIVERSE AND LIFE ON EARTH

Although humans are a relatively intelligent life form living on a tiny speck of stardust in the vastness of the universe, we are subject to the same forces that govern the rest of the universe, including the laws of thermodynamics. The second law states that entropy—a dispersion and loss of order (information) in matter and energy—always increases, as confirmed by observation of the universe since its beginning in the big bang almost fourteen billion years ago. Yet life temporarily resists the second law in small areas of space-time in its interaction with geochemical cycles (Kleidon 2010), as in James Lovelock's Gaia hypothesis (Lovelock 1986); it pumps order in the form of energy and matter from the outside environment into the living organism, extracting order to build and maintain life, and sheds degraded (higher-entropy, less-ordered) energy and matter as waste products. These waste products can have major negative impacts for future life—a prime example being our export of carbon dioxide to the atmosphere, which is driving global warming. Agriculture is a strategy for supporting human life, increasing the flow of information and resources through human bodies and human-managed systems. It is a strategy that has supported increasing levels of human population and consumption, with major impacts on the environment and society.

Of course, as part of the universe, life is an endgame—eventually the second law will conquer all. But we humans exist at some point in time between the beginning and the end of the universe, and this is where all the fun is—figuring out what to do with this amazing opportunity.

2.1. Biological and Sociocultural Evolution

Understanding the opportunity we humans have requires understanding biological and sociocultural evolution and the relationship between these processes, which do not always work in the same direction. Biological evolution may be defined as cumulative genetic change over generations resulting from the selection of phenotypic (physical) traits with some heritable basis. Biological fitness is a measure of the number of copies of genes passed on to the next generation—that is, of how successful genes, organisms, or populations are in reproducing themselves (Wilson and Wilson 2008). Beginning at the time it first comes into existence, a population (or species) must increase in order to survive. A population constantly decreasing in size will eventually become extinct. The population of any biologically successful population or species tends to increase until limited directly or indirectly by the environmental carrying capacity for that organism.

Biological fitness is nonteleological. In other words, it is not future or goal oriented and is not an absolute, but a relative, value; it is defined by the specific environment in which an allele (a form of a gene), organism, or population exists (Wilson and Wilson 2008). The environment includes physical components such as temperature, soil, and moisture; other species including food, competitors, hosts, and parasites; and the demographic and genetic structure of the population being considered, or to which the individual organism of interest belongs (Hedrick 2005:204ff.). In addition, fitness trends at different scales may be inconsistent, so that increasing fitness values at one level can be coupled with decreasing fitness values at an adjacent level (Wilson and Wilson 2008). For example, a population of organisms with high relative fitness may evolve in response to changing selection pressures of a particular microenvironment, resulting in reduced fitness of that population in comparison with others in the larger environment, leading to reduced size or extinction of that population (Bergelson and Purrington 2002, Endler 1986:43–44).

Ecologists have described the adaptation of population growth rates via evolution in terms of r and K selection (MacArthur and Wilson 1967:149ff.). Where resources are variable and abundant, r selection favors species with high, unregulated growth rates. In environments with less variable and scarcer resources, K selection favors species with growth rates that decrease as the population grows in response to feedback from the environment indicating resource scarcity. These species' growth rate is described as density dependent, as it slows with increasing population density, allowing the species to survive and persist over the longer term. Selection operates against species that do not limit growth under such conditions. This shows that the criteria of biological fitness can vary through time as the nature of the resources organisms depend on changes.[3]

However, the evolution of phenotypes via selection by their natural environment over generations depends entirely on existing information, not predictions for future conditions. This means that evolutionary biological success is a measure that can look only from the present backward, into the past. Whatever drives organisms to multiply can be modified only by *current* environmental conditions; there is no mechanism whereby probable future environmental conditions can affect the reproductive drive of present organisms. Current conditions themselves may be "predictors" of future conditions when they are part of a trend—for example, if there is a tendency over time for a decrease in a required nutrient, selection may favor an organism adapted to lower availability, which *may be* preadapted to even lower future availability. However, present conditions are not always good predictors of future conditions; for example, when a threshold is passed, triggering an acceleration in rate of change or a state change, adaptation to the present may result in reduced fitness under future conditions. In contrast, the human capacity to anticipate the future with some accuracy allows us unique influence over our future biological success, depending on the time frame we are interested in.

We humans are an extraordinary species in the extent to which we have devised strategies to successfully direct more and more of the Earth's resources to our consumption. In a sense, through our cognitive abilities, we have expanded the environmental space for r selection for *Homo sapiens,* creating what appear to be unlimited resources. Our evolutionary biological success has functioned as it does in other species—it has been based on current environmental conditions, not on likely future ones. The diversion of energy and matter for the expansion of our population has increased for generations, especially in some parts of the world, with most of this increase based on assessments of our environment *today,* not of the *future* environmental impacts of that diversion. Thus, to a great extent, our cognitive abilities have been applied to supporting human biological fitness in the present, *not* to overcoming the inherent shortsightedness of biological selection and fitness.

This leads to what I call the *fitness paradox*—humans have been successful in increasing our numbers and rates of consumption by dominating the Earth's resource cycles to a greater degree over a shorter time span than any other species with individuals of similar size. The paradox is that this past success threatens our future survival. That is, biological success as defined under r selection has a threshold beyond which it turns into biological failure, because of the conflict between an increasing number of organisms and their impact on the limited resources needed to support them. As that resource threshold is approached, success is increasingly determined by the logic of K selection. The human species, much more than any other, has the ability to predict the future based on knowledge of past events and to change our behaviors and institutions based on solid predictions. So the question is: Will humans be able to transition from an r-selection to a K-selection mode of growth via sociocultural evolution?

The economist Kenneth Boulding described a situation analogous to r and K selection in terms of sociocultural evolution: he contrasted unlimited or open systems, where growth-centered "cowboy economics" is successful in the short term, with limited or closed systems, where steady-state-centered "spaceman economics" is required to avoid disaster (Boulding 1968). "Spaceman economics" implies that humans can incorporate predictions about the future and about the effects of their present actions on that future in decision making. It also suggests that humans have goals for how they want the future to be, as well as ideas about what actions are needed to achieve those goals (see chapter 3 for more details). The critical questions are: What mechanisms are responsible for the transition from "cowboy" to "spaceman" economics? And what kind of life does that transition lead to—for example, a stable population size with short average life span or a stable population size with long average life span (Cohen 1995); a stable level of consumption equitably distributed or a stable level of consumption with a few living in extreme luxury and the rest impoverished?

Thus, for *Homo sapiens*, incorporated within the fitness paradox is a *cognitive paradox*—the human cognitive abilities that helped to create the fitness paradox also offer a solution—we can harness the cognitive paradox and *think* our way out of the fitness paradox! In other words, while evolution blindly evolved cognitive traits to increase short-term biological fitness, these traits included behavioral plasticity and the ability to consciously control thoughts and behaviors. Consciousness allows me to reflect on "myself" in this sentence, and it allows you to do the same when you read it (Hofstadter 2007), in order to make decisions about the future. For example, some tendencies—such as empathy, sociality, and altruism—can be consciously encouraged at the individual or group level, while other traits—such as territoriality, materialism, and greed—can be subdued, or vice versa. Our decisions, active or passive, about what characteristics should be and are promoted will affect the long-term fitness of *Homo sapiens* (Wilson et al. 2009). Our sociocultural evolution will have a major impact on the future of our biological evolution and sucess.

No matter how we choose to address our problems, it is helpful to recognize that humans will not persist indefinitely. Of course, there will be periods of relative homeostasis where forces tending to cause increase or decrease of a population or species in a given environment are balanced. However, over evolutionary time, homeostasis is not the rule—populations and species come into being and some time later become extinct. Evolutionary success is always temporary; no individual, population, or species lives forever, even those with the ability to evolve rapidly in response to changing conditions. One estimate of the average life span of a species based on the fossil record is about three million years (Gilinsky et al. 1989,

Gould et al. 1987). This is analogous to the situation within individual organisms where physiological systems tend to be homeostatic during the life of the individual, yet all individual organisms eventually die. It is not clear at this point what the balance will be between our ability to change our environment and our ability to change our behavior to avoid extinction. However, as is true for a finite individual organism, there are different ways for us as a species to approach the opportunity we have for life and different choices we can make about how we experience it.

2.2. Competition in Resource and Energy Cycles

Through time and space, different individual organisms, populations, and species on Earth cooperate and compete to obtain access to a limited quantity of matter and energy. For now, the matter is limited to the planet Earth, and energy is effectively limited to that derived directly from the Sun and converted to heat or chemical energy via photosynthesis by green plants, or to electrical energy via photovoltaic cells. Other sources of energy are those derived indirectly from the Sun (wind, waves, tides), or from nuclear fission (and potentially fusion) and geothermal sources, and all have associated costs—some quite high, as with nuclear fission, as seen in disastrous accidents such as those at Chernobyl on April 26, 1986, and Fukushima Daiichi on March 11, 2011. However, even if there were a source of unlimited "clean" energy, like all energy it could be useful to organisms only by energizing matter, so life on Earth is ultimately limited, both globally and locally, by the amount, structure, and distribution of physical matter of the planet itself. This means that there is no such thing as energy with no environmental impact, because even if it could be *generated* with no impact, energy can be *used* only by affecting the environment. Most energy controlled by humans is used to convert natural resources into humans or things humans use, which means a decrease in resources available to maintain natural ecosystems, on which humans, and all other life, ultimately depend.

Much of the history of life on Earth is the history of changing the locations of critical elements such as carbon, oxygen, hydrogen, nitrogen, calcium, and phosphorus (Raven, Andrews, et al. 2005, Raven, Handley, et al. 2004) across a wide spatial scale, from their position in chemical compounds to the distribution of their abundance on Earth, from the molten core to the upper atmosphere, from oceans to mud puddles. Nutritional biochemistry in living organisms—including humans—reflects the early geochemical evolution of the Earth (Fedonkin 2009). In other words, human evolution has been constrained by the evolution of chemical and biochemical pathways, and today the Earth's biogeochemistry is embedded in human physiology, with cellular metabolism repeating to some extent the "main events of the coevolution of geochemical and biotic processes in the early biosphere—life remembers the youth of the Earth" (Fedonkin 2009:1321).

At very local levels, the relative abundance of elements and compounds controls the growth and multiplication of organisms. For example, phosphate ions in the rhizosphere (the root zone) are required for the growth of crops such as sorghum, and sorghum is used to make porridge, which will determine the number of people that can live in a West African village. Populations and species have blindly evolved the ability to organize the matter and energy in their internal environments in ways that tend to optimize their growth and reproduction under current external conditions, while they also affect those external conditions. For example, photosynthesizing plants helped to increase the oxygen content of the atmosphere (see below). (In chapter 8 I discuss biogeochemical cycles in relation to climate change and diet.)

At the global level, too, resources become limited in relationship to the number of organisms and the demand they place on resources. For example, the more than seven billion humans living on Earth in 2013 were equivalent to about 350 billion kg of matter (at 50 kg per person)—matter that is temporarily unavailable to other organisms. And this does not include the resources in crops, domestic animals, and human-made capital, such as buildings, cars, and roads—but I don't know if this calculation has been attempted. In terms of the total photosynthetic output of green plants (net primary product, or NPP), Vitousek et al. have made a widely cited estimate that humans use 38 percent of our planet's NPP, of which 4 percent is used directly for food and an additional 34 percent is either co-opted for nonedible products or destroyed (Vitousek et al. 1986). Recent studies support the estimate of 38 percent of NPP used by humans, while pointing out that of the remaining 62 percent of NPP, only about 15 percent might be available in forms humans can use; given projected growth in demand, humans will deplete this available NPP within decades (Running 2012).

2.3. Does Life, and Agriculture, Have a Purpose?

This overview of humans in the context of the evolution of the universe helps provide the necessary context for understanding our agrifood system—its historical development, current accomplishments and problems, and possible futures. Many behaviors and institutions were selected for their survival value during the long course of human biological, and then sociocultural, evolution. But many of those were selected for during a time when human impact on the Earth was relatively small, and they seem maladaptive under current conditions. That is, they can lead to a reduction in human biological fitness in the environments we are creating on Earth. For our continued survival, humans need to rapidly evolve socioculturally to encourage the development of alternative ideas, behaviors, and social structures (Gowdy et al. 2010).

A key to solving the world food crisis and creating more sustainable agrifood systems is replacing the biological criteria of success based on blind, short-term biological selection for fitness in response to current conditions with sociocultural criteria that modify current trends to create alternative futures, while at the same time being adapted to those futures. Of course, agreeing on the vision for the future is critical and will not be easy (see chapter 3). However, empirical evidence from the natural sciences makes it clear that it will require replacing the biological criterion of increased physical growth in numbers with the sociocultural criterion of increased cognitive growth, and replacing unlimited per capita consumption with a focus on nonmaterial changes to improve our lives (after basic physical needs are met) (Jackson 2011).

So, what should be the goal of agriculture? As we saw in the introduction, the mainstream approach to the food crisis tends to assume that the goal of agriculture is to increase the food supply and decrease hunger—that is, to nourish human bodies. But this avoids going beyond the biological purpose of food. The ultimate goal of agriculture must be the same as that of human life, which in turn must be subjectively decided by humans—it cannot be simply to keep humans alive and multiplying. That is tautological nonsense.

The purpose of human life is defined by many different cultures and in different ways as what can be interpreted as subjective happiness. The Japanese advocate of natural farming Masanobu Fukuoka is often quoted: "The ultimate goal of farming is not the growing of crops, but the cultivation and perfection of human beings" (Fukuoka 1978:119). In his book *Eating Animals,* novelist Jonathan Safran Foer captures one example of this deeper meaning of food and human life in a story about his Jewish grandmother's struggle to survive while hiding from the Nazis in World War II. Even though she was starving, she refused to eat pork because it is prohibited by the Jewish religion: "If nothing matters, there's nothing to save" (Foer 2009:16–17).

Reaching at least a partial consensus on the purpose of human life is a key part of defining our goals for agrifood systems, as we shall see in chapters 3 and 4. The country of Bhutan has institutionalized the idea of happiness being a central goal in its Index of Gross National Happiness (Ura 2008). In an effort to reduce the subjectivity of such a goal, neuroscientist Sam Harris has suggested instead of "happiness" the term "human welfare," which he argues can be objectively measured in terms of electrical patterns in the brain (Harris 2010). Whatever we even partially agree upon as the purpose of human life, the unique human opportunity is the possibility of creating a better world in the dynamic tension between physically based biological evolution and cognitively based sociocultural evolution.

3. POPULATION GROWTH

Our success as a species at increasing our numbers and our per capita consumption, and the effects of this on society and the environment, are central to thinking about food and agriculture. In this section we will look further into population growth as a driver of human impact.

3.1. How Do Populations Grow?

Human population growth is similar to the growth of populations of other organisms. What determines population growth rate? On a local scale:

$r = BR - DR + (IR - ER)$, where
r = growth rate
BR = birth rate
DR = death rate
IR = immigration (migration in) rate
ER = emigration (migration out) rate (fig. 1.1)

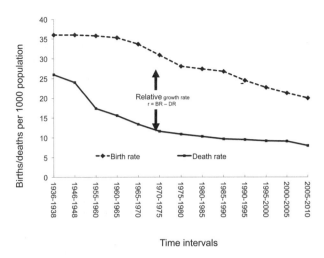

FIGURE 1.1. Global birth and death rates per 1,000 population. Data from (PRB 2010, UN 2012).

On a global scale, immigration and emigration are irrelevant, so that when BR is greater than DR, there is positive growth or natural increase, which means the potential for infinite population size, limited only by the resources needed to support it, and when BR is less than DR there is negative growth, leading eventually to extinction. It is obvious that for population size to be stable, BR and DR must, on average, equal each other.

So, depending on the relationship between BR and DR over time, the shape and properties of the growth curve of a population will be different. There are several kinds of growth that populations can have, and it is important to understand human population in terms of these alternatives. They differ in terms of the combination of relative growth rates (the numbers added during each time interval as a proportion of the population at the end of the previous interval) and the absolute growth rate (the number added during each time interval). Note that I discuss growth here as positive, but everything applies to negative growth as well.

Linear growth (also called *arithmetic growth*) is change by a constant amount (the absolute growth rate stays the same), which means that the relative growth rate (r) is decreasing, since the base population is growing, but the number of individuals added each year remains constant.

In *sigmoidal, logistic,* or *density-dependent* growth, growth is linked to carrying capacity (K), such that the absolute growth rate is determined by the difference between K and population size P at any given time. As P approaches K, both the absolute and relative growth rates decrease toward zero, so that P never exceeds K.

Superexponential growth is change by an increasing relative growth rate due to DRs decreasing or BRs increasing or both. Of course, the absolute rate will also increase.

Exponential growth is change by a constant relative rate, which means that absolute rates will increase. This has long been recognized as a special type of growth, and it is commonly used today as a model for growing populations. The following are different forms of the equation for exponential growth.

$$P_t = P_o \, (e^{rt})$$
$$P_o = (P_t / e^{rt})$$
$$r = (lnP_t - lnP_o) / t$$
$$t = (lnP_t - lnP_o) / r, \text{ where}$$

P_o = population at time o (i.e., the starting population size)
P_t = population at time t
e = the base of the natural logarithm (2.71828 . . .)
r = the relative growth rate
t = time between P_o and P_t
ln = natural logarithm

One convenient way to think about exponential growth is in terms of doubling time (DT), or the time it takes a population to double in size—and the time period during which doubling will occur will remain constant as long as the exponential growth rate remains constant. DT is often used by demographers as an intuitively appealing way to express the growth rate of a population and to compare populations. When a population doubles, P_t = $2P_o$, and therefore, by substitution in the equation for P_t, $2P_o = P_o \, (e^{rt})$, and $e^{rt} = 2$, and taking the natural log of both sides gives us rt = 0.693, and the time for the population to double = 0.693/r. Therefore, a quick way of estimating the DT of a population undergoing exponential growth is

DT ≈ 0.7/r, or when r is expressed as a percent, DT ≈ 70/r

There are many tales illustrating the astonishing result of continued doubling that is exponential growth during a given time period. For example, according to the legend of the Ambalapuzha Temple in Kerala in southern India, the Hindu deity Krishna once appeared in the form of a sage and challenged the king of the region to a game of chess.[4] For the prize if he won, the sage said he wished only a few grains of rice—one grain for the first square of the square chess board, two for the next, four for the next, and so on. The king was disappointed that the sage requested so little, but when he lost he soon discovered his dilemma. The sixty-four squares of the board would require 2^{63} grains of rice = 9,223,372,036,854,780,000, or more than 9,000 trillion grains of rice, equal to over 230 billion metric tons, nearly ninety times the size of the *total world cereal harvest* in 2011 (FAO 2013b)! Krishna then appeared in his true form and told the king he could pay off the debt over time in the form of *paal payasam*, a sweet rice dish, served for free to all pilgrims visiting the temple. This story gives us a feeling for the power of doubling to increase the original number rapidly, and for the consequences of not understanding what current trends mean for future conditions.

The next example gives a feeling for the dramatically increasing absolute rate of increase in exponential growth in relation to deciding how to deal with it. A pond has a single lily pad growing on it, which grows to two the second day and continues to double every day. After thirty days the pond will be completely covered, extinguishing all life living within it. The creatures living in the pond are aware of this, but given how busy they are with finding food, eating one another, settling disputes, and reproducing, they decide to wait until the pond is *half covered* before addressing the problem. What day will they finally meet to resolve the problem of the lily pads? The answer: the twenty-ninth day—too late!

3.2. How Does the Human Population Grow?

Understanding how the human population has grown in the past can help us to predict how it might grow in the future and to develop plans to deal with this growth. Joel Cohen, a population ecologist at Rockefeller University, has done a comprehensive review of the history of human population growth in relation to carrying capacity (Cohen 1995). He showed that the Earth's human population has grown at different rates—at very low exponential rates for foragers (gatherer-hunters), and then at increasing rates with the advent of agriculture about twelve thousand years ago, becoming super exponential. Since the mid-1960s, when the world's annual population growth rate peaked at 2.06 percent, the rate has been decreasing.

However, the absolute growth rate has remained fairly high and stable because the base population is so large. It has taken since the origin of our species, *Homo sapiens,* until about the year 1800 for our population to reach one billion, only 130 years to add the second billion, 30 years for the third, and only 12 to 14 years for the fourth, fifth, sixth, and seventh (in 2011) (PRB 2011). In 2012 the world population of 7.1 billion was growing at a rate of 1.2 percent per year, which means a net increase of about 231,000 more people every day and over 84 million per year, and it is estimated that another billion people will be added by 2025 (PRB 2012).

The absurdity of continuing our current rate of growth is shown by the following calculation. In 2012 there was an annual world population growth rate of 1.2 percent and there were about 4640 people per km² of land in the world. However, with a world land area of 1.53 × 10^{12} m², it would take only 448 years for there to be one million people per km²—that is, one person per every square meter of land in the world, which is obviously not sustainable, no matter how you define it.

$t = (ln P_t - ln P_0) / r$
$t = (ln\ 1.53 × 10^{12} - ln\ 7.1 × 10^9) / 0.012$
$t = 448$ years

In more personal terms, this presents humans with a dramatic choice—we cannot have a long average lifespan (low DR), a high BR, and a stable population size at same time (Cohen 1995:18). If we want long lives and a stable population ($r = 0$), then on average each individual can have only enough children to replace themselves—only one! Any growth rate greater than zero for the Earth's human population is ultimately unsustainable because the population would increase indefinitely, and so if birth rates don't fall, death rates will have to increase to maintain a stable population size.

3.3. What Is the Future of the Human Population?

As we can see, the current rate of population growth has to change, and population projection is the field of demography concerned with predicting how different factors affecting population growth—birth, death, and immigration rates, population age structures—will affect the growth rates and future sizes of human populations. Given past and present changes, most population growth in the next decades will occur in the Third World, but with continued slowing of the rate of growth there as birth rates continue decreasing faster than death rates. As a result, these Third World populations will begin to resemble more and more the age structure of industrial populations (fig. 1.2). The narrowly based age pyramids of populations with low birth and death rates mean higher proportions of people in less productive older years, increasing the overall dependency ratio.

Another important demographic trend is increasing urbanization. As of 2010 the majority of the human population lived in urban areas for the first time in human history. Thinking of the world food crisis in light of growing urbanization and dependency ratios emphasizes the necessity of switching from a short-term biological evolution-based criterion of success to one based on a combination of the most objective assessment of the biophysical world and agreement on the goals of the agrifood system to guide our sociocultural choices and changes.

4. THE EARTH'S HUMAN CARRYING CAPACITY

We have seen that we humans have a remarkable biological potential for reproducing ourselves and an equally remarkable potential for controlling the environment to support this growth. We have also seen that the potential for growth is limited—but if we are to use our cognitive abilities to anticipate and adapt to these limits, we also have to understand a bit more about those limits. In this section we focus on human carrying capacity as the constraint to population growth and other aspects of human impact, both locally and globally.

4.1. Malthus's Argument

Thomas R. Malthus was an English cleric, famous for bringing to the attention of nineteenth-century Europeans the conflict between the potential for humans to increase their numbers "geometrically" (i.e., exponentially) without limits, while the food supply on which they depend is limited by the limited supply of land and other agricultural inputs and can only increase "arithmetically" (i.e., linearly) (Malthus 1992 [1803]:17–19). He believed that the incompatibility of these two rates of growth inevitably resulted in checks on human pop-

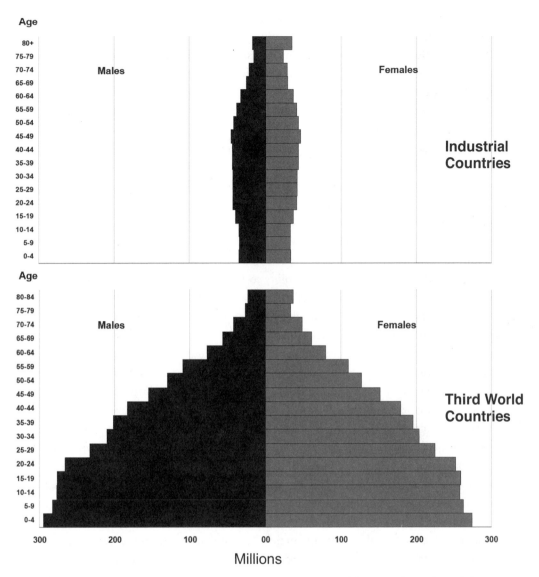

FIGURE 1.2. Population pyramids in the industrial and Third Worlds. Reproduced from slides 6 and 7 (population pyramids) from the 2010 WPDS slide show: www.prb.org/presentations /10worlddatasheet_presentation.pdf. © Population Reference Bureau. Used with permission.

ulation growth. In the first edition of *An Essay on the Principle of Population*, published in 1798, he emphasized two checks—positive and preventive (Malthus 1992 [1803]:15, Winch 1992:xiv, xvi). Malthus's *positive checks* are negative in terms of their effect on human happiness—they function to reduce impact via reduction in consumption and increase in death rates, causing misery and fear of misery. The possibility for preventive checks was not a very happy alternative either, in Malthus's view. For him, preventive checks function to prevent impact by reducing birth rates due to "vice," meaning sex outside of marriage and other unnamed "vicious" customs, including prostitution and contraception.

Malthus is often represented as a pessimist who believed that humans would unavoidably sink into misery as a result of population growth. What is not so often appreciated, however, is that beginning with the 1803 second edition of the *Essay,* Malthus emphasized the preventive check of moral restraint, by which he meant delayed or no marriage, and celibacy outside of marriage: "It is not the duty of man simply to propagate his species, but to propagate virtue and happiness; and that, if he has not a tolerably fair prospect of doing this, he is by no means called upon to leave descendants" (Malthus 1992 [1803]:271). He believed that humans were unique among animals in their ability to exercise this check due to their "distinctive superiority" in "reasoning faculties, which enables [them] to calculate distant consequences" (Malthus 1992 [1803]:21).

Thus, Malthus laid out the human predicament due to the conflict between food supply and a growing human population as one that would inevitably lead to pervasive misery and suffering, *except* that humans have the ability to foresee and plan for the future and to change their behavior accordingly, either in the manner he described, or through what is today more broadly available and socially acceptable safe birth control methods, an example of behavioral change arising from sociocultural change.[5]

As I briefly described in the introduction, food supply and human numbers have been increasing at similar overall rates, beginning with the Neolithic and continuing through the present. To some, this means that *Malthus was wrong* (e.g., Evans 1993). However, if you accept the idea that there are limits on the extent to which humans can divert energy and matter from natural ecosystems, and social limits on the extent to which humans can organize society to produce more food, then over the longer term *Malthus was right.* His fundamental observation seems incontrovertible—humans, like all organisms, tend to increase exponentially with no intrinsic biological limit to the ability to reproduce, while there are physical limits that prevent food supply from increasing exponentially except over relatively short periods of time. This means that the necessity of a sustainable agrifood system is Malthusian: food production, and therefore human numbers, are ultimately limited by the physical limits of the Earth, and sustainable agrifood systems must be adapted to this situation.

4.2. Human Carrying Capacity

I use the term human carrying capacity (HCC) to refer to the number of people a biophysical environment can support (culturally, socially, biophysically) without decreasing its ability to support the same number in the future. Human impact (HI) is the product not only of population size *(N)*, which I have emphasized so far in this chapter, but also per capita consumption *(C)*, and the technology *(T)* used by humans to provide what is consumed (HI = *NCT*) (following Daily and Ehrlich 1992:762), and a sustainable HI means that it does not reduce HCC over the long term (see section 5).

While there are many estimates of the Earth's HCC, most evidence suggests that with current growth rates it is finite in the relatively short term (i.e., in terms of the human evolutionary time scale), and it may be exceeded already (Cohen 1995, Giampietro et al. 1992), or will be very soon (Running 2012). Cohen reviewed more than sixty-five estimates of HCC made from the seventeenth century through 1994 and found a huge range, from one billion to a billion billion (10^{18}). While there was no trend in the average estimate, there was an increase in range, although more than 50 percent of the estimates (based on the upper bound when a range was given) were between 4 and 16 billion (Cohen 1995: 212 ff.). If only the short-term production potential of the planet is considered, very large estimates are possible—for example, 282 billion if all land area is used for crops, with the same yields as the highest yielding crops (Franck et al. 2011).[6] However, when you consider the longer-term limits of the biophysical environment of the Earth, subject to the laws of thermodynamics, then the estimates of HCC are much lower (Moran et al. 2009, Pereira 2009, Gilland 2006).

So far, we have assumed that HCC is defined in terms of the biophysical environment. However, it can also be defined as "social" carrying capacity (Daily and Ehrlich 1992:762) or, more correctly, socio*cultural* carrying capacity, to emphasize the role of subjective values in its definition and estimation. Biophysical and sociocultural dimensions cannot be separated, because biophysical limits can be determined only based on human values about the kinds of lives we want, or are willing to live, and the risks we are willing to take. Cohen concluded that because no estimates of HCC have included these factors, "taking into account the diversity of views about their answers in different societies and cultures, no scientific estimates of sustainable human population size can be said to exist" (Cohen 2005a). Sayre agrees, believing that carrying capacity has been used in an uncritical manner that does not acknowledge that "limits are rarely static or quantifiable, let alone predictable and controllable," often to advance ideological positions (Sayre 2008:132). His conclusion that the concept is not very useful, however, ignores the concept's potential for stimulating questions and research, as discussed in the rest of this chapter.

4.3. Organisms Affect Their Environments: Limits of Human Carrying Capacity

While some other species make use of technology to enhance their evolutionary success, humans alone have excelled at this, and have increased HCC dramatically through modifying the environment and other species to increase our food supply.

Ecologists have traditionally thought of causality between organisms and their environments as being unidirectional—namely, organisms change under the influence of other organisms and the environment. But organisms also affect their environments in ways that in turn affect their adaptation (biological and cultural) to those environments in an interlocking system (Corenblit et al. 2011; Fedonkin 2009). This has been modeled by Odling-Smee et al. (2003) as

$$dO/dt = f(O, E), \; dE/dt = g(O, E)$$

which states that changes in organisms over time (dO/dt) are a function of both the organisms and the environments ($f(O, E)$), and changes in the environment over time ($dE/$

dt) are a function of both organisms and the environment (*g(O, E)*) (Odling-Smee et al. 2003:18).

Thanks to humans' unique cognitive and technological capabilities, which are much greater than those of other organisms, humans can uniquely modify their environments with the *goal* of increasing their carrying capacity, and human activities have increased to a level that is now having dramatic effects on the Earth's atmosphere, climate, hydrology, soils, and other species. Graphs that track changes since the beginning of the industrial period show a dramatic similarity in the superexponential increase in human activity (e.g., population size, fertilizer consumption, and McDonald's fast food outlets), on the one hand, and the effects of this activity on the Earth's biogeochemical systems (e.g., global warming, loss of forests, and species extinctions), on the other hand (Steffen et al. 2011: Figs. 1, 3; see also Vince 2011). These changes will have a negative impact on the biophysical HCC, but also on the sociocultural HCC, because of increased conflict, negative psychological consequences, displacement from homelands, and loss of cultural identity (World Bank 2012b:55ff).

Many scientists believe that we are affecting the biophysical nature of the Earth so dramatically that the consequences of human impact today will be evident thousands or even millions of years in the future, and so they have suggested naming the period beginning with the industrial age about 150 years ago the Anthropocene epoch (Vince 2011). Such an impact raises the possibility that the Earth could be leaving behind the relatively stable biogeochemical cycles of the Holocene on a "one-way trip to an uncertain future" (Steffen et al. 2011:757); indeed, that these cycles might be approaching a tipping point that would change the Earth dramatically, beyond anything ever experienced by humans before (Barnosky et al. 2012; World Bank 2012b).

5. DEALING WITH "THE ZONE"

Joel Cohen has aptly described the space where human population and the Earth's HCC overlap as the "zone" (Cohen 1995), and I add consumption and technology to population in comprising human impact (HI) (fig. 1.3). Based on his exhaustive review of estimates of HCC described above, Cohen believes that humans have entered the zone.

Being in the zone means that there is an existing or imminent imbalance between the demands of humans for goods and services and the ability of the environment and society to meet these demands. As I have mentioned, the good news is that humans are unique compared with other species in that we have the ability of self reflection, so unlike nonhuman biological evolution, our sociocultural evolution can be teleological—that is, it can be consciously directed toward achieving goals in the future (Richerson and Boyd 2005). Therefore, humans have the ability to adapt as we approach the zone in ways that slow our approach and even reverse our direction. However, in order to adapt, several key ingredients are required—adequate knowledge of the zone and the consequences of being in it, the value-based social decision to do something, and the will and perseverance to actually do it. Similarly, based on his review of societal collapses and perseverance in the face of environmental limits exacer-

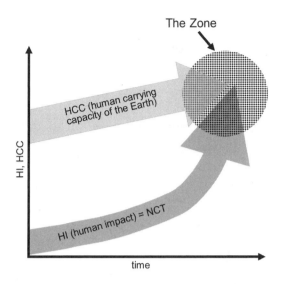

FIGURE 1.3. The zone where human impact approaches human carrying capacity (sizes of zone and HCC and HI arrows are arbitrary). © 2013 David A. Cleveland.

bated by human activity, Diamond sees the two biggest challenges to avoiding collapse as "long-term planning, and willingness to reconsider core values" (Diamond 2005:522).

Experience doesn't serve as a very good guide to dealing with the zone, since at the global scale we have never had to deal with it before. Biogeochemical changes (such as global warming) and their effects on agrifood systems (such as increased variability in crop yields) and human society (such as increased food insecurity and social unrest) are unpredictable. We can't simply extrapolate past trends into the future, especially because some of these changes will be nonlinear, and even pass biophysical thresholds, resulting in entirely new states we may have no knowledge of, as may be the case with climate change (see chapter 8). It's like anticipating an unknown guest coming for a meal that we must prepare using ingredients we've never seen before.

For example, if farmers in a village notice that the soil in their fields is eroding at the same time that yields are declining, they will expect further yield declines unless erosion is checked. However, if erosion has been going on for a long time, and yields have remained stable, how can farmers know that the topsoil in their fields will eventually pass a threshold beyond which yields will decline rapidly to very low levels? Thus, the challenge of teleological sociocultural evolution is to predict possible futures based on understanding current trends and predictions of potential dramatic deviation from trends, and to take action to direct the future toward desired outcomes. For this reason, the human ability to analyze the present as the basis for predicting the future is critical.

The question for sustainable agrifood systems is not only what biophysical mechanisms mediate between increasing demand for food and increasing agricultural production, but how we can detect when HCC and HI are going to approach or intersect each other and how we can respond. In the simplest terms, the choice is either increasing HCC or decreasing HI; once HCC is maximized under current knowledge and technology, decreasing HI is the only option. HI can be decreased by slowing or reversing population growth (by decreasing

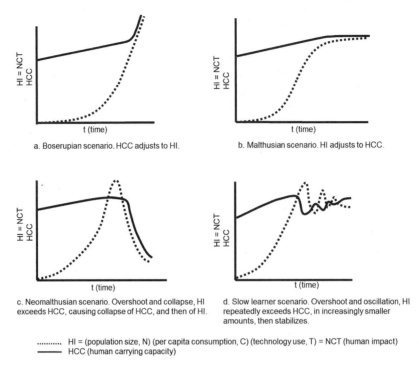

FIGURE 1.4. Future scenarios for human impact and human carrying capacity. Based on (Daily and Ehrlich 1992, Meadows et al. 1992).

birth rates and/or increasing death rates), decreasing consumption rates, or increasing the efficiency while decreasing the negative environmental impact of technology. Therefore, given our understanding of human population growth and the Earth's finite resources, the critical question is how we are going to reduce HI to below HCC.

Following Meadows et al. (Meadows et al. 1992), I discuss this in terms of four basic scenarios for human response to the zone (fig. 1.4). The first scenario assumes that HCC can be increased indefinitely in response to increasing impact. The other three assume that the ability to increase HCC is limited.

5.1. Human Ingenuity Increases Growth of Human Carrying Capacity: A Boserupian Scenario

Ester Boserup's first book on population and agriculture, *The Conditions of Agricultural Growth* (1965), was influential at the time it was published in 1965, and it remains so today (Malakoff 2011). It sounded a positive note about population growth at a time when there was growing concern. Following World War II there was a surge in population growth as death rates fell much faster than birth rates, and many were concerned about the possibility of food crises (Cohen 1995:67–68). Indeed, the 1960s was a period of unprecedented growth rates in the world population (peaking at 2.06 percent per year in 1965). It was also a period of widespread public concern about the effects of population growth, spurred by sensationalist predictions of the dire consequences of not slowing growth, including Paul Ehrlich's *Population Bomb,* which received wide publicity (Ehrlich 1968).

The Conditions of Agricultural Growth appeared just one year after Theodore Schultz's seminal book, *Transforming Traditional Agriculture* (Schultz 1964). Both Boserup and Schultz rejected the then-popular belief among agricultural economists and agricultural development professionals that small-scale farmers are irrational. Instead, they cited evidence suggesting that farmers are capable of responding in economically rational ways to the external forces of the marketplace, and in Boserup's work, of population pressure as well. They differed, however, in a key assumption. Schultz believed that increasing modernization of traditional agriculture was the only way of increasing output. Boserup believed that Schultz's conclusion failed to distinguish variation in intensity of land use in traditional farming systems (Boserup 1965:15–16, 1990:278–279). Boserup thought that even though intensification would be avoided when possible, because it leads initially to reduced production per unit labor input (diminishing returns), the pressure of population growth, and the stimulation of markets, would spur intensification and technological innovation that would lead to higher production, and eventually to higher returns to labor as well (fig. 1.4a).

There are many data supporting a positive correlation between population growth and increased HCC. Undoubtedly the best-known example of a Boserupian scenario is the World Bank–funded study in the 1980s of Machakos District, Kenya. This study showed that an agricultural environment that had been severely degraded as population grew, and that had been written off as irrecoverable, was reclaimed to productivity as a result of agricultural intensification, including extensive terracing and new market crops, transforming it into a poster child for Boserup's theory and the World Bank development policies (Mortimore and Tiffen 1994, Tiffen and Mortimore 1993, Tiffen et al. 1994). There have been many critiques of the study for its oversimplification and unfounded conclusions (e.g., Siedenburg 2006).

There is no doubt that humans can increase HCC, as they did dramatically with the invention of fire, tools, and agriculture. By looking selectively at that range of environments and time periods where carrying capacity can be increased, at least in the short to medium term, the Boserup hypothesis can be supported. Thus, a good deal of the controversy about the applicability of Boserupian views may be due to the failure to specify the range of conditions in terms of environmental HCC for each case study (Carr et al. 2009), precisely because it is assumed that an environmental HCC does not exist.

5.2. Negative Feedback Leads to Low Growth Rate of Human Impact: A Malthusian Scenario

As we have seen, Malthus's preferred strategy for dealing with the mismatch between exponential human population growth and arithmetic growth in food supply was preventive checks (moral customs) that reduce population growth when food is scarce, resulting in the logistic growth curve of HI in fig. 1.4b. In other words, awareness of the approach to or entrance into the zone can motivate humans to reduce HI and avoid the positive check of increasing death rates, analogous to *K* selection discussed in section 2.1.

The Kusasi of northeast Ghana before the advent of colonialism present us with an example of a Malthusian scenario, at least over the short term. I lived for eighteen months in Zorse, a village of Kusasi farmers, carrying out research for my dissertation, described in the preface (Cleveland 1980). The research was part of a project investigating the drought in the Sahelian region of West Africa in the 1970s. The Kusasi homeland, known as Kusaok, consists of Bawku District, Ghana, and neighboring areas of Burkina Faso and Togo. The Kusasi make their living in this environment with short-handled hoes, wooden flails, grinding stones, and an occasional ox plow. In good years, the rainy season between May and October can produce the illusion of prosperity, with thick green stands of millet and sorghum towering above the dispersed mud-and-thatch homesteads. The dry season, however, brings dramatic contrast, with the only relief from the barren, dusty fields being the gardens in dry streambeds watered by buckets carried from hand-dug wells (fig. 0.1).

Food production in Kusaok is labor intensive, so parents desire lots of children because children become net household producers by age ten, contributing directly to family welfare. Children also contribute to household food production indirectly—for example, by caring for younger siblings, herding goats, scaring birds from ripening crops, and gathering fuel and food—so that adults have more time for the heavier work in the fields and house of producing and processing food (fig. 1.5). Therefore, there is a high demand for children, which is rational in terms of optimizing household welfare, with fertility regulated not to maximize the number of live births, but the number of surviving children.

Traditionally, the high demand for children was balanced by fertility regulation embedded in community management of agricultural resources (see chapter 5), such as fields, pasture, and forests, and households helped one another in many tasks (Cleveland 1986b). Fertility was tied to the HCC of Kusaok via two social mechanisms. First was postpartum sexual abstinence—husbands and wives abstained from intercourse after the wife gives birth until the youngest child was old enough to remain healthy when her mother became pregnant again. Health is affected by nutrition (food supply) and disease (including water quality), so the period of abstinence increased as food supply and water quality decreased during periods of drought.[7] In addition, inter- and intratribal hostilities meant that the next-to-youngest child had to be big enough to run away if attacked while in the field with its mother. The net result was long birth intervals of three or more years.

The other mechanism controlling fertility was age at marriage, which was limited by the availability of bride-wealth cattle. The bride-wealth tradition requires that each young man give four cows to his future bride's family, and the would-be husband depended on his own extended family (father, uncles, grandfathers, and so on) for these cattle. But the supply of cattle is limited by their food supply, and by crop harvests, since good harvests allowed households to trade for cattle. Both of these, in turn, are controlled by the carrying capacity of the environment.

Thus, even though there was a high demand for children, Kusasis traditionally limited their fertility via feedback from the HCC of their environment by using Malthusian preventive moral checks—delayed marriage and postpartum abstinence.

FIGURE 1.5. Kusasi child gathering millet stalks for the cooking fire. Beginning at around age eight, Kusasi children transition from net household consumers to net producers. They do lighter tasks, like the one illustrated here, as well as child care, scaring away birds from the fields at harvest time, and herding animals; they also help older relatives in heavier tasks like pounding grain and hoeing. Photo © David A. Cleveland.

5.3. Human Impact Exceeds Human Carrying Capacity: A Neo-Malthusian Scenario

This is the scenario that Malthus dreaded, in which humans are either unable to understand that they are approaching or in the zone, unable to agree on a plan to do something about it, or unable to organize effectively to carry out the plan. The result is that HI overshoots HCC and causes environmental collapse, which in turn forces a reduction in HI through Malthus's positive checks—increasing death rates, misery, and the fear of misery, as food becomes scarcer (fig. 1.4c). As discussed above (section 4), there is much evidence that HI has already overshot HCC in many local situations, and perhaps globally as well.

The indigenous agricultural and demographic regime that created a Malthusian balance between HI and HCC in Zorse as described in the previous section changed dramatically and rapidly with the invasion and military dominance of European colonialists. Population grew despite substantial labor migration south to work in the colonial economy—between 1948 and 1970 the Kusasi population grew by nearly 50 percent. This growth can be partially explained by reduced mortality stemming from public health measures and improvements in transportation that permitted freer movement of people, cash, and food (Cleveland 1991). However, I found that it was also the result of a 44 percent increase in the total fertility rate (or TFR, the average number of births anticipated over the lifetime of a woman in a particular population) since 1943 due to a breakdown in traditional institutions of fertility control (Cleveland 1986b).

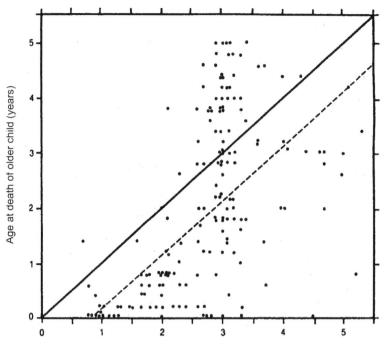

FIGURE 1.6. Birth intervals and child mortality. The figure shows the correlation between age at death of older child and birth interval between that child and the next youngest, with the solid line showing where these are equal. The dashed line marks where the age at death of the older sibling equals the interval between its birth and the conception of the next youngest child. Most of the deaths occur below this line—that is, before conception of the next youngest child. The correlation between age at death and birth interval here is high (0.72, P < 0.001) because a child's death leads to shorter periods of postpartum abstinence. The causation is reversed above the dashed line, when the older child dies after the next youngest is conceived—that is, shorter birth intervals increase the probability of the older child dying, and the correlation here is 0.78 (P < 0.001), mostly due to the strong relationship at younger death ages, when the influence of the conception of a younger sibling has the greatest effect on its older sibling. For deaths occurring above the dashed line only up to age 3.5 the correlation increases to 0.92 (P < 0.001). It is in the critical birth intervals of about three years that most changes in death rates have occurred through time. Adapted from (Cleveland 1986b). © 2013 David A. Cleveland.

The first cause of increased fertility was shorter birth intervals resulting from shorter periods of postpartum abstinence. One reason for this was improved child health, which was a consequence of increased transportation of food in time of famine and public health interventions, especially after the mid eighteenth century (vaccinations, disease vector control, cleaner drinking water, curative medicine) (fig. 1.6). The combined result was that post-partum abstinence was no longer dependent on the ability of the *local* environment to pro-

TABLE 1.1. Zorse Demographic Variables

	1948–1952	1973–1977
Total fertility rate (TFR)	5.4	6.7
Birth interval (months)	43	39
Age at marriage[a]	19.5	18
Mortality rate per hundred, 0–5 years	30	20
Migration rate per hundred males	37	76

Source: Cleveland 1986b.

[a] This is age at first birth, since marriage in Kusaok is a process that is difficult to date.

duce food and healthy conditions. The reduced infant and child mortality rates meant that parents could reduce birth intervals by decreasing the period of postpartum abstinence, thereby increasing the number of surviving children and improving household welfare (table 1.1, fig. 1.7).

Another reason for reduced periods of postpartum abstinence was European colonial rule—first by Germans, soon replaced by the British—in northern Ghana and Zorse beginning in the early twentieth century. The British military occupation led to decreased local hostilities among neighboring groups, diminishing the importance of mobility for mothers with young children, one reason for long birth intervals. Of course, the main purpose of the application of British colonial military power was extracting profit from farmers, which increased violence overall, due to British military dominance and coercive tactics such as forced labor.

The second cause of increased fertility rates was the decreasing age at which women married, a result of young men's becoming able to acquire bride-wealth cattle from outside of the village (Cleveland 1991). The colonial strategy was to break up largely subsistence-based agricultural communities, which were locally self-sufficient, in order to extract labor. At first this was done by physical coercion, and then by imposing a tax on each household, payable in British currency, which could be earned only by labor migration to work in commercial agriculture, mining, and public works in the south. These sectors were directly tied to the European economy for the benefit of Britain. Low wages and poor working conditions encouraged most migrants to return to their savanna villages when they were sick, injured, or too old to work. When Ghana gained its political independence from Britain, this new pattern of migration had become firmly established and was maintained by the changes in the social, economic, and transport systems.

The resulting increase in migration of young men to the south led to their economic independence from their families and community. They could now obtain bride wealth and marry without their family's help, without depending on the local environment to support the accumulation of bride-wealth cows. Along with their increasing exposure to outside values (e.g., more liberal attitudes toward sex), this led to the weakening of traditional values and the authority of elders. Therefore, marriage age (and thus fertility) was no longer

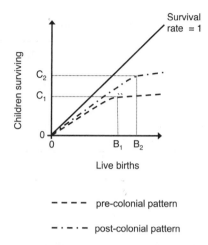

- - - - pre-colonial pattern

- · - · - post-colonial pattern

FIGURE 1.7. Value of children and fertility in Zorse, Ghana. The number of children surviving is positively correlated with the number of live births up to a point when birth intervals become so short that the health of the next-to-youngest child suffers and mortality increases. A survival rate = 1 is shown for reference. Because child mortality decreased after colonization (see text), birth intervals could be shortened and more children born (from B_1 to B_2) without increasing the mortality rate, making possible an increase in surviving children from C_1 to C_2. Therefore, parents chose to shorten the period of postpartum abstinence and have children closer together (shorter birth intervals and increased number of live births) to maximize the number of surviving children. A greater number of surviving children means more food available since children are net producers from about age ten. Adapted from (Cleveland 1986b). © 2013 David A. Cleveland.

connected to the local community and resource base (HCC), but to the international economy. The net result of this migration was a younger age at marriage, which in turn increased fertility (table 1.1).

At any one time about 50 percent of working-age males and 15 percent of working-age females from Zorse and the Upper Region were migrants in southern Ghana for a period of a year or more.[8] Significantly increased dependency ratios mean that as a result of this migration all remaining working-age adults had to support themselves plus four dependents, instead of supporting only three dependents, as would be the case without migration. Remittances by Zorse migrants are equal to only a small fraction of the value of their lost productive labor: 50 kg of grain is required per consumer per year, but the average remittance from migrants was enough to buy only 2.5 kg. Evidence suggests that the net effect of migration on Zorse has been negative; neither labor productivity nor land productivity is likely to compensate for the higher dependency ratio.

What my research in Zorse revealed was a classic neo-Malthusian scenario—HI exceeding the HCC, HCC in turn declining under increasing HI, but HI continuing to increase in terms of population growth (table 1.1). The incentive to have more children was reinforced because increasing population densities were accompanied by a decrease in land productivity—due to erosion and lack of organic matter, labor shortages during periods of critical farm activity, reduction in natural vegetation, and increasingly inadequate food supplies—which led to chronic malnutrition and further eroded the quality of the labor force.

Examples from Zorse and elsewhere, and theories of common property management (chapter 7), support the hypothesis that fertility behavior can be adjusted to HCC in traditional societies. The industrial and scientific revolution led to the illusion of unlimited resources, which has encouraged delinking HI and HCC, under Boserupian assumptions.

5.4. Human Carrying Capacity and Human Impact Gradually Reach Stability: A Slow Learner Scenario

In the slow learner or overshoot and oscillation scenario, HI and HCC fluctuate by increasingly smaller amounts, then stabilize with HI below HCC (fig. 1.4d). This scenario implies that humans learn to regulate HI only when the consequences of not doing so are experienced, as consecutive increases in HCC due to innovation are followed by surges in HI, which decrease HCC.

Norman Borlaug, who received the 1970 Nobel Peace Prize for his leading role in creating the Green Revolution, argued for more rapid learning. He stated in his acceptance address that any "breathing space" won by the huge increase in yields and production resulting from the high-yielding crop varieties, fertilizers, irrigation, and other inputs of the Green Revolution should be taken advantage of by human beings to lower their birth rates: "the frightening power of human reproduction must . . . be curbed; otherwise the success of the green revolution will be ephemeral only." (Borlaug n.d.:30–31).

It now seems clear to many that at the global scale the Boserupian scenario has reached its limits, and the Malthusian scenario has been left behind. We are in an Anthropocene neo-Malthusian scenario—there is overwhelming evidence that HI exceeds HCC. Therefore, we can strive only to successfully speed up the fourth scenario—to be faster learners. It seems that a key part of this learning will be to replace the dominant, empirically unsupported assumption that increasing HI based on unlimited economic growth is possible, as well as its supporting value-based assumption that doing so is a good thing (see chapter 3).

6. CALCULATING HUMAN CARRYING CAPACITY: WATER FOR RICE FOR ENERGY

In the previous sections we saw how different sets of assumptions about HI and HCC affect beliefs about the zone and about what human response should be. In this section I provide a simple illustration of how to calculate HCC by dividing available resources by the human requirement for those resources, using the example of water needed to grow irrigated rice to provide calories for human energy (fig. 1.8).[9] By highlighting the critical role of efficiencies and assumptions, this exercise helps demonstrate both the range of possibilities for dealing with the zone and the arbitrariness of any given estimate of HCC.

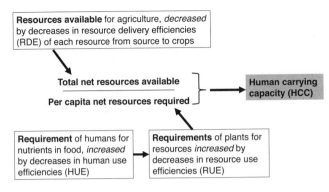

FIGURE 1.8. General method for estimating human carrying capacity. © 2013 David A. Cleveland.

In its most basic form, the estimate of HCC is:

HCC (number of people) =

$$\frac{\text{(Availability of resources for food production / time period)}}{\text{(Requirement per capita for food, in terms of agricultural resources needed to produce it / time period)}}$$

However, it would be very difficult to obtain the data needed for this formula. It is much easier to calculate HCC for a critical resource and a critical requirement.

HCC (number of people) =

$$\frac{\text{(Availability of the critical resource for producing the critical human-required nutrient / time period)}}{\text{(Requirement per capita for the critical nutrient in amount of critical resource needed to produce it / time period)}}$$

Figure 1.8 illustrates the variables used for calculating HCC.

6.1. Efficiencies

The HCC calculated here is the result of dividing the requirement for a critical resource into the availability of that resource, subject to three key efficiencies affecting the movement of the resource from its origin to its final human output. These efficiencies are: the resource-delivery efficiency (RDE) of delivering the resource to the crop (e.g., water from river to root zone of the crop plants), the resource-use efficiency (RUE) of the crop plants in converting the resource to required food (e.g. water into calories contained in rice), and the human-use efficiency (HUE) of humans converting food into the output desired by humans (e.g., calories in harvested rice to work energy) (fig. 1.9).

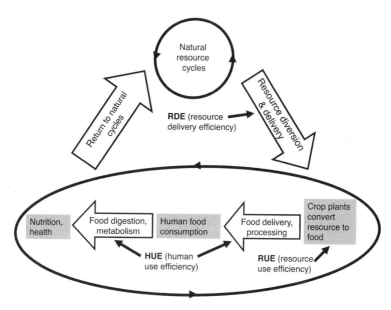

FIGURE 1.9. Three key efficiencies affecting human carrying capacity. © 2013 David A. Cleveland.

The base availability of the resource has to be adjusted by the RDE, and the base requirement has to be adjusted by the HUE and by the RUE:

$$HCC = \frac{\text{(Base availability of the critical resource per time period, for producing critical human required nutrient)} \times \text{(RDE)}}{[\text{(Base human requirement for the critical nutrient per time period)}/\text{HUE}]/\text{(RUE)}}$$

Efficiencies are ratios of output to input, and for RDE and HUE, because output and input are measured in the same units, the maximum efficiency is 100 percent—it is impossible to extract more of a given resource than you put in.[10] To use an estimate of efficiency in a formula it must be in decimal form (e.g., 70 percent = 0.7). However, when output and input are in different units, such as RUE, efficiencies cannot be expressed as a percent, but the ratios for different scenarios can be compared. HCC decreases as efficiencies decrease, either as a result of decreasing resource availability in the numerator or increasing requirement in the denominator.

6.2. Assumptions

Even though estimates of HCC can be useful in providing insights into the determinants of human-environment interactions, they are based on assumptions that make them quantitatively ambiguous:

- that the diversion of natural resources in the amounts used in the calculation do not negatively affect ecosystems or society in the short or long term, when in fact this is difficult to determine;

- that consumption of food is done in a way that does not negatively affect ecosystems or society—in other words, that it is equitably distributed and consumed in forms that are nutritious and satisfying; and

- that the resource chosen, the nutrient chosen, and the crop chosen (irrigation water, energy, and rice in the example) are representative of other resources, nutrients, and crops. However, we know that many resources are required for food production, that many nutrients are required for human health, and that crop species and varieties differ in the amount of nutrients they contain and their efficiency in converting resources into those nutrients.

6.3. An Example

In this and the following sections I show how to calculate the HCC illustrated with the example of water needed to grow irrigated rice to provide kilocalories for human energy (fig. 1.10) and how changing the values of key variables can dramatically change the final estimation of HCC.

$$\text{HCC (billions of people)} = \frac{\text{Availability of water (km}^3\text{) for irrigating rice year}^{-1}}{\text{Requirement of water (m}^3\text{) to grow rice providing kilocalories required person}^{-1}\text{ year}^{-1}{}^{11}}$$

6.4. Availability of Resources

The availability of resources for production is estimated based on those available given current technology, infrastructure, and social organization. Increasing their quantity in the future depends on changing one or more of these parameters. The movements and modifications of that resource to make it immediately available to plants will necessarily reduce its quantity or quality or both, in turn reducing the potential for future food production. This is measured as the resource-delivery efficiency (RDE):

$$\text{RDE} = \frac{\text{Amount of resource delivered to plant}}{\text{Amount of resource extracted from natural system}}$$

For irrigation, RDE is referred to as the irrigation efficiency (IE), and because it is always less than 1, it reduces availability from base (extracted) levels.

$$\text{Availability} = (\text{Base availability yr}^{-1}) \times (\text{IE}), \text{ where IE} = \frac{\text{Water delivered to root zone}}{\text{Water extracted from river, lake, or groundwater aquifer for irrigation}}$$

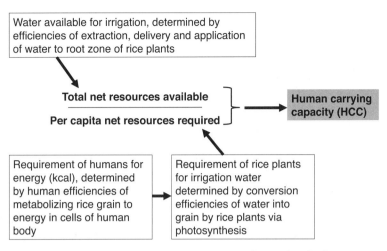

FIGURE 1.10. Estimating human carrying capacity for one key food production resource: irrigation water for growing rice for human energy requirements. © 2013 David A. Cleveland.

The total volume of water on Earth is about 1,400 million km³, of which only 2.5 percent, or about 35 million km³, is freshwater. For this example I will use an arbitrary estimate of 9,687 km³ of water per year available for withdrawal for agriculture globally. The two main components of IE typically used are conveyance efficiency (CE) and application efficiency (AE). Examples of factors that decrease CE are evaporation and leakage from canals, breakage of canals, and extraction from canals for other uses, such as household washing. We can assume for this example that CE also includes efficiency of extraction. Factors that decrease AE include unevenness of the field, so that there are high spots that don't get enough water and low spots that get too much, loss of water below the root zone, in animal burrows, and evaporation. Multiplying efficiencies results in lower overall efficiency—for example, if CE = 0.5 and AE = is 0.6 (typical for small-scale, traditionally based irrigation), then IE = 0.3.

$$\text{Availability} = (\text{Base Availability yr}^{-1}) \times [\text{IE} = (\text{CE}) \times (\text{AE})]$$
$$= (9,687 \text{ km}^3 \text{ yr}^{-1}) \times (0.5) \times (0.6)$$
$$= 2,906 \text{ km}^3 \text{ yr}^{-1}$$

6.5. Requirement for Resources: Human Nutritional Requirement

On the requirement side, we have to first determine what individual (per capita) requirements are for food in terms of the recommended dietary allowances (RDAs) for critical nutrients—most often energy (measured as kilocalories, or kcal) and protein (measured in grams). This is a scientifically complex topic with many political and ethical aspects, and hence it is very controversial.

For calculating HCC we will use the conservatively low round number of 2,000 kcal per person per day. Then we adjust the gross human requirement for the critical resource by the efficiency with which the harvest is transported from the field; processed, cooked, and eaten; and ultimately converted to usable nutrients by the human body. Since all of these steps reduce the overall quantity and sometimes the quality of the nutrient, this process will increase the effective requirement and ultimately decrease the HCC. We refer to this as the human-use efficiency (HUE):

$$HUE = \frac{\text{Amount of food (nutrient) metabolized by humans}}{\text{Amount of food (nutrient) harvested}}$$

Factors that contribute to reducing the amount of a nutrient that is metabolized, proceeding from harvest to human metabolism, include:

- harvest and postharvest losses: grain, tubers, and so on left in the field, eaten by insects, rodents, or microorganisms
- losses in transportation
- losses in processing and storage
- processing that reduces nutrient availability
- health-related reductions in digestive and metabolic efficiency
- conversion of plants to animal food

It is conceptually useful to divide HUE into two components: from field to consumption (field to mouth), and from consumption to final use by the body (metabolism). To estimate the effect of HUE on resource requirement, we divide the base requirement by HUE, which increases the requirement.

Requirement person^{-1} for food, in terms of critical nutrient
= (Base human requirement for the critical nutrient) / (HUE)

For our example, if we assume that the efficiency of human metabolism is 0.75 and the base field mouth efficiency is 0.80:

Requirement (kcal)
= [(2,000 kcal day^{-1} person^{-1}) × (365 days yr^{-1})] / [(0.75) × (0.80)]
= 1,216,667 kcal person^{-1} yr^{-1}

6.6. Requirement for Resources: Conversion of Resources to Food

The third major step before calculating the HCC is to estimate the amount of a resource required to produce a particular amount of human nutrient. We do this by working backward, converting the human requirement for food in terms of the critical nutrient into the amount of critical resource required to produce it—for example, by converting the requirement for energy in kcal into kg of water required to produce those kcal in harvested rice grain. Once the resource is spatially available to the plant—for example, water or nutrients in the root zone—it must be taken up and converted into the food required by humans, the resource-use efficiency (RUE).

RUE =

$$\frac{\text{Amount of food (nutrient) produced by the plant}}{\text{Amount of resource used by the plant}}$$

To estimate the RUE we typically begin by calculating the amount of the nutrient in a kg of total aboveground dry matter (TDM), then divide this by the amount of water required to produce 1 kg TDM, which we assume is 500 kg water. To convert TDM to grain we multiply by the harvest index (HI), a measure of the proportion of TDM that is grain (edible portion). For our example, if we assume that HI = 0.5, then 1 kg TDM = 0.5 kg grain. If there are 3,500 kcal per kg of grain, then 1 kg TDM = 1,750 kcal.

kcal (kg TDM)$^{-1}$ = [(1 kg TDM) (HI = 0.5) (3500 kcal kg^{-1})]
kcal (kg TDM)$^{-1}$ = 1750 kcal

Finally, we divide the kcal contained in 1 kg of TDM by the 500 kg of water required to grow 1 kg TDM:

RUE =
= [1750 kcal (kg TDM)$^{-1}$] / [500 kg water (kg TDM)$^{-1}$]
= 3.50 kcal kg^{-1} water

6.7. Calculating Human Carrying Capacity

Putting the three parts of the equation together:

HCC =

$$\frac{\text{Availability per year}}{\text{Requirement per year [(Human nutritional requirement) /(Conversion efficiency of resource to nutrient)]}}$$

For our example:

Availability
= 2,906 km^3 water yr^{-1}

TABLE 1.2. Estimating human carrying capacity

Human carrying capacity (HCC) variables	Units	Text example	Alternate scenarios: Boldfaced cells show modified assumptions			
			33% of calories consumed as animal products, with 25% efficiency	Increase conveyance efficiency by 50%	Increase HI by 10%	20% decrease in available water
HCC (billions) Availability of resource yr^{-1} / Requirement of resource person^{-1} yr^{-1}	Billion people	8.4	6.3	12.5	9.2	6.7
HCC (individuals)	People	8.4E+09	6.3E+09	1.3E+10	9.2E+09	6.7E+09
Base level availability	(km^3 water) yr^{-1}	9,687	9,687	9,687	9,687	7,750
RDE (resource delivery efficiency; irrigation efficiency)		0.30	0.30	0.45	0.30	0.30
Availability of critical resource — Conveyance efficiency		0.50	0.50	**0.75**	0.50	0.50
Application efficiency		0.60	0.60	0.60	0.60	0.60
Availability of water	(km^3 water) yr^{-1}	2,906	2,906	4,359	2,906	2,325
Availability of water	(m^3 water) yr^{-1}	2.9E+12	2.9E+12	4.4E+12	2.9E+12	2.3E+12

Base caloric requirement per person per year		$\text{kcal person}^{-1}\text{ yr}^{-1}$	730,000	730,000	730,000	730,000	730,000	730,000
Base caloric requirement per person per day		$\text{kcal person}^{-1}\text{ day}^{-1}$	2,000	2,000	2,000	2,000	2,000	2,000
HUE (human use efficiency) Food use in body efficiency			0.60	0.45	0.60	0.60	0.60	0.60
			0.75	0.75	0.75	0.75	0.75	0.75
Net field to mouth efficiency			0.80	0.60	0.80	0.80	0.80	0.80
Nutrient needed to be grown each year	Base field to mouth efficiency		0.80	0.80	0.80	0.80	0.80	0.80
	Animal field to mouth efficiency		1.00	0.75	1.00	1.00	1.00	1.00
Proportion of calories fed to animals			0.00	**0.33**	0.00	0.00	0.00	0.00
Grain-calorie to animal-calorie efficiency			0.00	**0.25**	0.00	0.00	0.00	0.00
Requirement of critical resource needed to supply requirement for critical nutrient (kcal) per person each year	Nutrient needed to be grown each year	$\text{kcal person}^{-1}\text{ yr}^{-1}$	1,216,667	1,616,833	1,216,667	1,216,667	1,216,667	1,216,667
	RUE (resource use efficiency, per kg water)	$\text{kcal (kg water)}^{-1}$	3.50	3.50	3.50	3.50	3.85	3.50
	Resource input for dry matter	$\text{(kg water) (kg TDM)}^{-1}$	500	500	500	500	500	500
Resource use efficiency (RUE)	Calories in 1 kg dry matter	$\text{kcal (kg TDM)}^{-1}$	1,750	1,750	1,750	1,750	1,925	1,750
	HI (harvest index, grain per total dry matter)	$\text{(kg grain) (kg TDM)}^{-1}$	0.50	0.50	0.50	0.50	**0.55**	0.50
	Calories per 1 kg grain	$\text{kcal (kg grain)}^{-1}$	3,500	3,500	3,500	3,500	3,500	3,500
	Resource use efficiency (per cubic meter of water)	$\text{kcal (m}^3\text{ water)}^{-1}$	3,500	3,500	3,500	3,500	3,850	3,500
Requirement of water		$\text{m}^3\text{ water person}^{-1}\text{ yr}^{-1}$	348	462	348	348	316	348

Requirement

$$= [(1{,}216{,}667 \text{ kcal person}^{-1} \text{ yr}^{-1}) / 3.5 \text{ kcal kg}^{-1} \text{ water})]$$

$$= 348 \text{ m}^3 \text{ water person}^{-1} \text{ yr}^{-1}$$

HCC

$$= (2{,}906 \text{ km}^3 \text{ water yr}^{-1} \text{ available}) / (348 \text{ m}^3 \text{ water person}^{-1} \text{ yr}^{-1} \text{ required})$$

$$= 8.4 \text{ billion people}$$

6.8. Sensitivity Analyses

Table 1.2 summarizes the above calculation in the "Text example" column, with "Alternate scenarios" columns illustrating the change in HCC due to modifying the assumptions of the text example for the highlighted variables. This sensitivity exercise shows that HCC varies widely with seemingly small differences in estimates of the values of variables in the equation. The first scenario assumes that 33 percent of the required calories are consumed as animal foods, which are produced by converting the calories in rice to calories in animal foods with only 25 percent efficiency. The large reduction in HCC points out the large contribution that animal foods make to total HI, and the potential for diet change to reduce impact, as discussed further in chapter 8. The second scenario shows the importance of water management and the need for effective physical and social means of efficiently delivering and using resources, as discussed further for irrigation in chapter 7. Increasing harvest index is the changed assumption in the third scenario, which has been a major factor in the ongoing domestication and breeding of crops, and especially important in the Green Revolution, as discussed in chapters 2 and 5. The last scenario shows the dramatic decrease in HCC that results from a 20 percent decrease in the availability of water and highlights a major challenge of global climate change for agrifood systems, as discussed in chapter 8.

These estimates of HCC and the way they change with differences in assumptions about the current and future realities are useful even though they are based on uncertain data and arbitrary assumptions. They provide us with some conceptual understanding of how to estimate the number of people the Earth can support, reinforcing the notion that there is a finite amount of resources available to support the human population. They also give us insight into the *efficiencies* in the relationships between variables determining the HCC, including how HCC varies with change in these variables. Finally, estimating HCC can help to increase awareness of the zone and stimulate discussion to figure out how we want to deal with it, given that the growth of HI is so great that just increasing efficiencies will probably not be sufficient to help us step back from the zone (see chapters 8 and 9).

TWO

Agricultural Revolutions

1. INTRODUCTION

Imagine waking in the morning, opening your eyes to see a sky beginning to lighten to a clear blue, the light of the rising sun filtered through the leaves of acacia trees towering over you, the smell of soil and plants and the remains of last night's fire in your nose. You rise, walk a short distance away to relieve yourself behind another acacia tree, and return to the camp where the members of your extended family are beginning to stir. You share bits of cooked tubers and meat from the evening's meal and begin to plan your day—the routes you will take, walking in the savanna searching for edible fruits and tubers and hunting small animals.[1]

This was a typical day in the life of our species, *Homo sapiens,* for the first 190,000 years or so since its appearance approximately 200,000 years ago. And then, after eight thousand generations of feeding ourselves by foraging, we slowly, hesitantly, and unconsciously started a revolution—we began to grow our food. We were becoming farmers, and as farming spread, foragers were pushed to the most marginal lands, and their numbers decreased until today they make up a fraction of one percent of the world population. Our growing dependence on cultivated food moved us into a revolutionary trajectory that we have followed with increasing speed and that has changed us and just about everything else on this planet. The way we obtain our food and what we do to it before we eat it has been a major part of the history of our species. As a result of agriculture, our lives are dramatically different than they were during the first 95 percent of our history spent as foragers.

The human invention of agriculture is called the Neolithic revolution, and it set off a race between the unlimited biological potential for human reproduction and consumption and the ability of humans to increase food supply for a growing population by continuing agricultural innovations. We are now living at the heart of the most revolutionary changes to our agrifood system since the beginning of agriculture—coming to terms with the realization that we need to reduce our global agrifood system's impact on the environment and to increase its contribution to happiness for all the world's still-growing population of more than seven billion people.

2. AGRICULTURAL REVOLUTIONS

Dividing up history is always an arbitrary and contentious endeavor, and like any modeling of reality, it is justified only if it ultimately leads to deeper understanding and more effective practice. Figure 2.1 is a timeline of six major revolutions that have occurred in the history of agriculture: Neolithic, global, industrial and scientific, Green, biotechnology, and sustainability (table 2.1). I define agricultural *revolutions* as periods of relatively major changes that mark the overall *evolution* of agriculture. Of course, they are present in different degrees in different places in the world. *All of these revolutions are still ongoing; they cannot be ranked in any absolute way; and many different forms of agriculture combining different aspects of these revolutions exist contemporaneously.*

Each of the first five revolutions are Boserupian ways of dealing with "the zone" (as described in chapter 1). That is, they are supply-side approaches, because they focus on increasing human carrying capacity (HCC) through sociocultural changes and more intensive management of growing environments and crop evolution to achieve higher rates of food production. The sustainability revolution is fundamentally different than the others. It is an example of scenario 4 for dealing with the zone—that is, it uses demand-side (and distribution) approaches, is based on the assumption that humans have overshot the Earth's HCC, and is focusing on finding ways to reduce human impact (HI). The sustainability revolution consciously questions the desirability of the overall effect of the other revolutions and is an important focus of this book, which I will cover in depth in chapters 3 and 4. In this chapter I briefly describe the first five revolutions.

2.1. Three Key Relationships in Agriculture

Agricultural revolutions affect the three key human relationships:

- *management of other species,* most dramatically by controlling their evolution through selection of planting materials, which tends to increase usefulness to humans, as well as the codependence of humans and domesticated plants, and today includes breeding and genetic engineering of plants;
- *management of ecosystems,* including biogeochemical, hydrological, and energy cycles, to create environments supportive of domesticated plants, accompanied by increasing dependence of humans on these managed ecosystems, and today

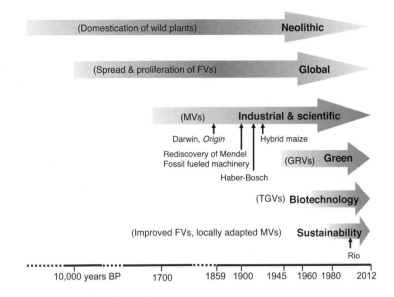

FIGURE 2.1. Major agricultural revolutions and crop varieties dominant in each. © 2013 David A. Cleveland.

includes massive diversion of water, fossil fuel, and other resources to the agrifood system; and

- *management of other humans,* to support the first two changes, including social organization of labor at the household, community and higher levels, and the production of technology; most often this eventually leads to increasing social hierarchy and changes in cultural values, population structure, nutrition, and health, and today includes structures and ideas that result in global trade in a wide variety of foods, as well as highly unequal access to food and high levels of malnutrition.

Changing ratios of inputs and outputs are key components of these three human relationships: *input/input*—for example, decrease in (human energy in/animal energy in); *output/output*—for example, increase in (grain harvested/total aboveground plant biomass), also known as harvest index; and *output/input*—for example, increase in (energy in harvested food out/energy in human labor in). However, the results are often mixed, and increased value to humans in one area from a change in ratios may be accompanied by decreased value in another area. For example, an increased harvest index means higher yields, but it also often means that humans have to spend more time and other resources nurturing the crop, which can result in decreased (harvest out/resources in).

While these ratios of outputs and inputs have evolved over time in different ways in different places, they usually result in changes that increase the direct benefit to humans, at least over the short term, and at least for those in control of the agrifood system. It is the

TABLE 2.1. Agricultural Revolutions

Revolution	Date began	Description
Neolithic	Up to 12,000 years ago	Domestication of plants and animals, increasing management of agroecosystems for domesticates and supporting social and cultural changes in humans. Supported increasing human fertility and population growth rates, but also poorer health. First domestications were variable in time and space.
Global	Soon after Neolithic	Local and global spread of domesticates and farming knowledge and technologies. Led to great proliferation of locally adapted farmer crop varieties (FVs) and conversion of native vegetation to cropland and pasture. European global diaspora and colonization beginning in sixteenth century was major episode. Current globalization led by multinational corporations and industrial country governments seeking new markets and raw materials, especially in the Third World.
Industrial and scientific	Late 18th c. C.E.	Use of increased power of science and technology for discovery and manufacturing of inputs, including the Haber-Bosch process for nitrogen fertilizers, and machinery using fossil fuels. Application of biological sciences to plant breeding and agronomy. Led to great increases in efficiency of labor and decreases in efficiency of energy, with fewer people needed for growing food.
Green	1960s	The promotion of industrial agriculture in the Third World, including the introduction of modern crop varieties (MVs) with high yields due mainly to high harvest index under optimal conditions created by increased inputs (irrigation water, inorganic fertilizers and pesticides, fossil fuel). Dwarf, disease-resistant wheat and rice were first widely adopted Green Revolution varieties. Especially effective in southern Asia and Latin America; did not have much effect in Africa.
Biotechnology	1980s	Direct and targeted intervention in the genetic makeup of plants (and animals), bypassing natural reproductive processes. Major effort beginning in the 1980s to engineer transgenic crop varieties (TGVs); first major commercialized TGVs were introduced in 1993 and widely planted in the United States in 1996. Currently being aggressively promoted globally and often strongly resisted, especially in the Third World and Europe; AGRA (Alliance for a Green Revolution in Africa) is a major effort promoting TGVs in Africa.
Sustainability	1980s	Search for distribution and demand-side alternatives to previous revolutions (focused on increasing equity and decreasing demand), which were supply-side (focused on increasing production) and increased yields and total food supply in tandem with the growing population. Response to the growing evidence for the environmental, sociocultural, and economic unsustainability of past revolutions.

unintended, and often initially unnoticed, longer-term negative effects of agrifood systems on the environment and on people that have triggered the sustainability revolution.

2.2. Is Agricultural Evolution Unilineal or Multilineal?

The changes that have occurred between the advent of farming and the contemporary industrial world's genetically engineered crops, immense laser-leveled fields and cities of millions depending on food grown by a very few, may not represent an inevitable or unique path. In fact, the agriculture practiced by the greatest number of farmers today—that is, small-scale agriculture in the Third World—is very different than the industrial world agriculture that is commonly thought of by agricultural scientists and development professionals as the logical and inevitable end product of agricultural evolution. As mentioned in the introduction, Third World agriculture covers more area and involves many more people directly in food production than industrial world agriculture; more than two billion people live on almost five hundred million small-scale (less than 2-ha) farms, including half of the world's undernourished people and the majority of people living in absolute poverty, and produces up to 80 percent of the food consumed in the Third World (IFAD 2011:1,62). In addition, a growing number of farmers in the industrial world are doing things differently—growing organically and on a small scale, and selling directly—part of a growing localization movement that has sprung up as an alternative to the mainstream (see the introduction and chapter 9).

This suggests that agricultural evolution is a multilineal process producing a bush with many coexisting forms, not a unilineal process producing a tree with one terminal endpoint, with the side branches eventually dying off. In the nineteenth century, many evolutionists thought of biological evolution as a tree with Europeans as the terminal branch and other human races as inferior species destined to either evolve into white Europeans or die off (Desmond and Moore 2009). Indeed, one of Charles Darwin's great contributions was to propose the theory, which he supported with a massive amount of data, that evolution was not unilineal, and the only illustration in his groundbreaking *Origin of Species* is a bushy tree of life that emphasizes multilineal evolution (Darwin 1859:116 ff.).

The simplistic ideas of unilineal biological evolution popular in nineteenth-century Europe were replaced by the ideas of Darwin and others that more accurately reflected the theory and data and were influenced by emerging values about egalitarian human social relations (Desmond and Moore 2009).[2] As discussed in chapter 1, biological evolution is not teleological or unilineal, but sociocultural evolution can be both (though not necessarily either), as I will discuss below. Yet, ironically, the mainstream agriculture that promotes the most novel technologies, such as genetic engineering, remains dominated by antiquated nineteenth-century ideas of social evolution. These ideas rationalize and justify the domination of commercial agrifood systems in the industrial world and their extension to the rest of the world.

A central hypothesis of this book is that there are many alternatives to the mainstream agrifood system. As mentioned in the introduction, these alternatives may promote building on traditional knowledge and technologies as the best way to address the food crisis. Which of these options are "best" will depend on changing environments and societies, and on the value-based criteria for judging them, as discussed further in chapters 3 and 4. We should be open to thinking about industrial agriculture as coequal evolutionarily with other forms, including continuously adapting, small-scale, traditional agriculture, and not as superior to and destined to replace it.

3. THE BEGINNING OF AGRICULTURE

Why, when and *where* did agriculture begin, *what* did it entail, and *who* was responsible? To put this topic in broad evolutionary perspective, agriculture began not with humans, but with single-cell organisms, and later insects. Humans invented agriculture fairly recently, after spending most of our life as a species foraging for plants and animals.

3.1. Agriculture Before Humans

Looking at the evolution of agriculture in nonhuman species can shed light on the general principles underlying this rare phenomenon and help us to understand the goals of more sustainable human agrifood systems. If we define farming broadly as the management of one species by another that uses the former for food, probably the earliest type of what could be called farming was by unicellular organisms. This practice has been documented in a contemporary species of slime mold, the social amoeba *Dictyostelium discoideum* (Brock et al. 2011). While farming clones of these molds do not domesticate their bacterial food supply, they have evolved to stop feeding before they exhaust this food supply, and they place some of the species of bacteria they feed on in their spore sacs, which can then be used to colonize a new location devoid of the preferred edible species. Nonfarming clones do not do this. Farming and nonfarming clones coexist sympatrically, in part because the bacteria that they both feed on are not domesticated, which limits the farming amoeba clones' potential to evolve greater fitness relative to nonfarming clones. In contrast, human farmers and their domesticated crops replaced foragers relatively rapidly, in part because the crops were specifically adapted for human use by domestication, as I will discuss below.

The most sophisticated and evolutionarily successful farming evolved among the insects, beginning about fifty million years ago—once in ants, once in termites, and seven times in beetles (Mueller et al. 2005). This involved cultivation of another species, as opposed to the transportation done by the amoeba. Out of the first ant farming, four additional types of farming evolved, the most successful in terms of numbers being the leaf-cutter ants. Starting about eight to twelve million years ago, leaf-cutter ants began to carry pieces of leaves to their nests, which they inoculated with a domesticated fungal clone and then cultivated by removing diseased growth. The leaf-cutters were the first farming ants to domesticate another species—the fungi they cultivate cannot live freely—and this feat allowed leaf-cutters to be evolutionarily very successful, becoming the top herbivores in the New World tropics (Schultz and Brady 2008:5438).

Some researchers have suggested a parallel between humans and ants in terms of the challenges of growing genetically narrowed domesticates in monoculture—modern potato varieties by humans and varieties of fungi, which have probably become inbred, by the most specialized leaf-cutters (Schultz et al. 2005:178–181). Both the potato and fungus are propagated as clones and are therefore very genetically uniform, making them especially vulnerable to disease. To cope with this, the ants protect their fungal gardens from the environment, monitor them to control pathogens, manage a range of microbes that help suppress disease, and even introduce other, nonclonal, more genetically diverse cultivars to increase their ability to respond to diseases affecting their food clones (Mueller et al. 2005). Thus, a complex of four symbionts appear to have coevolved: (a) the leaf-cutter ants, (b) the group of fungi that the ants cultivate in gardens and that are their only source of food, (c) a parasitic fungus that infects the fungal gardens, and (d) a symbiotic bacteria that grows on the ants and at least partially controls the fungal parasite (Schultz and Brady 2008).

The critical difference in the way ants and humans have dealt with pathogens of their cultivated species and other similar challenges of farming is that ants and their symbionts had five million years to evolve biological ways of adapting, more than four thousand times longer than humans. The most specialized, the leaf-cutter ants, have evolved adaptation to their cultivated fungal diet, both losing the genetic ability needed as foragers and acquiring genetic adaptation to the new diet (Nygaard et al. 2011). Humans have not had time to evolve major biological adaptations to their symbionts but have had to rely on sociocultural ways of adapting, using sophisticated cognitive abilities that ants lack. And they have had to adapt over a much shorter time period: about twelve thousand years for agriculture as a whole, only several decades for large areas of genetically uniform monocropped fields, and even less time still for some of the specific genetic novelties used in recent transgenic crop varieties (TGVs).

3.2. Foragers

Humans were foragers until very recently in history. If the genus *Homo* originated about 2.5 million years ago, and the species *H. sapiens*, the only surviving species in the genus, originated about two hundred thousand years ago (Garrigan and Hammer 2006; Schwartz 2012), and agriculture first began about twelve thousand years ago, then humans have been foragers for all but about 0.5 percent of our history as a genus and 6 percent of our history as a species. To get an intuitive feeling for what this means, the period during which we as a genus and species have been farmers is proportional to 17 seconds in one hour, and 3.6 minutes in one hour, respectively.

Humans evolved biologically during the period before agriculture, known as the Paleolithic age, which began about 2.5 million years ago with the first stone tool making by human's *Homo* ancestors, such as *Homo habilis* (e.g., Schwartz, 2012). A combination of factors—including meat consumption, cooking of plant and animal foods that makes more energy and nutrients available and digestible, and cooperative social organization allowing greater energy efficiency of foraging—are now hypothesized to have contributed to the evolution of *H. sapiens*' large brains (Carmody and Wrangham 2009, Pennisi 1999).

Foragers have been pushed out of the more favorable environments by farmers, so today foragers live in relatively inhospitable environments and are in contact with farming groups.

I therefore use the past tense when referring to them, even though much of our understanding of foraging comes from studying relict surviving groups, because in general foragers were probably much better off before the spread of agriculture.

Foragers were very mobile, following the abundance of resources as affected primarily by the seasons, so in order to be successful they had to understand environmental patterns and cycles and to anticipate changes. In fact, some researchers suggest that it was the working knowledge of plants and ecosystems acquired by foragers through their gathering and hunting and landscape management that established the basis for agriculture (Harlan 1992). There is evidence that foragers have had major effects on ecosystems—for example, using fire to manage vegetation for food and fiber, as in California, United States (Anderson 2005), and New Zealand, where foragers who arrived there about eight hundred years ago affected vegetation assemblages to the extent that they lacked the adaptive capacity to recuperate, resulting in significant changes in dominance of preferred species, freshwater biochemistry, and patterns of water and soil movement (McWethy et al. 2010).[3]

Forager group sizes were flexible, alternating between congregation and dispersal, especially in response to seasonal variation in the concentration of resources. Maintaining mobility meant having few material goods and allowed movement away from social problems. Forager societies had strong social enforcement of egalitarianism, including a strong emphasis on sharing, which helped keep differentiation in status (wealth, power, and prestige) to a minimum. These same forces and others were used to minimize violence or preempt it when conflict arose, although homicide rates in forager groups have been observed to be the same as those in industrialized cities (Boehm 2012). All of these behaviors were evolutionarily advantageous and enabled foraging groups to function successfully in their environments with the technologies at hand.

Foraging was formerly thought to be a very difficult way of life, involving lots of work for meager food and a short lifespan—with Hobbes's famous description in the *Leviathan* of human life in a state of nature as "solitary, poor, nasty, brutish and short" often assumed to apply—and therefore seen to provide a logical explanation for the beginning of farming as an escape from foraging. For example, in one of most widely viewed anthropological films—*The Hunters*, made by John Marshall in the 1950s about the Ju/'Hoansi (formerly known as the !Kung San, or the "Bushmen") of the Kalahari desert in southern Africa—phrases like "far from generous land" and "ceaseless labors of women" are common.

In the 1960s anthropologists started measuring and counting foragers' activities, and their data led to a major shift in thinking (Lee 2012). It began to be accepted that Ju/'Hoansi and other foragers worked less and had a better diet than many farming groups, even though Richard Lee's seminal observations were made in the 1960s, a time of severe drought in the relatively marginal environment of the Kalahari desert. In comparison, farmers eat less desirable food, work longer at more monotonous and harder jobs, experience a higher rate of disease, and have a less secure food supply. Because of these observations, some researchers even view foragers as the "original affluent society" (Sahlins 1972).

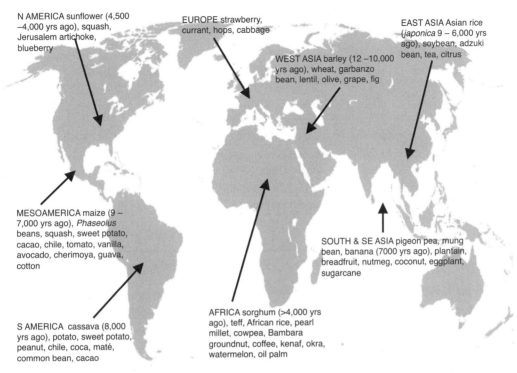

FIGURE 2.2. Areas of origin of some important crops. Based on data from (Simmonds and Smartt 1999, Purugganan and Fuller 2009, Meyer et al. 2012); base map (http://en.wikipedia.org/wiki/File:BlankMap-World-large-noborders.png) from the Wikimedia Commons (http://commons.wikimedia.org/wiki/Main_Page), a freely licensed media file repository.

This lends support to the idea that the transition to agriculture in many cases may have been a back-and-forth process. There are a number of documented examples of forager or pastoral groups who plant seeds and return later for the harvest, but do not do this consistently or depend on it for the majority of their food, and do not give up their foraging way of life, returning to foraging when they are able to, either as a supplement in normal times or when crops fail (Lee and Daly 2000).

3.3. The Neolithic

Because foragers were managing plants and ecosystems to increase production of preferred plant products well before the beginning of agriculture, it seems unlikely that a lack of basic knowledge prevented humans from taking up agriculture earlier. So what caused humans to change a mode of subsistence that had supported their spread over thousands of generations, from east Africa into much of the rest of the world? Because agriculture began in many locations in the world, in some places as early as twelve thousand years ago, with different domestication patterns for different crops (Simmonds and Smartt 1999, Smith 2001) (fig. 2.2), there are surely unique factors affecting the time in each case. However, the fact

TABLE 2.2. Human Population Growth Rates and Agricultural Revolutions

Period	Annual growth rate or average of estimated range (r)	Change in r since last period
250,000 BP–12,000 BP, foraging	0.0009	
12,000–2,000 BP, Neolithic, and global agricultural revolutions	0.07	↑ 75.5 times
2,000BP–17th c., Global and industrial-scientific revolutions	0.07	stable
17th c.–1950, Industrial-scientific to Green Revolution (and public health revolution beginning around 1950)	0.6	↑ 7.6 times
1950–1975, Green Revolution and public health revolution to fertility revolution (world population growth rate peaked at r=2.1 in 1965–1970)	1.9	↑ 2.8 times
2012	1.2	↓-0.4

Source: Averages calculated from data in (Cohen 1995), 2012 from (PRB 2012).

that the Neolithic occurred in these different places in such a relatively narrow time period, compared with the previous period of foraging, strongly suggests common factors as well.

Although I will not discuss in detail the causes for the beginning of agriculture, a key variable in the opinion of many researchers is the relationship between population and food availability, the latter influenced strongly by climate. There seems to be an association between major increases in food availability and increases in population growth rates (table 2.2), yet it is difficult to determine which one leads to the other, and it seems likely that there has been much variability through space and time, that causal arrows probably often point in both directions, and that other variables also come into play. The relationship between the ability to obtain food from the environment and human impact (including population, chapter 1), is key to developing a more sustainable future agrifood system, as discussed in chapters 3 and 4.

It has been suggested that agriculture was not able to develop until the onset of the relatively stable climate of the Holocene epoch (beginning about twelve thousand years ago and continuing until the current Anthropocene epoch) (Atahan et al. 2008). The preceding Pleistocene epoch (about 2.5 million to twelve thousand years ago, and thus roughly coterminous with the Paleolithic age) was marked by intense variability of climate, with repeated glacial advances and retreats. For example, between 50,000 and 11,600 years ago there were no two-thousand-year spans relatively free of large climatic changes on the scale of one hundred years (Feynman and Ruzmaikin 2007).

In contrast, other scholars have argued that agriculture developed as an adaptation to a pattern of short-term climatic variability—for example, in the Near East, where it may have evolved as an adaptation to alternating runs of good and bad years (Abbo et al. 2010). According to this theory, slow climate change would not result in agriculture that includes strategies for both kinds of years, and rapid climate change would destroy agriculture's buffering effect. Indeed, there is increasing evidence from a number of locations for a scenario of fluctuating levels of dependence between farming and foraging that may have gone on for extended periods. One key indicator of the rate of domestication of cereals is the rate at which the nonbrittle rachis (nonshattering) trait increases (Zohary and Hopf 2000). The rachis is the axis on which seeds develop; a brittle rachis is strongly selected for in wild populations because this disperses the seed. When humans started gathering grain, they unconsciously favored the small number of plants with nonbrittle rachises, since the seeds of these plants did not detach as easily. When they planted and harvested this grain year after year, this unconscious selection increased the proportion of plants with nonbrittle rachises (see fig. 2.4 below). The slow rate of increase in frequencies of domestication traits supports this scenario. For example, some estimates based on the archaeological record are that nonbrittle rachises in domesticated barley, einkorn wheat, and rice increased at a rate of 0.03 to 0.04 percent per year, suggesting continued gene flow from wild populations with brittle rachises as humans continued to gather wild grain (Purugganan and Fuller 2009:845). In the lower Yangtze River valley on the east coast of China, rice cultivation began about 7,700 years ago but stopped about two hundred years later, when the site was flooded by the ocean (Zong et al. 2007). Another excavation in the same area also found that farming began at least seven thousand years ago but was limited by frequent flooding and combined with foraging and cultivation of wild rice populations. The environmental changes assumed to result from a greater dependence on agriculture didn't occur until much later, during "a period associated in the lower Yangtze with technological advances in agriculture, and relatively stable hydro-geomorphological conditions" (Atahan et al. 2008:567). Continuing foraging well after the start of farming is also seen as having been a rational strategy in terms of energy efficiency. A study comparing early agriculture with foraging in diverse regions of the world found foraging to be more productive (in terms of net kcal per hour of labor) (Bowles 2011). One possible explanation for this is that eventually the shift to farming over foraging occurred for socioeconomic reasons, including sedentarism, shared labor, and military consolidation.

However, there are other sites where the data suggest that agriculture was taken up when the supply of foraged wild foods became limited by changing climate, as on the coast of Portugal (Dean et al. 2012). It seems possible that the adoption of agriculture occurred in very different ways in different locations and times. Foragers undoubtedly manipulated plants and their environments, as well as other humans, and the Neolithic was based on the knowledge they acquired through long experience. However, agriculture marked a substantial increase in these processes, and *these profound transformations in human relationships with other species, the environment, and other humans began a trajectory of change that continues to this day and helps define our choices for the future.* In the next three sections I describe in more detail the three major changes that marked the Neolithic.

4. HUMAN MANAGEMENT OF PLANT EVOLUTION: DOMESTICATION AND VARIETAL DIVERSIFICATION

Domestication involves genetic changes that adapt plants to the house, or *domus*; through this process a plant becomes dependent on humans for survival (Harlan 1992). Simmonds defines domestication as follows: "A plant population has been *domesticated* when it has been substantially altered [genetically] from the wild state and certainly when it has been so altered as to be unable to survive in the wild" (Simmonds and Smartt 1999:1). That is, full domestication means that a plant cannot survive without humans.[4]

Why does selection for increased yield and good taste often make domesticated plants no longer able to survive on their own and, therefore, dependent on people to manage their growth and reproduction? It is because the genetic information that directed the available resources to enable plants to survive on their own has been replaced via selection with information that redirects resources to the survival of farmers. For example, domesticated plants whose leaves are used for food have often been selected for good taste and nutrition, and through this process they have lost the bad-tasting chemicals that served to repel insects and other herbivores and are thus more dependent on humans to protect them.

Domestication can also be looked at from the perspective of the plants—humans have been "domesticated" by plants in the sense that we can no longer survive without our crops. This is especially apparent at the global scale, where only a small fraction of our present population of more than seven billion could survive as foragers. A key result of domestication is codependence between humans and domesticated plants.

However, whereas humans can select plants with goals in mind, plants cannot do the same with humans, although biologists often use a stylistic shorthand that ascribes evolutionary goals to nonhuman organisms or to the process of natural selection. When popular writers dramatize this shorthand it reinforces misunderstandings about our relationship with our crop plants. For example, when Michael Pollan demonizes maize for its domination of the U.S. agrifood system (Pollan 2006:28–31), it can mislead readers into ascribing malevolent attributes or even intentions to a plant species and seeing some humans and their institutions as somehow less responsible for the negative consequences of this dominance.

4.1. Elements of Domestication

With domestication, humans began to be directly involved in the evolution of plant (and other animal) species—artificial, human selection changed the genetic makeup (allele frequencies) of plant populations over time. This sometimes resulted in the evolution of new, domesticated species, although as already mentioned, many of these species remained capable of hybridizing with their wild ancestors (Ellstrand 2003). Domestication of new species is ongoing (Simmonds and Smartt 1999), and domestication via genetic engineering is being promoted (Gressel 2008). Plants were likely domesticated earlier than animals and were more important to early farmers in terms of total nutrients supplied. In this book I focus on plants, though much of what I say applies to animal agriculture as well.

Three elements necessary for plant domestication, and evolution in general, are (fig. 2.3):[5]

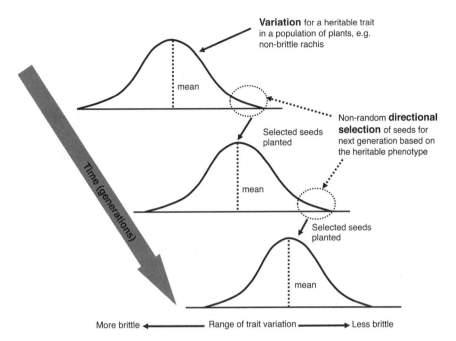

FIGURE 2.3. Cumulative, directional phenotypic selection for a heritable trait changes the population mean for that trait; this can lead to macroevolution if the change is great enough to result in a new species. © 2013 Daniela Soleri. Used with permission.

- *Genetic diversity*—that is, plant variation in heritable phenotypes. *Phenotypes* are all of the physical *traits* of an organism—for example, size, color, or number of seeds produced; drought resistance; or the chemical compounds produced that give the food made from a plant its flavors. Two important phenotypic traits for domestication in many species are the brittleness of the rachis (described earlier) and reduced seed dormancy. Phenotypes are the result of both the genotype and the environment. The *genotype* is the genetic makeup of the plant and is *heritable*—that is, it is passed from generation to generation. The genetic makeup consists of the genes and their different versions, or *alleles*.

- *Phenotypic selection*—the selection of seeds or other propagules that will produce the next generation based on the parental phenotype. To the extent that the selected phenotypic traits are based on the genotype, the phenotype will be inherited; the aspects of the phenotype that are a result of environmental influence will not usually be inherited. Humans are an important selective force in domestication, either directly by selecting plants, or indirectly by managing environments.

- *Time* sufficient for a genetically distinct population (or species) to evolve. While the unit of selection is the individual plant, the result in terms of evolution is changed allele frequency in the population. *Directional selection* can result in cumulative, directional change of allele frequencies through generations. Any change in allele frequencies between generations is *microevolution,* and when this cumulative change over generations leads to a new species it is *macroevolution.*[6]

Figure 2.3 illustrates the situation for sexually propagated crops. The domestication of clonal crops—for example, most fruit trees—is also a process of selecting individuals from within a population, but it may not continue every generation because there is usually no creation of new heritable variation. Instead, propagation is vegetative (clonal)—for example, using root suckers of olives or offshoots of dates. Another kind of vegetative propagation is grafting, which requires more skill. The earliest evidence of grafting is the importing of grape budwood, which occurred in Mesopotamia in 1800 B.C.E., in the Mediterranean area by the first century C.E., and in China by the third century C.E. (Janick 2005:262).

4.2. Genetic Diversity

Genetic diversity is the raw material of evolution, including domestication. Yet the domestication process typically involves a loss of diversity because the first domesticates are derived from a very small sample of the individual plants in their wild relative species. Domestication usually involves the creation of new sister species or subspecies through reproductive isolation from its wild relative.

Genetic *differentiation* is the result of reproductive separation of populations in different environments leading to those populations having different allele frequencies in response to different selection pressures, or selection for unique alleles resulting from mutations in the separate populations. But this reproductive isolation does not usually result in physiological or genetic barriers, so that when the domesticated species and its wild relative have the opportunity, they are often capable of cross-fertilization, a process called *hybridization,* resulting in hybrid plants and potentially hybrid populations (Ellstrand 2003). The result is increased genetic variation, as evidence suggests for the history of maize (Hufford et al. 2012), which is then available for selection (natural and human) to act upon to produce new populations and varieties (this process is explained in more detail in chapter 5).

Cycles of differentiation-hybridization, both between domesticated species and their wild relatives and among domesticated varieties themselves, enhance the potential for the evolution of new populations and species in response to changing selection pressures (Harlan 1992). Farmers have sped up these cycles by moving plants between different biophysical environments and creating new environments through cultivation practices (Harlan 1992). For example, maize was domesticated in Mesoamerica but was subsequently moved out from this center to both North and South America by early farmers. Later, when differentiated populations, such as South American and Mesoamerican varieties, were brought together again, new races (groups of related varieties) of maize resulted (Matsuoka et al. 2002).

Farmers also learned how to promote the creation of diversity by facilitating hybridization through interplanting or encouraging sympatry, as has been suggested that some Mexican farmers do with maize and its wild relative, teosinte (Benz et al. 1990). Farmers also encourage hybridization between different varieties, including between their own local *farmer varieties* (FVs) and *modern varieties* (MVs) created by plant breeders, as documented for pearl millet farmers in Rajasthan, India (Vom Brocke, Presterl, et al. 2002).[7]

4.3. Selection

Selection means the identification, based on phenotypic traits, of some individuals within a population that will be used to propagate the next generation; this means that when there is phenotypic variation the contribution of individuals of one generation to the next generation will not be random, and the selection differential (S) will be greater than zero. Phenotypic selection can have genetic consequences only if variation in the phenotypic trait selected is at least in part genetically determined and can therefore be inherited. If phenotypic variation in the trait selected is not genetically determined, meaning that it is not heritable, the result will be that even though selected seeds or other propagules are phenotypically different than those not selected ($S>0$), there will be no genetic change. This can be the case when farmers select large seed for planting (Cleveland and Soleri 2007a).

If the selected trait is heritable, then selection can result in genetic change between generations (R), and if this selection is directional and continues over generations, it is microevolution (mEv), and over many generations can result in macroevolution of new species (Ev). Figure 2.4 illustrates macroevolution for the brittle rachis trait in the domestication of cereals (compare fig. 2.3, see section 3.3). I will discuss S, R, and mEv in more detail in chapter 5.

It is generally assumed that selection by farmers—in combination with natural selection, gene flow, recombination, and mutation—contributed to the large amount of intraspecific diversity that evolved during the global revolution following domestication. As Norman Simmonds stated, "Probably, the total genetic change achieved by farmers over the millennia was far greater than that achieved by the last hundred or two years of more systematic science-based effort" (Simmonds and Smartt 1999:12), an insight verified by a genome-wide review of maize wild relatives, FVs, and MVs (Hufford et al. 2012). This process was likely unintentional at first, but farmers learned how to intentionally select the plants in a population whose characteristics they most desired.

Phenotypic selection can be classified in terms of the selection agent, and when the agent is a farmer or plant breeder, their intentions in selecting (fig. 2.5). Each of the resulting types of phenotypic selection can have a range of outcomes in terms of S, R, and mEv. *Natural selection*, acting on diversity created by mutation and recombination, accounts for all of the biological diversity that evolved since the beginning of life on Earth until the Neolithic. The genetic changes that define crop domestication and subsequent evolution of diversity in crop species are inextricably linked with changes in selection pressures due to human behaviors in a process called *artificial selection*—that is, humans directing the evolution of plants by selection. *Artificial selection* is both *direct*, a result of human selection of

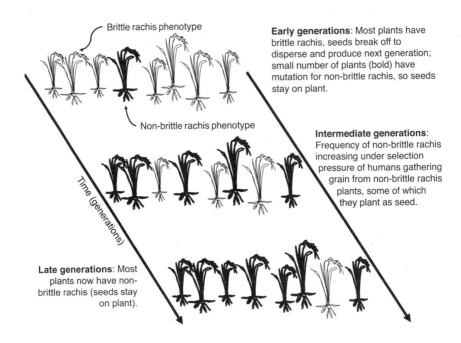

Brittle rachis phenotype

Early generations: Most plants have brittle rachis, seeds break off to disperse and produce next generation; small number of plants (bold) have mutation for non-brittle rachis, so seeds stay on plant.

Non-brittle rachis phenotype

Intermediate generations: Frequency of non-brittle rachis increasing under selection pressure of humans gathering grain from non-brittle rachis plants, some of which they plant as seed.

Time (generations)

Late generations: Most plants now have non-brittle rachis (seeds stay on plant).

FIGURE 2.4. Selection for nonbrittle rachis in domestication of grain crops. © 2013 Daniela Soleri. Used with permission.

planting material, and *indirect,* a result of the environments created in farmers' and plant breeders' fields and storerooms.[8]

It is impossible to determine the type of selection that resulted in past crop evolution, and experts differ on the type they believe was most important. Selection early in the Neolithic may have been mostly indirect or unintentional direct selection. Simmonds and Smartt emphasize indirect selection—"the art of cultivation is perhaps the peasant's most potent contribution" (1999:13). As already mentioned, in southeast China the earliest cultivation of both wild and domestic rice happened about 7,700 years ago, and the evidence suggests that this occurred where farmers were intensively managing coastal wetlands with fire, to control vegetation, and bunds, to control flooding and increase nutrient concentration in fields (Zong et al. 2007). Bringing wild plants into human-modified environments, such as compost heaps near houses, as well as exchanging seeds and other propagules, also facilitated domestication via hybridization, as with *Luecaena* in southern Mexico, and probably with two other important domesticates from that region, agave (*Agave* spp.) and prickly pear cactus (*Opuntia* spp.) (Hughes et al. 2007).

Because many of the early domesticates were grain, legume, and other species whose seeds were eaten, and because selection occurred via the harvesting and replanting of these seeds, direct selection undoubtedly also played a part. Direct selection can be either *intentional* (conscious, based on explicit criteria)—as when farmers select plants to produce a next generation that better fits their criteria (production of edible parts, taste, aesthetics, and

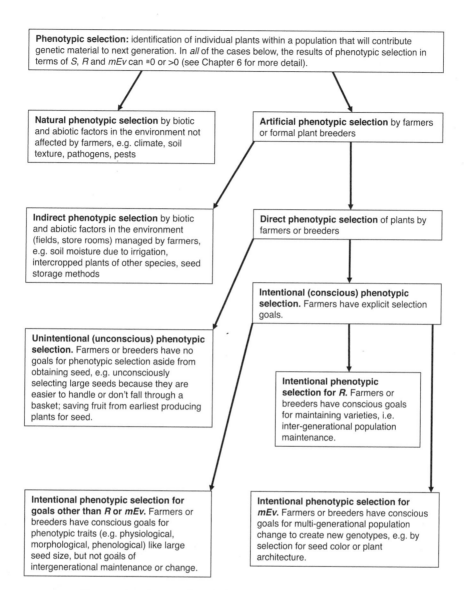

FIGURE 2.5. Phenotypic selection classified according to the agent of selection and intention of the farmer or plant breeder as agent. See text for key to symbols. © 2013 Daniela Soleri and David A. Cleveland. Used with permission.

so on)—or *unintentional* (unconscious, based on implicit criteria)—as when no conscious decision is made about the trait selected for, such as when large seeds are automatically selected because they are easier to handle (Harlan 1992). Intentional selection can be to increase the frequency of desirable nonheritable phenotypic traits in seed to be used for planting (with S the desired outcome); to maintain desired traits, or eliminate undesired ones that arise between generations (with R, genetic change or gain, the desired outcome); or to make a change in the variety (with microevolution, mEv, the desired outcome). The plant breeder Robert Allard has emphasized direct, intentional selection: "The consensus is that even the earliest farmers were competent biologists who carefully selected as parents those individuals . . . with the ability to live and reproduce in the local environment, as well as with superior usefulness to local consumers" (Allard 1999).

4.4. Time

The third ingredient for evolution is time—with enough time, selection on heritable variation in a population results in new traits. Estimates of how long domestication took have often been based on when the domestication syndrome appeared. The crop *domestication syndrome* is a group of shared traits that characterize many domesticated plants in contrast with nondomesticated plants (Gepts 2004, Harlan 1992, Simmonds and Smartt 1999). Its key features include:

- lack of seed dispersal by shattering—for example, via loss of brittle rachis— facilitating harvesting;
- reduced or no seed dormancy, facilitating uniform germination, seedling establishment, and maturity;
- compact growth habit, increasing harvest index and thus increasing yield per plant and decreasing interplant competition;
- reduction of toxins, increasing nutritional value and palatability; and
- coevolution of host plant and beneficial microorganisms.

Genetic evidence for a number of crops suggests that domestication could have occurred in one to two hundred generations, short periods of time relative to the twelve thousand or so years that crop plants have been cultivated (Gepts 2004). Domestication syndrome traits often appear to be determined by a small number of genes with large effects, suggesting that domestication could proceed relatively rapidly. For example, Paterson et al. (1995) found a small number of quantitative trait loci (QTLs) coding for the domestication syndrome traits of seed size, photoperiod sensitivity of flowering, and brittle rachis in taxonomically distinct cereals with diverse centers of origin (sorghum, rice, and maize). In the common bean (*Phaseolus vulgaris*), control of the domestication syndrome involves genes that have a large effect (>25–30 percent of phenotypic variation for a particular trait) and these genes together account for a substantial part of the total variation observed (>40–50 percent) (Koinange et al. 1996). Simulations based on sequence variations at loci coding for biochemical or structural phenotypes in maize and its close and distant relatives have estimated that domestication could have taken from 10 (Eyre-Walker et al. 1998) to between 315 and 1023 generations (Wang, Stec, et al. 1999).

Yet while there may be potential for rapid domestication, the actual environments and practices of foragers and farmers suggest that this may not have been the case. Recently, integration of genetic and archaeological data indicates that changes associated with the domestication syndrome did not occur simultaneously, and domestication often took several thousand years (Purugganan and Fuller 2009) (see section 3.2). For example, large grain size appears to have evolved early under cultivation, but fixation of the loss of shattering occurred later and more slowly, perhaps due to different intensities of selection. The management of *japonica* rice in some parts of China began about 9,000 years ago, but approximately 6,500 years ago half of those grains retained the shattering phenotype. Another reason for rethinking rapid domestication is new genetic data. For example, recent research on the maize genome has found that hundreds of genes, many more than originally hypothesized, were involved in the domestication of that crop, suggesting that the domestication syndrome may include many phenotypic traits that have yet to be identified (Hufford et al. 2012). Archaeobotanical data also suggest longer domestication processes, including evidence that both farming and foraging were practiced by the same or neighboring human populations for a long time, reducing selection pressures for domestication traits and allowing gene flow between domesticates and their wild relatives (Fuller et al. 2012).

In addition to selecting for characteristics of the domestication syndrome (Harlan 1992), domestication in sexually propagated crops may have resulted in increased autogamy (self-pollination) and therefore homozygosity (a state in which all alleles for a gene are the same), especially in cereals and small pulses, expressed phenotypically in greater trueness to type in a population over generations. In contrast, some vegetative propagation practices may have selected for heterozygosity (a state in which alleles for a gene are different), by incorporating large seedlings into clonal populations, the result of heterosis, and therefore for allogamy (cross-pollination), as contemporary evidence suggests for cassava (Pujol et al. 2005).

Most traits in the domestication syndrome are related to increased yield under cultivation, but farmer selection has been a powerful force for evolutionary change in crops based on food preferences as well. For example, three major genes involved in starch metabolism in maize were found to have unusually low genetic diversity as compared to that of its closest wild relative, teosinte (*Zea mays* ssp. *parviglumis*)—strong evidence of selection for specific processing and culinary qualities important for the primary manner in which maize has been consumed in its regions of origin and highest diversity (Whitt et al. 2002). Three other loci contributing to the sweetness of the maize grain in certain varieties in particular locations exhibit evidence of strong directional selection, further evidence of selection for culinary traits by farmers (Whitt et al. 2002). Similarly, evidence of strong selection by early rice farmers for sticky, glutinous grain quality resulted in the sticky rice favored by upland northeast Asian peoples in contrast to the nonglutinous rice varieties used by other Asian groups, and presumably would be among their fundamental choice criteria, perhaps as an adaptation for eating with fingers or chopsticks (Olsen et al. 2006).

It is, of course, impossible to know for sure the relative roles of hunger, cuisine, and other factors that motivated foragers and incipient farmers to domesticate and pamper these first crops, but it is likely that many factors were important to varying degrees in different places at different times.

5. ECOSYSTEMS MANAGEMENT

In order for domesticated crops to produce a harvest, humans had to become involved in manipulating ecosystems to create improved growing conditions for the crops—for example, by protecting crops from herbivores (from insect caterpillars to elephants), removing weeds to reduce competition, or planting crops together in polycultures that benefited one another (chapter 6). In addition, the effects on ecosystems of crop production and direct manipulation mean increased diversion of nutrients, water, and energy from ecosystem function to human use (chapter 8). Today, agricultural production has a profound impact on ecosystems, both positive and negative (Power 2010).

5.1. Managing Environmental Variables

Nutrients. In general, agriculture diverts resources from natural- to human-controlled cycles and speeds up the cycling within agricultural ecosystems, which in turn become increasingly dependent on farmers to maintain them. In agriculture, nutrients and organic matter are removed from the soil by burning vegetation to clear fields, cultivating fields which increases soil erosion and decomposition by soil organisms, and harvesting crops. These nutrients need to be replaced by applying compost, manure, rock dust, or concentrated commercial fertilizers; by growing cover crops; or by fallowing. In a fallow, no crops are grown, and wild and weedy plants growing in the fallow can be turned into the soil, allowing the buildup of organic matter and nutrients.

Energy. Solar energy drives an ecosystem through photosynthesis, which converts that energy into chemical energy in organic compounds (the net primary productivity, or NPP) that living organisms, including humans, can use. During the course of human history, from foraging through the Neolithic to the present, the trend has been to increase the proportion of the NPP appropriated, and we are fast approaching the limit to this trend (Running 2012). More recently, fossilized NPP (coal, oil, natural gas) has become a mainstay of agriculture—but we are also approaching limits in the amount of this we could use without pushing climate change to extremes (Helm 2011, Rogelj et al. 2013) (see chapters 1 and 8).

Water. The first farmers probably relied completely on choosing locations to plant where rain and naturally occurring soil moisture appeared adequate for maturing their crops, like the Hopi and Zuni farmers described in the preface. Later, farmers also harvested and concentrated rainwater where crops could use it. At least as early as five thousand years ago irrigation systems in present-day Yemen were importing surface water to fields (Harrower 2008), and today about 30 percent of the world's crops are irrigated.

Pests and diseases. Humans reduce damage from these by manipulating cropping patterns and environmental conditions—for example, by removing trees whose shade may encourage fungal disease. They also exert direct control, by physically excluding or removing pests and using biological and chemical controls.

Plants. One of the biggest demands of management is controlling weeds, plants that compete with crops for sun, water, and nutrients. Weeds put more of their resources into competing, whereas crops have been selected to direct resources to edible parts, and so are often incapable of outcompeting weeds. Farmers also learned that different combinations of crops gave more satisfactory yields.

5.2. Intensification

Agricultural intensification usually refers to the process of increasing intensity of land use, described in terms of amount of inputs or number of crops per unit of land, in time or space. Inputs that can be increased per unit of land include human capital (e.g., time, labor, management, cropping cycles), human-made capital (e.g., hoes, machinery, irrigation pipes, computers), and natural capital (e.g., irrigation water, green manure, crop species and varieties).

Cropping intensity can vary in time or space. Increasing cropping intensity over time, also called *cropping ratio,* involves increasing the number of crops grown in a given period. Intensity over time = cropping ratio (CR) = crop cycles per year, e.g.:

- 1 crop per year: CR = 1 / 1 = 1
- 2 years of one crop per year, followed by 8 years of fallow: CR = 2 / 10 = 0.2
- 3 crops per year (as in rice in southeast Asia): CR = 3 / 1 = 3.0

Increasing cropping intensity in space involves increasing the number of varieties or species grown in a given area at the same time, known as *polyculture.*[9] Small-scale traditional agriculture often involves polyculture, which can increase biodiversity, resulting in increased yield and yield stability.

Intensity in space = number of different varieties or species at a given time:

- *mixed cropping:* more than one crop in a field at the same time
- *intercropping:* a type of multiple cropping where crops are grown in alternate rows
- *alley cropping:* a type of intercropping, with perennial and annual crops in alternating rows

6. SOCIOCULTURAL AND BIOLOGICAL CHANGES IN HUMANS

Although humans initiated the revolutionary move from foraging to agriculture, agriculture has affected humans as much as it has affected ecosystems and other species.

6.1. Biology and Demography

The move from foraging to agriculture involved a great reduction in the diversity of food plant species. For example, in central Africa, Harlan (1992) estimates that foragers using about 1,410 species were replaced by farmers using an order of magnitude fewer species. This loss of species diversity decreased diversity of the diet, increased the proportion of starchy foods eaten, and increased sedentarism due to the demands of tending crops, which likely led to an overall reduction in quality of life (Larsen 1995, 2006). As mentioned in section 3.2, the move to farming may have significantly increased the time and labor needed to procure food.

This suggests that there were external pressures, such as increasing population (Richerson et al. 2001) or coercion by emerging social elites, that forced foragers to give up greater dietary diversity and better health for a more abundant supply of food. Further evidence for this is that intensive farmers revert to more extensive methods when resource constraints are reduced, as in the case of the Kofyar in northern Nigeria (Netting 1993) (see section 3.2). However, based on contemporary human behavior and food preferences, it is also conceivable that agriculture may have made particular prized foods or food qualities available that in themselves were incentives.

Many researchers believe the evidence shows that the increased food supply and sedentarism resulted in rising birth and death rates, with the former dominating, leading to greater rates of population growth. There is some support for increased birth rates in skeletal remains—the beginning of the Neolithic is correlated with a rise in the ratio of five- to nineteen-year-old skeletons to all skeletons five years or older in 133 cemeteries across the northern hemisphere (Bocquet-Appel 2011). Analysis of maternally inherited mitochondrial genomes from separate farming and forager populations from three regions (Europe, southeastern Asia, and sub-Saharan Africa) showed rapid population growth strongly correlated with archaeological dates for the beginning of farming, and this occurred about five times faster among farmers than foragers (Gignoux et al. 2011:6048).

6.2. Culture and Society

In order to discuss the changes in human culture and society that began with the Neolithic, we need to define some key terms that I will use throughout the book and will focus on in more depth in chapter 3 (section 2). These are terms commonly used in English, but here I will give them more precise social science definitions.

Knowledge is what is inside our brains (mental constructs, including value-based and empirically based assumptions, conscious and unconscious, all of which have a physical basis in the brain), and a *culture* or cultural group comprises people who share a large proportion of this knowledge (Romney et al. 1996)—for example, regarding the value of children, how to plant cassava, or the cause of thunderstorms. A culture may change as a result of experience or information, including agriculture. The sum of knowledge of a culture acquired over generations is referred to as *cumulative culture*. Note that this definition of culture does not include physical artifacts, as it does in common usage.

Cumulative culture and cooperation, prime factors in human biological success, may both be based on the social structure of foragers, the social structure for the vast majority of human history. One study of thirty-two contemporary foraging groups found that their social structure is comprised of extended interacting networks of people, the majority of whom are not related to one another. This suggests that cooperation is not the result of people increasing their individual fitness by helping relatives who share their genes. These large social networks may have provided the selective pressures that led to the evolution of the capacity for "social learning that resulted in cumulative culture" (Hill, Walker, et al. 2011).

Behavior is determined by knowledge and the external environment, and it consists of the things people do, including speaking and writing. Behavior of individuals results in patterns of behavior in social groups referred to as *society*. This includes the organization and activities of work groups for weeding in a village in Mali, or those of farmers markets or the county cooperative extension program in a U.S. county. Sociocultural changes supported the processes of domestication and ecosystem management that increased food production. For example, the creation and sharing of knowledge of plants and environments, and the mobilization of labor to clear and tend fields or build irrigation systems, were organized around household and kinship structures and religious ceremonies. The social networks described in the preceding paragraph created groups larger and more spatially extended than is often possible with direct kinship, and so may have encouraged the spread of innovations, technologies, and materials, and the growth of cumulative culture (Hill, Walker, et al. 2011).

Technology refers to things and processes humans create to mediate with and control their environments, including other organisms from soil nematodes to maize plants to people. The name Neolithic means "new stone," which refers to the change in lithic (stone) technology that marked the transition to farming. Some stone, and later metal, tools have been preserved in the archeological record and provide clues to developments and changes in technology over time. However, farmers also developed many other kinds of technology that may not have left traces in the archeological record. For example, we know through written records that farmers in China have been using citrus ants as a biological control of herbaceous insect pests in citrus and other fruit trees since at least the beginning of the fourth century C.E. (Huang and Yang 1987), but this practice may have existed long before this date. Farmers in Mesoamerica invented stone manos and metates to grind maize seed into flour and clay griddles to cook tortillas (Flannery and Sabloff 2009), and they also invented the nixtamalization process—cooking maize with calcium carbonate, which increases its nutritional value (Katz et al. 1974).

As human population sizes grew and groups became more dependent on agriculture, competition for resources increased, both within and between groups. With the rise of states and empires, agricultural resources and even food were increasingly controlled by the elites who held power. Water control began fairly early, with simple structures to direct water to crops and to hold it for later use. With the rise of more hierarchical social systems, and larger populations to feed, irrigation often became a major activity, with much time, labor, and other resources devoted to constructing elaborate irrigation systems. The archeological evidence for the simultaneous occurrence of increasingly complex irrigation and social organization has most often been interpreted through Karl Wittfogel's theory of hydraulic despotism in the Middle East (Wittfogel 1957)—namely that the organization necessary for large irrigation systems to function requires a centralized social hierarchy topped by a few powerful elites. However, observations of contemporary agricultural societies reliant on irrigation indicate that, while centralized control may be necessary for irrigation systems to function, it does not necessarily need to be in the form of a social hierarchy (Davies 2009). Based on research from East Africa, Davies has shown that groups of users working cooperatively can successfully fulfill the same function, an example of common property management, which I will discuss in chapter 7.

THREE

Thinking Critically about Sustainable Agrifood Systems

1. INTRODUCTION

What is *your* assessment of the current state of our agrifood system? What is it based on? There are lots of scientific reports on the subject, but their results often conflict, and interpretations of these results can be even more at odds with one another. So how do you choose which reports to use in your assessment? What parts of the system would you like to change? How would you like to change them? What do you think you and others can do to make those changes? How would you work with those who have different ideas about the changes needed?

These are difficult questions to answer—questions that some people devote much of their lives to. Underlying these questions, however, are more fundamental ones that need to be addressed. How do we know what we know? Why do we think we know it? For many who have thought deeply about these questions, the answer is that it is impossible to have true knowledge, to know reality in an absolute way.

This is an idea that has been at the core of many philosophical traditions for centuries, and it is also key for understanding the concept of sustainable agriculture. It underlies the view of science as having the goal of building models most useful for a given need, in contrast to the idea of science as a discovery of truth. To paraphrase a statement by Lao Tzu, "To know when one *does not* know, and *when* and *what* one *does* know, is best" (Waley 1958). It is best in the sense that more accurate, reliable knowledge supports more effective action to reach desired objectives.

But in a book about agrifood systems, why am I talking about the philosophy of knowledge? I believe that addressing such fundamental issues will help us to put our questions about agrifood systems sustainability into perspective. "Sustainable agriculture" has been used to mean everything from giant, laser-leveled fields of genetically engineered soybean to tiny hill-side plots growing tumbles of traditional maize, bean, squash, and herbs, cultivated by hand. Does the proliferation of incompatible definitions, this definitional slipperiness, mean that the concept should be discarded? I don't think so; I believe that analyzing the persistent use of the term to mean different things provides an entry into discussing the values underlying the fundamental concept of sustainable agrifood systems, and therefore a way to move forward.

There is something about "sustainability" that seems to have broad, perennial appeal, in spite of the many criticisms that have been leveled against it. I think this is because sustainability in its most basic sense captures the idea that most humans do not want to suffer or disappear, and being "sustainable" implies a long, relatively comfortable future for us and our descendants. The goal of our perseverance through time, along with that of the things that support us, is appealing to most humans—we don't want to suffer as a result of our support systems disappearing. We want to continue living our individual and social lives, and we want the environment that supports us, including by food production, to continue as well. Given the environmental and social problems created by our agrifood systems, discussed in the introduction to this book, achieving sustainability implies the need for thinking in broad terms across time and space.

This fundamentally simple idea was expressed in the oft-quoted Brundtland definition of sustainable development: "development that meets the needs of the present without compromising the ability of future generations to meet their own needs," with priority given to the "needs of the world's poor." In this view, economic growth is seen as essential for increasing sustainability, and limits to growth are not "absolute" but imposed by "the state of technology and social organization on the environment's ability to meet present and future needs."[1]

The Brundtland report was an important step in raising awareness of the need to increase the sustainability of human impact. Although it is hard to disagree with the general definition of sustainability given in the report, such definitions are "opaque": they don't analyze the values and assumptions about the way the world works that underlie them, and they seldom give concrete indicators that can serve as a guide to recognizing when an agrifood system is "sustainable" or "unsustainable." For example, although the report seems to endorse the Boserupian assumption of the potential of technology to increase human carrying capacity, it does not justify why this key assumption should be accepted. I see the key reason for the frustrating disutility of such definitions as their failure to address the fundamental nature of the concept—*sustainability is a subjective statement of goals, and so entirely dependent on the values of those defining it.*[2] Indeed, the term "sustainability" is often used as an adjective to describe anything to which the person or organization wants to give a patina of acceptability—evidence that sustainability has garnered considerable cachet, as Lele (1991) pointed out in an influential paper a year before the Rio Earth Summit in 1992, an event that firmly established the concept of sustainability in global consciousness.

In order to make the concept of sustainability useful for analyzing agrifood systems, we can think of it as comprising a four-part process:

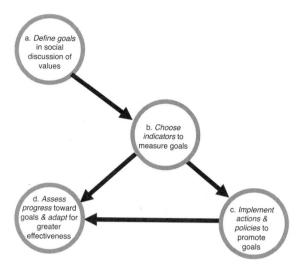

FIGURE 3.1. The process of analysis for increasing sustainability in agrifood systems. © 2013 David A. Cleveland.

a. defining sustainable agrifood systems as goals for changing the current situation through social discussion of values;

b. choosing and using indicators to objectively measure and assess current sustainability as defined;

c. creating and implementing specific actions and policies to achieve goals and increase sustainability; and

d. measuring indicators to assess progress toward agreed-upon goals and to adapt the process for greater effectiveness (fig. 3.1).

However, because of a failure to recognize the inherently subjective nature of the concept of sustainable agrifood systems as a whole, many efforts conflate subjective definitions, objective indicators, and policies, leading to debates and programs in which it is difficult to achieve any meaningful progress. In chapter 9 I discuss in detail the localization movement as an example of this. In the present chapter I analyze the concept of sustainable agrifood systems, and in chapter 4 discuss the assumptions underlying three of that concept's major emphases: economic, environmental, and social.

2. A MODEL OF REALITY, KNOWLEDGE, AND BEHAVIOR

In this section I discuss the fundamental question of how we know what we know, which is critical for understanding agriculture and food. Greater insight into how to answer this question can help to make sustainability a useful concept in creating "better" agrifood systems.

2.1. Constructivism versus Positivism

Much of the discussion about the nature of knowledge is polarized between constructivist and positivist camps (e.g., Harding 1998, Hull 1988). The *constructivist* view is that knowledge is dominated by social forces, including power relationships, and is historically and culturally particular; the process of knowledge creation is dominated by preexisting knowledge, including values, acquired through participation in a particular institutional or social setting, often mediated by the social control of technology and information.

The *positivist* view is that more and more universal and accurate knowledge of biophysical reality is a valid goal; knowledge acquisition is dominated by empirical methods meant to identify and eliminate social and cultural influences in order to ascertain the true nature of the world outside the individual mind. Positivism is often closely linked with the mainstream economic view of the world. For example, I, along with tens of thousands of other members of the American Association for the Advancement of Science, received this letter in 2010 (emphasis added):

Wed, 20 Oct 2010

Dear Dr. Cleveland,

Increasingly, unfounded public misgivings about science—the *mistaken belief that scientific findings have a point of view or are somehow partisan*—are becoming a serious barrier to progress both in science and in society's use of our work. . . .
What concerns me about this skepticism is that it could have serious consequences for many issues we care about, unless we work together to reverse it . . . most of the public is probably unaware that since the 1940s, *science and technology have driven over half of America's economic growth*. And in today's economic climate, we can't afford to turn away from this *vital national resource*. . . . I hope you can see the immediate need for our work—and help us restore trust in science so that it can be of maximum service to our nation and our world.

With thanks,
Alan I. Leshner
Chief Executive Officer

Leshner's letter epitomizes the problems that many see with the positivist approach. It is inconsistent in that it claims to have a view of science that is free of value assumptions, but at the same time claims that science is an integral part of the economic system of the United States and industrial world, implying that this economic system and its emphasis on growth is free of partisanship and values and therefore not to be questioned. Many objective observers would find this position problematic, based on examples such as that of agricultural biotechnology (chapter 5, section 6), commerciogenic malnutrition (chapter 8, section 7.2), or localwashing (chapter 9, section 4.1).

While positivist science often accepts the assumptions of the mainstream economic-political system, the reverse is also true. The mainstream economic-political system often accepts a positivist view of science, stating that decisions must be based on "sound science," which is assumed to provide one correct answer, even though this same system often attempts to manipulate science for political and economic gain (UCS 2012). For example, proponents of transgenic crop varieties (TGVs) often assume that scientific research has demonstrated these to be an essential (or even the key) component for solving the world food crisis, without any significant risks, and therefore deem any questioning of this assumption as bad science.[3]

2.2. A Holistic Approach: Knowledge about Reality as Models

The idea that we can know reality objectively, but that this knowledge is always indirect and incomplete, is a move away from both the extreme positivist view that science discovers ultimate truth and the extreme constructivist view that all knowledge is subjective and equally valid (Cleveland 2001). The holistic view is a middle way between these two extremes, based on the idea that knowledge comprises models; there is an objective reality outside of the mind, but we don't have direct access to it—that is, to the "truth" about it—only to abstract models of small parts of the world. A holistic view sees generating scientific knowledge both as an important way of increasing objective knowledge of reality *and* as a social process that reflects the particular cultural and social contexts in which it operates. It is a position that "rejects both epistemic absolutism and irrationalist relativism" (Bourdieu 2000:111).

Both holistic and positivist approaches work by testing ideas, but the holistic view realizes that tests are rarely conclusive (especially in the life and social sciences) and that models are inherently limited, so the goal is useful models. While a positivist view of science may persist among nonscientists, including politicians, and among many scientists when engaged in partisan issues, many scientists also agree that science is about finding the model of reality that is most robust in a given situation, for a specific goal.

Failure to understand that knowledge about biophysical and social reality is always in the form of models means that arguments can devolve into debates about which side has the "truth" or the "facts," rather than about the assumptions under which the knowledge was generated and how it is to be used, which requires a discussion of values. This confusion can keep our local and global communities in chaos, unable to reach the consensus necessary to move forward with effective policy to reach goals (Costanza 2001).

Models are theories, which I define as generalizable concepts about the way the world is, including causal relationships, and models can provide the basis for predictions and actions (Hull 1988:485, Medin and Atran 1999:9). According to some philosophers of science, there is no such thing as a purely observational term, since our descriptions of our observations are necessarily affected by theory, and apparently nontheoretical terms such as "animal" and '"dorsal" are theory-laden, and, therefore, "theory-free observation languages and classifications are impossible" (Hull 1988:8, 485).

The idea that we can know only models of the world is supported by the fact that it is logically impossible to fit the outside world into our brains. This is a conundrum that has interested not only scientists but also philosophers and artists—for example, in the (very) short story "Del rigor en la ciencia" by Jorge Luis Borges and Adolfo Bioy Casares (Borges 1960). This story tells of mapmakers whose maps of their province became more and more accurate until they were the same scale as the province itself, at which point they became useless. The 2008 film *Synecdoche, New York,* explored some of the same ideas about relationships between models and reality.

Models are necessary to make meaningful observations and conclusions about reality, but models are only "correct" in the context of the available knowledge about objective reality and subjective goals that the models are meant to serve. Therefore, when these goals later change, models become inadequate and can inhibit new ways of seeing things, until the accumulation of inconsistent observations force a change—or "paradigm shift," in the phrase made famous by Thomas Kuhn (1970 [1962]). A well-known example of this from biological science was the belief of pioneering maize cytologist and cytogeneticist Barbara McClintock that resistance to her observations and conclusions on genetic transposition ("jumping genes") in maize was the result of "an implicit adherence to models that prevents people from looking at data with a fresh mind"—what McClintock called "tacit assumptions" (Keller 1983:178).

2.3. A Holistic Model of Reality, Knowledge, and Practice

Because all knowledge of the world is in the form of models that are inherently incomplete, we need to understand what influences the creation of this knowledge in order to be able to judge its accuracy, reliability, and utility. The study of different kinds of knowledge, including how knowledge is created and used, is called *epistemology.*

Figure 3.2 presents a simple holistic model of reality, knowledge, and practice, based on current research in philosophy, neuroscience, physiology, psychology, and the social sciences. It provides a structure for analyzing data and conflicting interpretations of data. This model is critical for understanding and using a dynamic concept of sustainability. It is based on the assumption that there is an objective reality outside of the mind, but that humans can know this reality only subjectively—that is, as models of that reality—and as such, each of us constructs our subjective realities from our perceptions of objective reality. Science, in the sense of testing ideas about reality through observation and experiment, seems to many to be the best method yet devised for creating shared models of our shared reality. Yet science is inevitably based on scientists' ideas about the way the world *should* be and *is,* based on subjective values that provide the basis for systematic analysis (Oreskes 2004). "Good" science makes its values—what Joseph Schumpeter called "preanalytic vision" (Costanza 2001)—explicit, and systematic scientific analysis is consistent with this vision.

Input from biophysical and social reality comes in the form of physical stimuli (e.g., light waves) sensed by receptors and processed by the neuroendocrine system to become new "knowledge." This is not a deterministic process—while human perceptual systems are evolutionarily adapted to the physical nature of stimuli, the stimuli perceived are probabilistically related to their source in the world (Faisal et al. 2008, Geisler 2008). In addition, the path of a stimulus from its origin to the receptor in the human body is often enhanced,

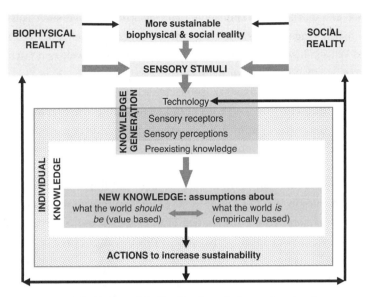

FIGURE 3.2. A holistic model of reality, knowledge, and action for sustainable agrifood systems. Shaded arrows represent knowledge generation, and solid black arrows represent effect of knowledge (via action) on reality and knowledge generation. © 2013 David A. Cleveland.

recorded, or calibrated by human behavior and technology (e.g., by a microscope or digital camera). Next, the stimuli are sensed by the body (e.g., in the case of light waves, by photoreceptors in the retina) and interpreted through body and brain structures, which include preexisting knowledge and imagined stimuli (Berger and Ehrsson 2013). This results in new knowledge—for example, the identification of microorganisms seen through the microscope or social interactions as recorded by the digital camera. Such knowledge is the basis for actions that can then change the biophysical and social reality of agrifood systems to make them more sustainable both in terms of our definition of sustainability and our perceptions of it.

Knowledge generation can be unconscious as well as conscious, and emotions and intuitions can function to consolidate unconscious perceptions in ways that facilitate adaptive behavior (Adolphs 2009). This is illustrated clearly in Evelyn Fox Keller's well-known biography of Barbara McClintock, whose "feeling for the organism" was a way of "seeing the patterns," an ability to empathize with the organism, to "get down in that cell and look around," until the objects become subjects, because organisms are "fantastically beyond our wildest expectations" (Keller 1983:6, 149, 200). This way of understanding is not uncommon in "objective science," or even physics; Einstein, for example, stated, "To these elementary laws there leads no logical path, but only intuition supported by being sympathetically in touch with experience" (quoted in Keller 1983:145). Darwin's passionate hatred of slavery influenced his belief that humans were descended from a common ancestor and was an important motivator in his development of the theory of evolution (Desmond and Moore 2009).

While emotion and intuition also contribute to unscientific theories, such as those of creationists or climate change deniers, the critical difference is that the theories of McClintock, Einstein, and Darwin have been subject to and supported by, as well as revised in light of, years of diverse, independent empirical tests.

What these observations point to is that objective science is one component of the way we create knowledge about the world—the deepest understanding also requires subjective science, allowing the unconscious mind to process the huge amount of perceptual information coming in through the body's sensory organs, since the conscious mind, and therefore objective science, cannot comprehend it.

The holistic model illustrates how our knowledge, and thus the formal and informal research that generates it, is subject to context-specific, contingent variation at all steps in the process, and therefore requires evaluation of this process and of the assumptions it is necessarily based on in each specific instance. In this way it can provide a framework for analyzing conflicting statements about the world food crisis and sustainable agrifood systems, and it demonstrates the need to bring together natural sciences, social sciences, and the humanities in understanding these and other critical concepts.

3. FARMERS' KNOWLEDGE AND SCIENTISTS' KNOWLEDGE

A key question in discussions about the role of knowledge in solving the world food crisis and how to create more sustainable agrifood systems is whether modern scientific knowledge (SK) and traditional farmer knowledge (FK) complement or conflict with each other.[4] How do we know which system is better in a given situation, or which components of each could be combined to create a body of knowledge better than either one alone? How we approach this question is an example of how our models of knowledge and reality affect our understanding of the world, of one another, and of how to work together to improve our agrifood system.

FK has been growing dramatically in importance in development policy and practice and in academic research, especially since the Rio Earth Summit in 1992, which highlighted the value of traditional knowledge. However, FK is rapidly disappearing as cultural groups become exterminated or absorbed into larger regional and national groups, as industrial agriculture continues to displace traditional agriculture, and as rural people move to cities. Is FK worth saving—is it essential for a more sustainable agrifood system? Or has FK become useless as the result of massive environmental and social changes, and should it therefore be replaced by SK? Or is there a way to combine FK and SK that would be better than either alone?

As I saw in Zorse and described in the preface, the mainstream view for quite a while has been that SK is key for solving the food crisis and FK is inferior and irrelevant, as in a recent conference organized by the International Food Policy Research Institute (IFPRI) on African agricultural development that emphasized technology transfer and modern science, with no mention of FK (Lynam et al. 2012).[5] From this perspective, FK is unscientific and inadequate for meeting food requirements, and small-scale, traditional farmers either need to leave farming or be integrated into modern, commercial agriculture (Collier 2008). The Alliance for a Green Revolution in Africa (AGRA) project described in chapter 5 is a prime example of this view.

The alternative view is that FK is key and SK is inferior or irrelevant, as in a report on local farmer knowledge in India (Kumar 2010). From this perspective, SK is rationalistic, reductionist, theoretical, generalizable, objectively verifiable, abstract, and imperialistic, in sharp contrast to FK, which is organic, holistic, intuitive, local, socially constructed, practical, and egalitarian (Escobar 1999, Scott 1998:340).

These perspectives agree about many of the ways in which SK and FK are different, but they value these differences in opposite ways, which means that defining sustainable agrifood systems would necessitate choosing between them. However, other social scientists, agrifood system practitioners, and farmers view FK and SK as fundamentally similar in a number of ways (Agrawal 1995). In contrast to the view that SK is universal and FK is local, research supports a more holistic view that both are universal at a basic level. Understanding both their similarities and differences is important for understanding how traditional farmers and others in the agrifood system can reach their goals for increased sustainability (Cleveland and Soleri 2002b). I present here the holistic view that FK and SK are both similar *and* different.

FK and SK are both universal. The universal as opposed to local nature of FK is supported by the many studies showing a common basis for local indigenous and scientific taxonomies of plants and animals (Berlin 1992). Boster's research with Aguaruna farmers in the Amazon found that their cassava classification tends to classify the smallest distinct taxonomic unit in patterns similar to those of scientists (Boster 1985). Based on a review of research on rain forest peoples, Ellen concluded that many individual observations lead these peoples to deductive models of how the natural world works that are then privileged over "accumulated inductive knowledge" when knowledge is passed between people (Ellen 1999:106).

In some cases, FK may even be more complete than SK—Malawian farmers' taxonomy of cassava varieties based on plant morphology visually distinguishes varieties between which scientists can see no differences, but whose distinctness is supported by molecular analyses for cyanogenic glucoside levels and by genetic analysis (Mkumbira et al. 2003). A study of farmer-managed *Phaseolus* (common bean) diversity in Oaxaca, Mexico, found that farmers were distinguishing inter- and intraspecific phenotypic variation and using this knowledge to tailor their planting of that diversity across heterogeneous growing environments (Worthington et al. 2012). These farmers make use of intraspecific variation that is often phenotypically subtle, but verifiable using genetic analyses, and includes divisions into subracial taxonomic groupings.

FK and SK are both local, culturally constructed, and embedded in specific contexts. While FK may be dismissed as local and nongeneralizable, SK can also be culturally relative, "local" knowledge. Agrawal points out that science-based approaches to Third World development have been shown to have failed because they ignored local contexts, indicating that SK is also

FIGURE 3.3. Eliciting farmer knowledge using scenarios in Oaxaca, Mexico.
Crop genetic resources specialist and plant breeder Flavio Aragón-Cuevas
interviews Lucilia Martínez using props (photographs and small bags of
maize seed) to illustrate a scenario. Photo © Daniela Soleri. Used with
permission of photographer and subjects.

culturally constructed and contextual (Agrawal 1995:425). Work by social scientists, histori-
ans, and philosophers on the nature of scientific knowledge since the 1920s has explicitly
explored the role of personal psychology, historical contingencies, and social context in its
production (Giere 1999). Dove suggests that indigenous knowledge and scientific knowledge may be similarly limited in their ability to comprehend complex natural phenomena
(Dove 1996b).

Understanding differences and similarities between FK and SK requires a baseline or
reference point. Because most who have attempted this comparison are scientists, it is SK

that has provided the methodological base for comparison. Therefore, attempts to understand FK include the following assumptions (Soleri and Cleveland 2005): (a) that there is an objective reality that both FK and SK are based on; and (b) that SK can provide a description of this reality that can serve as an *ontological comparator*. The ontological comparator chosen should not include knowledge that is unique to the experiences of the outside scientist or the local farmer. This is especially important when outside researchers or development personnel are the ones in charge of making comparisons, as is usually the case, because there is a strong tendency for them to assume that levels of knowledge that are linked to the specific experience of the outsiders are universal, a point I will discuss further in chapter 5 regarding collaboration in plant breeding.

Creating an ontological comparator requires determining the most basic level of reality with which both farmers and scientists interact. This can then be used to develop methods, such as constructing hypothetical scenarios for farmers to evaluate. Farmers' responses allow outsiders to compare how interpretation of this basic model in more complex contexts differs among farmers, among scientists, and between farmers and scientists (Soleri and Cleveland 2005) (fig. 3.3). Using such scenarios, Daniela Soleri and colleagues found patterns in FK across different crops and countries, as well as patterns between FK and SK that supported the hypothesis that empirical and theoretical FK and SK consistently reflect similar environmental patterns and relationships (Soleri, Cleveland, Smith, et al. 2002). However, they also found differences among farmers, among scientists, and between FK and SK that could often be explained in terms of differences in crop varieties, environments, or cultural values.

4. GUIDELINES FOR CRITICAL ANALYSIS

The guidelines for critical analysis in table 3.1 operationalize the model of reality, knowledge, and practice described in section 2.3, and depicted in figure 3.2. They facilitate critical thinking about the information received from lectures, readings, discussion, and direct observations about the world, as well as research to generate new knowledge, and even new values, although not all components of the guidelines are applicable in all situations. My goal in using these guidelines is to generate better models. I call them "better" based on the assumption that more accurate, reliable models support more effective action to reach desired objectives. I also call them "better" in the sense that they provide an explicit and shared basis for creating those models in specific contexts and so can be more useful than the gridlock of conflict that often handicaps discussions about our agrifood systems.

Making our different types of assumptions explicit and analyzing them is the first step in these guidelines, which I will cover in detail in the following section. Identification of the particular *problem* that is to be addressed and then the *theory* or *theories* relevant to our understanding of that problem is the next step in this process. A theory describes the larger mechanism that we believe is at work in the problem situation; it is the way in which we frame the problem we are looking at. To understand a problem in ways that we can use to

TABLE 3.1. Critical Analysis Guidelines

Components and definitions	Questions to ask
1. **Assumptions:** Empirically based assumptions about the way the world *is*, including how it works, and value-based assumptions about the way it *should be* and *should* work; assumptions include theories, and hypotheses that both have and have not been tested. All conclusions based on testing hypotheses are also empirically based assumptions about the way the world is and works. See fig. 3.4.	What are the major assumptions about the way the world *is* and the way it *should be* present in each of the critical analysis components identified below?
2. **Problems:** The main issues being addressed.	What problems are addressed? What data are used to describe them? What disagreements, if any, about those descriptive data are discussed? What values underlie classifying the issue as a problem?
3. **Theories or models:** Statements about the way the world is or works. Theories or models may be widely accepted by the research community, or accepted by only a small minority.	What relevant theory or theories are used to frame the problem?
4. **Hypotheses:** In their most elegant form, are logically falsifiable statements about reality based on theory—that is, about the way the world is or works. Hypotheses are often stated informally as questions. Often a general hypothesis is followed by more specific hypotheses. A good hypothesis clearly indicates the data required to test it.	What specific hypotheses (ideas) are tested? How are they related deductively to theory or inductively to previously existing data?
5. **Data:** Information used to test the hypotheses.	What data are used? Do you think some important data are omitted?
6. **Methods for data collection:** The ways in which the data are collected.	What methods are used to collect the data? Do you think they are appropriate—that is, do they result in data that are precise, accurate, and reliable enough?
a. Social context: Social variables that could influence research.	How might authors' backgrounds, affiliations, funding sources, and other intellectual or socioeconomic links influence the research?
b. Research design: At a basic level, the design could be a case study, a survey or an experiment in the laboratory, a test plot, or in the field.	What is the research design?
c. Site selection	How is the site for the research or project chosen? How representative is the site of the universe to which results will be extrapolated?

TABLE 3.1. *(continued)*

Components and definitions	Questions to ask
d. Sample selection	How is the sample selected? How representative is the sample of the site?
e. Variables	What variables are measured? What units are used and how are they chosen? What epistemological approaches are used? How are they chosen?
f. Data collection	How are the data collected? What techniques and technologies are used?
g. Human subjects	If people are involved, how are they involved (as subjects, participants, collaborators)? How are data gathered (as personal interviews, group interviews, mailed forms, web forms, observations)? If there is direct contact between researchers and people, what language is used? What evidence is there for the extent of rapport and trust?
h. Ethics	If humans, other organisms, or other special entities (e.g., ecologically or culturally sensitive locations) are involved, are required protocols and ethical norms being followed?
7. **Analyses:** The process of using data to test hypotheses.	How do the researchers use their data to test their hypotheses? What qualitative or quantitative methods (e.g., statistics, modeling) are used? Are they appropriate for the data? Are the assumptions required by the statistical and other tests satisfied?
8. **Results:** The outcome of testing the hypotheses.	Are the hypotheses accepted or rejected? Do the authors remain objectively open to considering alternative hypotheses to the ones tested, or are they constrained by their assumptions and theories?
9. **Conclusions:** The application of the results to address larger issues contained in the theory and problem definition.	Are the conclusions based directly on results? If not, what are they based on? Are they internally logical? Are they parsimonious? Do the authors explain their interpretations? What changes do the conclusions suggest to the theory the hypotheses are based on? What changes in the problem definition do they suggest? What interventions and actions in the world do they support? Are some conclusions not adequately based on results, and instead are actually speculations?
10. **Speculation:** Statements about the implications of the results that are not strictly based on the results.	Do the authors speculate about other conclusions that could have been drawn if different assumptions were made, hypotheses tested, data collected, analyses done, or results obtained? What new hypotheses are suggested?

take action and make improvements we develop *hypotheses*—clearly testable statements that delimit the variables and processes that we are investigating. Good hypotheses are designed so that the result of testing them is information that helps us make decisions about actions that address the problem. The process of testing our hypothesis includes identification of the *data* needed, the *methods* required, and the implications of our methods for the type and quality of our data. Once the data are collected they are *analyzed* to obtain the results of our hypothesis testing, from which we may reach *conclusions* that respond directly to the problem we are working on. Finally, we may wish to *speculate* about the meaning of our results, or the experience of testing our hypothesis, considering ideas that we acknowledge are not directly supported by the research but that we believe may be useful for future work and problem solving.

A key element in critical thinking is the testing of some assumptions as hypotheses (#1, 4, 7, 8 in table 3.1). Some may think hypothesis testing is limited to science and have a negative opinion of it because they think it is too reductionist. However, hypotheses are not only used by scientists but are a necessary part of human mental life, both consciously and unconsciously, including the simple heuristics that each of us uses daily (Gigerenzer and Todd 1999), as well as the more difficult, conscious analysis that can correct faulty heuristics (Kahneman 2011). Indeed, the process of biological evolution can be considered a passive (unconscious, nonteleological) process of interactive hypotheses testing of genotypes by their environments (mostly) via phenotypes, based on the criterion of fitness.

5. THE KEY ROLE OF ASSUMPTIONS

Better knowledge has three components: (1) empirically based knowledge of how the world *is,* including how it works, through analysis that is as objective as possible of the reality external to the mind; (2) subjective, value-based knowledge of how the world *should be;* and (3) understanding of how these two types of knowledge interact. Assumptions play a key role because *all* knowledge is composed of different kinds of assumptions.

A model of knowledge such as the one presented in figure 3.2 provides a framework for analyzing statements about the sustainability of agrifood systems in terms of epistemological processes, focusing on the assumptions that play a key role. One of the themes of my research—and of this book—is that it is necessary to question the assumptions in any theory, research design, interpretation of empirical data, or statement of values.

Appreciation of the contingency and contextuality of knowledge generation as outlined above in the holistic model, for both subjective values and empirically based observations, helps clarify that assumptions are an essential component of this process—assumptions are detrimental only when failure to identify them unduly biases our methods, results, or conclusions.

A nonrigorous approach that fails to examine assumptions dominates much of the discussion of sustainable agrifood systems. For example, the authors of an article calling for a data collection network to measure the sustainability of agriculture at various locations around the world using a set of indicators (Sachs et al. 2010) ignore the need to first state

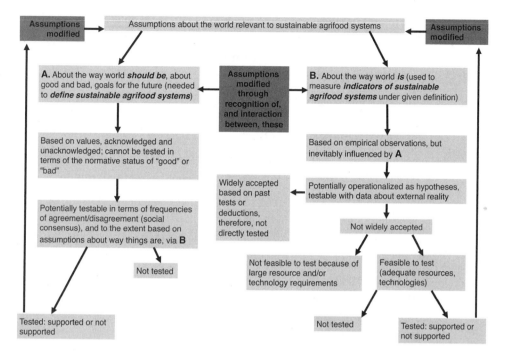

FIGURE 3.4. A taxonomy of assumptions for analyzing statements about the world relevant to defining and measuring sustainable agrifood systems. © 2013 David A. Cleveland.

their definition of "sustainable agriculture," which the indicators are to measure, and the values that their definition is based on. The implication is that there are no alternatives to their unstated definition, which in this case is implicitly that of conventional agriculture. Therefore, data collection is likely to reinforce this implied definition rather than to promote wider discussion of how to define sustainable agrifood systems as the goal which indicators are meant to measure via data collection.

Assumptions can be made about the way the world should be, or about the way the world actually is. Analyzing assumptions using the taxonomy in figure 3.4 is an important step in understanding different emphases and perspectives in sustainable agrifood systems.

A. Assumptions about the way the world *should be* that are necessarily based on values about what is good and what is bad. We can classify potential tests into three categories:

1. cannot be tested in terms of the normative status of "good" or "bad"—for example, ethanol made from corn is bad because edible food should not be used for biofuel;

2. can be tested with data about biophysical and social reality, to the extent that values are contingent on reality—for example, increase in biofuel production is bad because it is correlated with increase in food prices;

3. can be tested with data (from observations or interviews) on degree of social consensus, to the extent that values are judged by the number of people holding them—for example, biofuels made from maize are bad because surveys show that most people believe they are bad;

B. Assumptions about the way the world *is,* which are necessarily based on empirical assessment and are potentially testable with data about reality. We can also classify these potential tests into three categories:

1. do not need to be tested, to the extent that they are self-evident or are widely accepted based on past tests—for example, humans require a minimum amount of food in order to be able to function;

2. not feasible to test, because of the large amount of resources required for data collection and analysis, or because data do not exist—for example, the average number of calories consumed in each community in India in 2012 was greater than in 2008;

3. can be tested as hypotheses with data that can be reasonably gathered—for example, the average greenhouse gas emissions per unit of protein from cereal grain is less than from red meat.

The two main types of assumption also influence each other. Testing of empirically based assumptions is critical not only to assess their empirical validity but because the results can also influence value-based assumptions (see A.2 in list above), and vice versa. Both intuition and social conformance can reinforce empirically incorrect ideas about the way the world is (Bloom and Weisberg 2007, Kahneman 2011). While it is probably human nature to use "rational" thinking to justify value-based preferences, it is also via interaction with others that conclusions based on these preferences can be modified by more objective, scientific information (Haidt 2007), especially when the assumptions are identified. In other words, to the extent that value-based assumptions also "draw on factual presumptions, often made in an implicit way," value assumptions remain "subject to revision in the light of more knowledge" (Sen 2000:942). Therefore, while many value-based assumptions cannot be directly tested with data about the objective world, making them explicit and including them in discussion could improve science and policy development. In other words, we need to discuss values openly, find areas of agreement, and then evaluate the evidence or "facts" that those values are in part based on. Facts can change values, but we should try to let values change facts as little as possible, and instead use them to help us determine how to use facts to achieve our value-based goals.

6. DEFINING, MEASURING, AND IMPLEMENTING SUSTAINABLE AGRIFOOD SYSTEMS

Now we can build on the preceding discussion of knowledge to go into more detail about the four steps in creating more sustainable agrifood systems (fig. 3.1). It is especially important to identify and analyze assumptions, because they influence all of the steps. For example, because the economic aspects of the agrifood system are assumed to be the most important in the mainstream economic emphasis, the positive, direct economic effects (such as amount and value of production and agricultural employment) and some positive economic externalities (such as nonagricultural employment generated, foreign export earnings, and ecosystem services) are tracked by businesses, governments, and organizations such as the FAO. Because other aspects are assumed not to be very important, they are not monitored. This means that the relationship between agrifood systems and their negative environmental and social externalities (such as greenhouse gas emissions, water pollution, biodiversity loss, unjust labor practices, increased cost of food, and malnutrition) are often poorly documented or even invisible because they are of little interest to the mainstream economic system, and relatively few data have been systematically collected and valued monetarily (Eshel 2010).

6.1. Defining Agrifood System Sustainability

It is obvious that there are problems with the current agrifood system and that we need to produce, process, transport, and consume our food more "sustainably." But how? Since all of the components of any definition (the variables, their levels and rates of change, the boundaries in space and time, and so on) can be defined only subjectively, doing so requires individual reflection and social discussion; all definitions must be cognitively and socially constructed.

To be useful, a definition must also be shared by a critical number of people. Because the world is globalizing at an increasing rate (i.e., super exponentially), consensus for many key areas of sustainability, such as climate change, needs to be reached on a global scale. This in turn implies the need for some sort of mechanism to manage this social process.

However, once a definition is agreed on, an agrifood system can be evaluated objectively, by choosing and measuring indicators of that definition. There are some general aspects of definitions of sustainable agrifood systems that are widely accepted—such as that they should include environmental, economic, and sociocultural criteria. Sustainable agriculture, for example, is often defined as agriculture that conserves natural resources and ecosystems for future generations, is economically viable, and is socially equitable (cf. Francis and Callaway 1993:4, Thompson 1995). However, this "triple bottom line," as it is often referred to in the business community (Savitz 2006), has become a cliché hiding disparate assumptions, potentially leading to greenwashing or drinking green Kool-Aid, as discussed later in this chapter. I will discuss in detail the assumptions underlying environmental, economic, and social emphases in definitions of sustainable agrifood systems in chapter 4.

Reaching agreement on a subjective definition of sustainability is made more difficult to the extent that our world is increasingly one of specialized and compartmentalized information that is difficult for laypersons to understand and that privileges "experts" who often disagree, leading to inconsistent and unstable public opinion (Costanza 2001). One approach to getting a critical number of people to agree on a definition is to formulate alternate scenarios of future states for stakeholders to evaluate (Peterson et al. 2003, Rouquette et al. 2009). Another approach is to review the literature for commonality of relevant "sustainability issues" (Walter and Stützel 2009:1277). Whatever process is used, defining sustainable agrifood systems includes answering at least five interconnected key questions, all of which depend on values.

1. *What should be sustained?* What is currently judged "good" and worthy of sustaining versus "bad" and needing to change? Because agrifood systems have such a broad array of effects, resources are limited, and conditions are constantly changing, difficult choices need to be made about what to sustain. For example, if farmers in a community are adopting new crop varieties that are better adapted to a changing climate, which traditional crop varieties should they continue to maintain, if they can't maintain all of them because the time needed competes with time required to maintain traditional pest management practices, food preparation techniques, or agricultural ceremonies?

2. *For whom should it be sustained?* What should be the sociocultural boundaries of the agrifood system? This often has a lot to do with who is doing the defining. For example, people who have access to a large variety of foods they enjoy would probably like to sustain this situation, while those who are hungry certainly do not want to sustain their situation. There can be conflicts between groups, especially when resources are limited, so increasing some aspects of equity may be seen as bad by those with more (Anand and Sen 2000)—for example, if decreasing hunger in Third World countries means reducing access to food for the well-off in industrial countries.

3. *Where should it be sustained?* What should be the spatial boundaries of the agrifood system? How should spatial heterogeneity be dealt with? For example, high soil erosion rates can reduce crop yields for a farming community living upslope while increasing crop yields for a community downslope, which benefits from the deposit of good soil in its fields. Indeed, erosion is an essential part of the geological dynamics of our planet—it's what created soil in the first place—so while it is generally considered something to be reduced to make agriculture more sustainable, over the long term, agriculture could not exist without erosion. Intermediate spatial boundaries on the scale of ecosystems or watersheds are often used in definitions of sustainable agriculture, but as mentioned above, for many aspects of the agrifood system, the most relevant spatial boundary is becoming the entire planet.

4. *How long should it be sustained?* What should be the temporal boundaries of the agricultural system? If it weren't for the meteorite that hit Earth about sixty-five million years ago, triggering the great Cretaceous-Tertiary boundary die-off, in

which about 70 percent of species became extinct, it may be that dinosaurs would still dominate Earth's ecosystems, mammals would be much less common and diverse, and humans would not have evolved (Gould 2002). And eventually our Earth's star, the Sun, will die, and all life on Earth will die along with it. These extreme limits on temporal boundaries illustrate that "sustainability"—in the sense of preserving a given state—is meaningless over the really long run because it defies the change that is fundamental to life and to the inanimate universe. At the other end of the temporal scale, the situation is also easy to resolve—most organisms most of the time would like to continue existing and would like their supporting social and biophysical environments to persist with them—at least for the next few minutes! Of course, it is the intermediate scales that are the most difficult to define—time periods on the scale of years or generations.

5. *How should it be sustained?* This includes choices about social structures (e.g., regulations, laws, and behavioral norms) and biophysical structures (e.g., tools, buildings, fences, and dams). For example, as I discuss in chapter 7, sustaining resources by reducing the negative effects of the agrifood system can be done via markets, government taxes and regulations, or community social organization—alternatives that can have very different overall impacts. A key issue for how any entity should be sustained is determining the balance between preservation (stasis) and conservation (which includes change)—for example, crop genetic resources can be preserved *ex situ* in cold storage in seed banks, or conserved *in situ* in farmers' fields. Seed banks will preserve a larger amount of the original genetic diversity, but conservation by farmers will result in evolutionary adaptation to changing conditions that might make the results more useful in the future, but also might result in loss of some diversity.

6.2. Choosing and Measuring Indicators

Objective measures of environmental, economic, or social variables tell us nothing about sustainability without a subjective definition. Therefore, the statement that environmental (or social or economic) criteria for sustainability are objectively prior to social (or economic or environmental) ones is not logical because any such ranking is part of defining sustainability—which is subjective! Once there is agreement on a definition of a sustainable agrifood system, then variables defined as indicators can be chosen, which will allow the agricultural sustainability of any given system to be evaluated, and suggestions made for increasing sustainability.

Therefore, it would be a mistake to think that the indicators are objective in an absolute way, because they too can be chosen and measured only via decisions based on subjective values. For example:

- *Choice of indicator variables.* Which of the often many different indicators of a given goal will be chosen? For example, if part of the definition is that everyone is well nourished, then objective indicators could include the level of satisfaction with the diet, the quantity of nutrients in the food normally eaten determined by laboratory analysis as a proportion of nutrient requirements determined experimentally, or direct assessments of nutritional status such as anthropometric measures (e.g., weight-for-age, triceps skinfold) or physiological measures (e.g., red blood cell count, blood cholesterol levels).

- *Choice of critical values for the indicator variables.* What levels, direction, and rates of change of the indicator variables will be assumed to indicate sustainability? If the *direction of change* is used instead of the value of the indicator variable, then some difficulty is eliminated, since it can be easier to agree on the direction a variable like nutritional status is moving in, as opposed to a critical value. For example, there is often disagreement about what levels of nutritional indicators like weight-for-age are optimal, but for undernourished, low weight-for-age children, increasing weight is not controversial. However, directional change such as increasing efficiency may not be a good indicator if the goal of sustainability is decreasing an absolute amount—for example, increasing the efficiency of greenhouse gas emissions (GHGE) (unit of GHGE per unit GDP) when the concentration of greenhouse gases (GHG) in the atmosphere needs to be reduced (Jackson 2009b) (I discuss this further in chapter 8).

- *Sampling.* Where and on what units are observations made? The choice of sample can have a strong influence on the data collected and thus on the results and conclusions. For example, to test the hypothesis that transgenes have unintentionally flowed into farmer crop variety (FV) populations in a given area, those plants most likely to have transgenes should be sampled, whereas to test the hypothesis that transgenes are present at a threshold frequency in a population, the whole population should be randomly sampled (Cleveland, Soleri, Aragón Cuevas, et al. 2005).

- *Observation and measurement.* What epistemological variables (technology, language, preexisting knowledge) could bias the results? The way interview questions are constructed and presented can have large effects on the way interviewees respond. Asking farmers what they know about crop selection using scientists' terminology, for example, will likely result in the conclusion that farmers' and scientists' knowledge are completely different (Soleri and Cleveland 2005). In contrast, asking them in terms relevant to their own experiences can elicit answers that suggest that farmers' and scientists' knowledge is very similar in some important ways.

- *Statistical analysis.* Statistics can quantify the probability that what we think we know about reality is "true"—that is, that it has a high probability of being a true

		Objective reality (which we cannot perceive directly)	
		H$_0$ true	H$_0$ false
Creation of new knowledge about reality leads to decision to:	Accept H$_0$	**Accept H$_0$ when true.** E.g., concluding TGVs are substantially equivalent to conventional MVs, i.e. have no important risks, when this is true Result: benefits increased	**Type II error, false negative. Accept H$_0$ when false.** Probability of *not* committing type II error = β. Decreases with increasing sample size. E.g., concluding TGVs are substantially equivalent to conventional MVs, i.e. have no important risks, when they are *not* substantially equivalent and *do* have important risks Result: costs increased
	Reject H$_0$	**Type I error, false positive. Reject H$_0$ when true.** Probability of type I error = α = significance level. Set ahead of time by determining P level (usually 0.05). E.g., concluding TGVs are substantially different & have important risks when they do not Result: benefits decreased	**Reject H$_0$ when false**; = power of the test = 1- β E.g., concluding TGVs are substantially different & have important risks when they actually do Result: costs avoided

FIGURE 3.5. A model of hypothesis testing. H$_0$ = null hypothesis. © 2013 David A. Cleveland.

statement about reality (fig. 3.5). Some of the assumptions about the way the world is can be formulated as hypotheses and formally tested using data, providing information useful for confirming or adjusting research and policy frameworks, as well as revising the value-based assumptions themselves (see section 4, this chapter). Statistics can provide the basis for inferences from samples to the universe of things sampled. However, because they are always tests of a model using limited information, hypotheses cannot be proven or disproven; they can only be supported or not supported, given the results of the statistical analysis (Karsai and Kampis 2010). It is also important to be aware that there are statistical alternatives to the dominant frequentist approach of null-hypothesis significance testing to making inferences about what the world is (Stephens et al. 2007).

6.3. Actions and Policies

Implementing actions and policies that will promote the goals of a chosen definition of sustainable agrifood systems includes:

· establishing baseline measurements of the chosen indicators—for example, the nutritional status of children five to fifteen years old;

· determining the variables that will move the chosen indicators in the direction defined as more sustainable—for example, testing the hypothesis that increased access to fruits and vegetables improves nutritional status;

- establishing the social and biophysical environments and tools that will likely move the indicators in the defined direction—for example, if it is concluded that increased access to fruits and vegetables improves nutritional status, then programs to encourage household and community gardens and promote sales of fruit and vegetables in corner stores could be established; and

- remeasuring the indicators to evaluate the success of the actions and policies and adjusting the policies (evaluation, monitoring)—for example, if the programs above were implemented, then the nutritional status of children in the community should be remeasured after implementation.

6.4. The Importance of Not Mistaking Indicators for Goals

Indicators are chosen in part because they can be measured more easily than goals, but the danger is their potential to displace or even undermine the original goals. For example, because research has shown that fresh fruits and vegetables contain nutrients such as vitamins and minerals that are important for good nutrition, it is reasonable to hypothesize that increased consumption of fresh fruit and vegetables would improve the nutrition of malnourished children in low-income neighborhoods, and to promote projects to increase the number of stores in these neighborhoods that sell fresh produce. However, directly measuring the consumption of fresh fruits and vegetables and the nutritional status of children would take much time and resources, so indicators of the success of projects to improve child nutritional status—such as the number of stores in or near the project neighborhoods selling fresh fruits and vegetables—are usually chosen. This is based on the assumptions that increased presence of fruits and vegetables will result in increased consumption which in turn will result in improved nutritional status. But there are a number of steps between improved nutrition and number of stores selling fresh produce—including the purchase of produce, which would depend on how expensive it is for the families in the neighborhood and people's buying habits; the preparation of the produce; and whether the malnourished children are eating it, along with what other foods these children are eating and their physical activity and other characteristics. In other words, it is not valid to assume that more stores results in better nutrition and declare the project a success based on an increase in number of stores selling fresh produce. To avoid confusing indicators and goals when the sustainability of a given situation is measured, and when actions and policies to increase agrifood systems sustainability are evaluated, we need to frequently check that the indicators actually do measure those goals as much as possible in each instance and that changes in indicators reflect advancement toward goals (fig. 3.6).

In the example of crop transgene flow, the indicator of presence or absence, or of a threshold frequency of transgenes in a nontransgenic crop, frequently becomes the goal, when the underlying definition of sustainability is often much broader. Sustainability goals could be empowering local farming communities to maintain crop genetic diversity (common for TGV critics) or helping farmers to increase production and market sales (common for TGV

FIGURE 3.6. Guarding against mistaking indicators for goals.
© 2013 David A. Cleveland.

supporters). A frequently chosen indicator of the former is elimination of any transgene presence or flow (González 2005), while for the latter transgene flow might not matter, or even be seen as positive (Raven 2005). The objective evaluation of the relationship between transgene flow and TGV adoption and either of these goals is often obscured or forgotten in the ensuing debate about indicators (Cleveland, Soleri, Aragón Cuevas, et al. 2005, Soleri, Cleveland, and Aragón Cuevas, 2006) (chapter 6).

Another example is the frequent use of food miles (distance from farm to retail) as an indicator of the sustainability of agrifood systems, which is based on the assumption that food miles are a major contributor to GHGE in the agrifood system and that their reduction is therefore correlated with reduced GHGE (Hill 2008) (chapter 9). However, because food miles account for a very small proportion of the GHGE of the conventional agrifood system, reducing food miles may result in only very small decreases in GHGE unless the relationship of food miles to GHGE is specifically addressed in localization—for example, by creating local food distribution hubs (Cleveland et al. 2011b).[6]

In climate change mitigation negotiations, the UN Collaborative Programme on Reducing Emissions from Deforestation and Forest Degradation in Developing Countries (UN-REDD) has emerged as one of the centerpieces, and it provides another example of confusion regarding indicators and goals. UN-REDD is based on the stocks and flows conceptualization of ecosystem services often favored by environmentalists as a useful basis for selecting indicators of sustainability that can help preserve ecosystem function. However, Richard Norgaard shows how this concept is being used to substitute the indicator of carbon credits, in the form of carbon sequestered in forests, for the sustainability goal of mitigating global warming. Moreover, he suggests that the concept of ecosystem services in general, and UN-REDD in particular, is being used by mainstream economics to justify continued economic growth (Norgaard 2010).

6.5. Drinking Green Kool-Aid

One form of substituting indicators for goals, or accepting claims based on little or no evidence, is done by those who believe in the goals embodied in a given definition of sustainable agrifood systems. I refer to this as *drinking green Kool-Aid,* which happens at a range of levels of awareness.[7]

When governments and expert bodies drink green Kool-Aid, they delude themselves and the public because of political "realities," often because of pressure from the powerful players in the mainstream economic system, such as multinational corporations, who want to maintain the status quo. For example, the U.S. National Research Council's report on limiting GHGE recognizes that "curbing U.S. population growth . . . or deliberately curbing U.S. economic growth, almost certainly would reduce energy demand and GHGE. Because of consideration of practical acceptability, however, this report does not attempt to examine strategies for manipulating either of these factors." (NRC 2010:39). A statement by the Ecological Society of America on the ecological impacts of economic growth highlights the concept of sustainability, yet it offers no substantive definition of that concept and sees economic growth as compatible with conserving ecological function: "The problem is not economic growth, *per se,* but the ways in which it is implemented" (ESA 2009:1).

In academic settings, critical analysis can be the victim of the wave of enthusiasm for sustainability that is sweeping college campuses. Conflation of the often-underspecified goals of "sustainability" with indicators is frequently used to justify physical and economic growth and fails to engage intellectually with the widely accepted scientific understanding of the need to reduce human impact (see chapter 8, section 1.0). This can be seen in the web pages of the largest U.S. organization promoting sustainability in academia—for example, in the assumption that an increase in "green" buildings makes campuses more sustainable without considering the overall impact of physical growth (AASHE 2010). Academia seems also to have failed to engage fully with social sustainability, including the issue of inequity (Breen 2010), and with the value basis of sustainability generally, as evident in the pattern of university hiring in sustainability (Vucetich and Nelson 2010).

6.6. Greenwashing

When indicators are intentionally substituted for goals, or when claims are made with little or no evidence in order to protect short-term advantages in the unsustainable status quo, the result is commonly dubbed *greenwash.* The charge of greenwashing is most frequently leveled at environmental policies of corporations and others who have an economic emphasis in sustainability (GreenpeaceUSA 2011). But there are many variations on the greenwash theme. *Fairwash* occurs when companies promote their food as produced and traded in ways that treat farmers and workers fairly when there is little support for this claim (Clark and

Walsh 2011). *Localwash* occurs when corporations such as Walmart advertise their food as "local" (Roberts 2011), based on indicators such as food miles, implying that their "local" food is also promoting goals of localization like stronger local communities (see chapter 9). Because the relationship with those goals is never explicitly measured, however, little increase in local control is likely to result (DeLind 2010), and in cases such as that of Walmart, the opposite may be more likely (see section 4.1, chapter 9).

According to the NGO Food and Water Watch, the Walmart pledge to increase by 50 percent the amount of local produce it sells by 2015 means that, locally grown foods will comprise 9 percent instead of 6 percent of its produce (Food and Water Watch 2012:8). Moreover, Walmart's definition of "local" means from within the same state, with the 9 percent goal calculated across all 3,804 stores in the United States. But even this modest move toward sustainability in one area is likely to have negative effects on sustainability in other areas. In order to meet this pledge and retain low consumer prices and high corporate profits, Walmart is driving consolidation of producers and suppliers, and controls over 50 percent of grocery supplies in twenty-nine U.S. metropolitan areas. There are studies indicating that because of the low wages it pays, the presence of a Walmart in a community actually increases poverty compared to matched communities without the store (Goetz and Swaminathan 2006).

Therefore, sustainable agriculture is seen primarily (in the words of multinational agricultural biotechnology and pharmaceutical company Bayer) as opening up "fresh business opportunities" to satisfy the "demand for sustainably produced goods [which] will increase, just as consumer awareness of sustainability issues will also rise" (Bayer 2009). Even more explicitly, a senior executive of Unilever, a major food and cosmetic multinational corporation, has stated that he "considered genuine achievements in sustainability and social responsibility as essential to [the corporation's] 'license to grow,' particularly in emerging markets" (Hamilton 2010). This approach substitutes the short-term indicator of well-being—measured in terms of consumption of goods and services—for the longer-term goals of well-being, which many define in terms of ecosystem functionality, biodiversity conservation, equitable social relations, subjective happiness, and physical health. This conflation often obscures fundamental ecological principles maintaining the production base, including natural ecosystem functions.

Critics of greenwash believe that "sustainable growth" and "green growth" are oxymorons, and they have proposed that to be truly sustainable means moving to steady state economies that do not grow (Daly and Cobb 1989, Jackson 2009a), or even degrow (Sorman and Giampietro 2013). However, this critique is not part of the public discussion of sustainability, in part because of concerted pushback from mainstream economics (e.g., Bartelmus 2013) and business (see chapter 8, section 1 for further discussion).

6.7. Co-optation

The term *co-optation* has been used to describe the efforts of corporations (Campbell 2001), states, or government entities to neutralize the effectiveness of movements for social change

(Gamson 1968) by controlling and/or redefining them, and so maintaining the status quo. Co-optation seems to be a common strategy of the few who control the dominant agrifood system to counter efforts by the many to gain more control.

Daniel Jaffee and Phillip Howard delineated five methods used by agrifood corporations to co-opt organic and fair trade alternatives in the United States in order to maximize their profits: removing rules, changing rules, altering the content of standards, eroding price premiums, and soliciting the assistance of the government (Jaffee and Howard 2010:389). Julie Guthman illustrates many of these tactics in her analysis of the evolution of organic agriculture in California, the place where the US organic movement began, showing how its meaning has been coopted (and greenwashed) by mainstream agribusiness (Guthman 2004). At the national level in the United States, the U.S. Congress voted in 2005 to approve the "Organic Trade Association Rider," a legislative clause heavily lobbied for by agribusiness interests, which overturned a federal court ruling and significantly weakened the four-year-old organic standards set by the U.S. Department of Agriculture (USDA) (DuPuis and Gillon 2009). The Cornucopia Institute, an organic industry watchdog organization, has detailed the extensive corporate influence on the U.S. National Organic Standards Board (NOSB) (Cornucopia Institute 2012), and the journalist Stephanie Strom has documented the consolidation of the organic industry through acquisition by mainstream corporations (Strom 2012).

FOUR

Sustainable Agrifood Systems

Three Emphases

1. INTRODUCTION

The heads of state of the world's leading industrial nations, members of the G8 (Group of Eight), met in Toyako, Hokkaido, Japan, on July 6, 2008.[1] They ate luxurious multicourse meals and chatted over drinks; in addition to an extensive selection of wines, the menu for *one* dinner comprised

> Corn and caviar; Smoked salmon, sea urchin; Hot onion tart; Winter lily bulb and summer savoury; Kelp-flavoured beef and asparagus; Diced tuna, avocado and soy sauce jelly, herbs; Boiled clam, tomato, shizo in jellied clam soup; Water shield and pink conger with soy sauce vinegar; Boiled prawn with tosazu vinegar jelly; Grilled eel and burdock; Fried goby fish with soy sauce and sugar; Hairy crab bisque soup; Grilled bighand thornyhead fish with pepper sauce; Milk-fed lamb flavoured with herbs and mustard, and roast lamb with crepes and black truffle; and Cheese, lavender honey and caramelised nuts. (Chapman 2008)

Is it surprising that one of the main issues they were dealing with was the world food crisis, which had erupted in 2007–2008 in the form of dramatic spikes in food prices, triggering widespread protests and riots? The G8's ideas about the food crisis are dramatically embodied by the meals they ate at the 2008 summit—meals that were featured in the media and became a cause célèbre for G8s critics, who contrasted these extravagant feasts with the chronic hunger faced by almost one billion people worldwide.[2]

High food prices have continued since 2008, accompanied by a plethora of international conferences, publications, and Internet blogs. While there is wide agreement that there is indeed a "crisis" in the agrifood system, there is also a lot of disagreement about the underlying causes of and solutions for this crisis, based on differing assumptions about how the system *does* and *should* work. How do we make sense of this? Which view of the problems, causes, and solutions to the food crisis is "right"?

One way to answer these questions is to evaluate the available information and the statements being made in terms of the assumptions that underlie the three different emphases in sustainability: the *economic* emphasis of the mainstream perspective and the *environmental* and *social* emphases of alternative perspectives (in some contexts I will refer to these latter two in the singular). The mainstream perspective, emphasizing economic concerns, dominates public discourse and is shared by most national governments, especially those of the industrial world, and many international organizations, such as the World Trade Organization, the World Bank, and the UN Food and Agriculture Organization (FAO). The alternative perspective, emphasizing environmental and social concerns, is that of many NGOs and community, farmer, and grassroots organizations. This classification is of course a simplification (as are all classification schemes), but it can help us to see broad outlines of the issues more clearly, and it provides the framework for moving toward more sophisticated analyses. For each emphasis I will examine those underlying assumptions about the way the world works and should work in terms of natural resources, human nature, internalizing externalities, and risk management.

2. THREE EMPHASES IN AGRIFOOD SYSTEM SUSTAINABILITY

Most definitions of sustainable agrifood systems include economic profitability, conservation of natural resources and ecosystems, and social equity (chapter 3) and can sound superficially similar. In practice, however, emphasis on any one of these components usually results in different policies. It is only by investigating the details of these definitions and proposals for achieving their goals, including the specifics of projects and organizations being supported, that it becomes evident to the neophyte that the same or similar words can have very different meanings.

Table 4.1 presents some of the key value- and empirically based assumptions underlying the definitions of sustainability from economic, environmental, and social emphases. Explicitly identifying and analyzing these contrasting assumptions can help move us further beyond the Brundtland Commission's opaque definition of sustainability (WCED 1987) introduced in chapter 3. It allows us to see the real challenge more clearly—to agree on a definition of sustainable agriculture that combines aspects of the three emphases through discussion of their assumptions. Once we agree on a definition, we can measure the sustainability of an agrifood system and develop policies and actions to increase that sustainability.

2.1. Economic Emphasis

The logic of sustainability from an economic emphasis has been termed *weak sustainability* (Neumayer 2004, Norton 2005). Weak sustainability is based on the assumption that economic growth is a prerequisite for social and environmental sustainability, that human-made and natural resources are substitutable, and therefore that the goal should be increasing *efficiency* of human activity. According to this logic, increasing growth in the size of the human economy is compatible with increasing sustainability, if the unit of negative impact, such as greenhouse gas emissions (GHGE), per unit of economic activity, such as gross domestic product (GDP), is decreasing (Bartelmus 2013:26–28,65–66). This is also referred to as *relative decoupling* of economic growth from negative environmental impacts (Jackson 2009b). It assumes that human impact (HI) is comfortably below environmental and social human carrying capacity (HCC), and therefore that growth in consumption is good because it will spur technological innovation, leading to continual growth in HCC (Norton 2005). To some economists, such as the late Julian Simon (1983), this means that people are the ultimate resource, and that population growth is good because it produces more ideas about how to increase HCC, as in a Boserupian scenario for dealing with the zone (chapter 1).

The economic emphasis dominates global and most national agrifood systems and defines sustainability as continual growth in yields, production, and profits, with solutions to the world food crisis achieved by continued industrialization, commercialization, and technological innovation. It is a unilineal vision that sees small-scale, traditional agrifood systems everywhere evolving into large-scale, modern industrial systems, a process driven by the rational decisions of profit-maximizing farmers (see Collier 2008). For example, in his popular text on economic development, Michael Todaro contrasts the "inefficient and low-productivity of agriculture in developing countries" with the "highly efficient agriculture of the developed countries," where the "specialized farm represents the final and most advanced stage of individual holding in a mixed market economy" (Todaro 1994:288, 310).

Since the economic emphasis is the core of the mainstream perspective on sustainability and dominates corporate, government, and development agency policy, the other emphases are often defined and discussed in contrast to it. Its critics see it as frequently undermining sustainability—an "opulence-oriented" approach with a goal of aggregate "wealth maximization" that ignores "social justice and human development" (Anand and Sen 2000:2031), thereby contrasting strongly with the social emphasis; and a technologically oriented, production-maximizing approach that ignores physical limits to growth, thereby contrasting strongly with the environmental emphasis (Gowdy 2007, Jackson 2009b). A growing body of data on the environmental and social impact of mainstream, industrial agrifood systems challenges the economic emphasis, and its environmental and social problems have been enumerated by many critics (e.g., McMichael 2009, Vandermeer 2009, Weis 2010, IAASTD 2009).

Some economists have suggested that there is a lack of evidence for key empirically based assumptions of the mainstream economic emphasis. Most mainstream theorists have abandoned the two theoretical pillars of mainstream economics—the notion that humans

TABLE 4.1. Common Assumptions Made by Three Contrasting Emphases about Agrifood System Sustainability

	Emphasis		
Variable	*Economic*	*Environmental*	*Social*
Value-based assumptions about the way the world should be for key variables, including in terms of HI = NCT (human impact = population × consumption × technology)			
Definition of sustainability (its goals)	Continuous growth of agrifood economy to provide wealth for future generations. Focus on economic (human-made) capital, allocative efficiency.	Conservation of natural ecosystems and biodiversity for their own sake (ecocentric) or to provide resources for people (anthropocentric). Focus on natural capital, ecosystem health.	Social justice via empowerment of indigenous peoples, women, and minorities, including equitable access to agricultural resources and food. Focus on social (moral) and human capital, fair distribution.
Human population (N)	Should increase.	Should stabilize or decrease.	Should be dealt with by education and development.
Human consumption (C)	Should increase.	Should decrease.	Should be more equitable.
Technology (T)	Should be more productive.	Should be more efficient.	Should be more accessible.
Decision making	Should be by economists to promote economic growth.	Should be by scientists to protect ecosystem functioning.	Should be by communities to promote equity.
Empirically based assumptions about the way the world is for key variables			
Human impact (HI) and human carrying capacity (HCC)	No limit to HCC, can be increased to stay above HI for a very long time.	HCC surpassed by HI (or will be soon).	HCC and HI best dealt with after social issues.
Markets	Translate self-interest into social good to achieve Pareto optimality. Trickle down to address inequity. High discount rate.	Do not adequately value natural resources; destroy environment. Must define limits ecologically. Low discount rate.	Do not adequately value social good; destroy community. Society should distribute goods equitably. Low discount rate.
Natural resources	Substitutable; demand and technology drive supply.	Complementary; finite biophysical (source and sink) resources limit supply.	Adequate, but unequally distributed.

TABLE 4.1. *(continued)*

| Variable | Emphasis | | |
	Economic	Environmental	Social
Human nature's dominant trait	Concern for self, selfishness.	Concern for natural environment, biophilia.	Concern for others, altruism.
Discount rates	Should be high to reflect human short term self-interest.	Should be low to reflect limits of key resources.	Should be low to reflect importance of future generations.
Negative externalities: most efficient way to internalize	Market rationalization.	Regulations based on understanding of ecosystems.	Community management and extension of compassion in space and time.
Risk management process and policy making	Ex post trial and error; policy best made by economists.	Precautionary principle; policy best made by scientists.	Precautionary principle; policy best made by communities.

Source: Based on (Cleveland and Soleri 2007).

are rational actors motivated by self-interest (the *Homo economicus* model) and the possibility of perfect competition in markets—in the face of much research (Gowdy 2007:3). Nobel prizes in economics have been awarded to a number of researchers for work challenging mainstream assumptions, including Daniel Kahneman's work showing how people are often "irrational" (Kahneman 2011) and Amartya Sen's work showing that mainstream economics does not adequately deal with social choice and equity issues (Sen 1999). The mainstream economic emphasis was heavily criticized from a number of directions after the collapse of the free market boom in 2008, which helped to bring on the food crisis (e.g., Skidelsky 2009).

Yet there seems to be a disjunct between research on economic theory and its application "in the field," as Gowdy and Erickson (2005) have pointed out. The policy recommendations of economists continue to be based on outmoded assumptions and are defended on the grounds that critics have not offered a viable alternative to the existing economic system (Gowdy 2005:209). Critics respond that alternatives are impossible to define in detail in advance because we have never faced the challenge of replacing an economic system of this magnitude, but that this does not justify ignoring the likelihood that if we don't work on creating an alternative, a disaster much more difficult to get out of will likely occur (Jackson 2009b).

2.2. Environmental Emphasis

An environmental emphasis views economic and social functions of the agrifood system as dependent on environmental functions and therefore maintains that environmental scientists' advice should have priority in making policy to increase sustainability. It defines sustainability in terms of conservation of ecosystems and biodiversity, both natural and domesticated (Arrow et al. 1995). It assumes that HI is exceeding the environmental HCC, and that human population and consumption should not grow—or should even decrease—in order to restore and maintain biodiversity and ecosystem functions that will otherwise disappear in spite of, or because of, technological innovation. In other words, it assumes that growth in HI has already carried us into the zone, and it supports *strong sustainability*, with the goal of decreasing the *absolute level* of HI (as opposed to decreasing the rate at which HI increases, as in the weak sustainability of the economic emphasis) (Bartelmus 2013:26–28,65–66, Jackson 2009a).[3] That is, instead of a relative decoupling of economic growth from negative environmental impacts, it assumes that *absolute* decoupling is required, which means a transition to a steady state, no-growth economy (Jackson 2009b). The environmental emphasis includes alternative approaches from within economics, including *ecological economics* (Gowdy and Erickson 2005, Daly and Farley 2004).[4]

An environmental emphasis sees economic growth as constrained by natural physical and biological limits. It rejects the consumption of goods and services as an indicator of well-being, as promoted by the economic emphasis. Instead it measures well-being in relation to longer-term outcomes, including ecosystem functionality and biodiversity conservation. For some, this implies that nonindustrial traditional farming is more sustainable because they assume that indigenous farmers are more in harmony with natural systems and their limits (e.g., O'Neal et al. 1995, Thompson 1995, Vandermeer 2009). There is much evidence for the failure of modern models imposed on local communities, whose traditional agrifood systems are more sustainable (Vandermeer 2009). For example, Norgaard's analysis of industrial-style agricultural development in the Amazon concludes that it has been a failure, whereas tradition-based local systems are more reliable and successful (Norgaard 1994:121).

A major challenge for environmental sustainability is the complexity of ecosystems, including agrifood ecosystems, about which scientists still have much to learn. This makes it difficult to set limits on the amount of impact humans can have without undermining ecosystems and to determine what actions are needed to restore negatively impacted systems—for example, in the case of nutrient influx abatement to reverse eutrophication in coastal ecosystems (Duarte 2009). A major example I will discuss in chapter 8 is anthropogenic climate change.

2.3. Social Emphasis

A social emphasis views economic and environmental functions of an agrifood system as secondary to its social functions and therefore maintains that communities should have priority in making policy to increase sustainability. It defines sustainability in terms of social justice for the less powerful—including Third World farmers, women, and indigenous and minority groups—and equitable access to food and agricultural production resources such as land and water. While HI may be exceeding the environmental HCC, more importantly, it is also exceeding the *social* HCC, in terms of the huge and growing inequity in access to food and resources, with the resulting social, cultural and psychological damage. In this emphasis, population may be considered a problem, but a problem best dealt with by improving access to technology and equitable consumption. Such improvement can occur by increasing the consumption of the poor, which may need to be achieved by reducing the consumption of the rich.

One of the key concepts of social sustainability is *cultural relativity,* a value-based assumption that the knowledge and values of each individual and cultural group is valid for that person or group, regardless of how outsiders might judge it. This contrasts with *universal ethnocentrism,* which is present among all cultural groups and means that people assume that their own culture is superior to all others. Ethnocentrism is a major problem especially with the mainstream economic emphasis, which, because of its overwhelming power and influence, is able to impose its assumptions on the rest of the world.

In his famous novel *Things Fall Apart,* Chinua Achebe captures the tragedy of ethnocentrism in the clash between cultural groups. The protagonist, Okonkwo, the chief of a Nigerian village, describes the conflict with the British colonial administrator: "We cannot leave the matter in his hands because he does not understand our customs, just as we do not understand his. We say he is foolish because he does not know our ways, and perhaps he says we are foolish because we do not know his" (1959:175).

However, while the concept of cultural relativity as promoted by the social emphasis is important for supporting social equality and justice, it also has some potential difficulties. *First,* the social emphasis often assumes that local knowledge has coevolved with the local environment to be the most efficacious, and therefore that no other knowledge, including modern scientific knowledge, could be superior or useful (see chapter 3, section 3). Yet if local people want to change their farming, is it socially just for outsiders to discourage them from evaluating alternatives, or from testing the empirical validity and efficacy of local knowledge? *Second,* upholding the right of one group to continue its traditional behaviors may infringe on the rights of other groups to continue theirs. It also needs to be balanced by the criterion of intergroup equity in negotiating conflicting claims of individuals or groups to resources based on "traditional" knowledge or practice. *Third,* there is always variation in knowledge among individuals within any cultural group, and cultural relativity does not provide an unambiguous way to resolve resulting conflicts, or to reach consensus within the group.

So far, I have described how the three emphases differ regarding the concepts involved in defining, measuring, and implementing agrifood system sustainability. In the rest of this chapter I will compare the emphases in terms of other key variables (table 4.1).

3. NATURAL AND OTHER RESOURCES

Assumptions about the external world—in terms of natural and other resources needed to produce, transport, process, and prepare food—and the effect these assumptions have in turn on resources are central to sustainability. The production function is a way of modeling the relationship between inputs and outputs of resources and is fundamental to thinking about many different processes in the agrifood system, including the relationship between stocks and flows of resources.

3.1. The Production Function

The production function plots outputs on the y axis as the dependent variable, whose value is determined by inputs, the independent variable on the x axis (fig. 4.1). It illustrates basic assumptions of the economic emphasis (Ellis 1993) and the way assumptions of the environmental and social emphases differ. It helps us to conceptualize a basic problem: What is the appropriate level of human activity for transforming resources (inputs) into food and, ultimately, into human happiness (outputs)? What are the specific outputs we want? And what is the level of inputs we are willing to invest to get them? Of course, this depends on many different factors, from how we subjectively value inputs and outputs, the physical scarcity of inputs, and the technology used to transform inputs to outputs.

Economists usually think about the output-input relationship in terms of monetary profit—profit being the excess in value of outputs over value of inputs. But output is defined differently from different perspectives. For example, Lyson and Welsh critique the mainstream economic assumption that production should be measured in terms of on-farm profits and propose instead measuring it in terms of crop diversity (1993). Their analysis of U.S. census data showed low levels of crop diversity associated with large corporate farms and high levels associated with smaller farms spending more on labor.

Inputs are typically categorized into land, labor, or capital, with "land" including all other physical inputs in addition to land, such as seed, water, and fertilizer. The total value product (TVP) curve tracks total output. The change in value of product (output) with a change in unit of input is the marginal value product (MVP), which decreases with each additional unit of input, with its rate of change equal to the slope of the tangent to TVP at a given point. Because the MVP tends to decrease for all combinations of inputs and outputs, this relationship is known as the *law of diminishing marginal returns,* which applies to many biological and physical, as well as economic, relationships—for example, the output of harvest in response to inputs of fertilizers.

The upper graph in figure 4.1 shows the cumulative increase in cost of inputs, or total factor cost (TFC). Maximum profit is at the point of economic optimum where the vertical distance between these two curves is at its maximum. The lower graph shows that the change in marginal factor cost (MFC) is zero (i.e., the curve is a straight line, showing that the cost per unit input remains the same as inputs increase), while the MVP is decreasing (i.e., the value added with each additional unit of input is decreasing), intersecting MFC at the point of maximum profit. This is the point of maximum profit because to the left of it MVP is greater than MFC, so the cost of each additional unit of input is *less than* the value of the output it produces, and there is additional profit to be made; to the right of this point MVP is less than MFC, so the cost of each additional unit of input is *more than* the value of the output it produces, so profit is decreasing. MVP continues decreasing until it becomes zero at the point of technical maximum, the highest output (TVP) possible with a given pro-

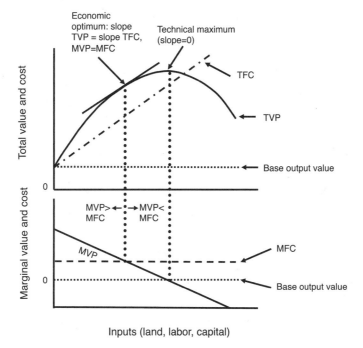

FIGURE 4.1. The production function. TFC = total factor cost; TVP = total value product; MFC = marginal factor cost = cost of each additional unit of input; MVP = marginal value product = value of additional output produced with each additional unit of the input = slope of TVP curve at any point; base output value = that which would occur without any of the input under consideration. Based on (Ellis 1993:23–26).

duction technology and mix of inputs. After that point, the MVP becomes increasingly negative for each additional unit of input, reducing the TVP.

The concept of diminishing marginal returns makes intuitive sense if you think about it in concrete terms. For example, imagine you have a large garden plot that needs weeding before the weeds choke out the young crop plants. To solve the problem you could invite your friends to come weed in the garden with you the next day, promising to feed them a big meal at the end of their work. The first few friends that show up really increase the rate at which the weeds are pulled and thrown on the compost pile, but after a dozen arrive, you realize that the additional output each one can produce is beginning to be less than what it will cost you to feed each additional helper (in other words, you have passed the economic optimum). After a couple of dozen friends arrive, things really slow down, because people start getting in each other's way. Worse than that, they end up trampling the crop plants! The result is that the increase in your total returns, or output, in terms of what you will eventually be able to harvest from the garden (TVP), decreases with each additional friend who shows up to help (you have passed the technical maximum).

3.2. Stocks and Flows of Natural Resources

A major point of disagreement about the production function between economic and environmental emphases is about stocks and flows of natural resources in the agrifood system, including land, water, soil, air, and genetic diversity. An economic emphasis assumes that resource flows determine resource stocks and, therefore, that increased consumption or population growth will lead to increased production. For example, Tim Dyson, while acknowledging the importance of environmental limits in a general and distant way (1996:19, 208), stated that "cropland can usually be rehabilitated if it is judged to be economically or politically worthwhile" (1996:149). This implies that it is economics and politics that determine supply of land (and other natural resource inputs to agriculture).

An economic emphasis also assumes that scarcity of a natural resource input can be remedied by substituting another input, such as technology or labor. This means that *natural capital* (natural resources) is not indispensible and that it can be substituted for by more and improved inputs of *human-made capital* (technology, infrastructure) or *human capital* (labor, management). As mentioned above, a focus of environmental economics has been monetizing ecosystem services, which facilitates their substitution, helping to rationalize natural resource degradation, including via agricultural production (Norgaard 2010).

These assumptions translate into policies for sustainable agriculture focused on producing more, but doing so "sustainably," so that sustainable agriculture may be defined as increasing the slope of the TVP curve in the production function (e.g., Lynam and Herdt 1992). A corollary assumption is that when the limits to the economic efficiency of any given production function are reached, production can always be increased by the invention of new, more efficient technology, generating a TVP curve with higher values. This implies that any "yield plateaus" in agricultural production can be overcome with new technologies, and new technologies are a key element of the mainstream solution for the world food crisis. The crop physiologist Lloyd Evans, in his authoritative book on crop yield, makes an analogy between the potential for increase in yield and the historical increases in accelerator beam energy as a result of successive design innovations that created new production functions and sustained its exponential growth (Evans 1993:29–30).

In sharp contrast, an environmental emphasis on the use of natural resources focuses on the physical limits of external reality and especially on the laws of thermodynamics (Walker 1992). It criticizes the assumption of the economic emphasis that production inputs are completely substitutable among themselves, because this denies the first law of thermodynamics, the law of conservation of energy (and matter)—that is, the necessity of increasing matter-energy (physical) inputs in order to increase outputs. For example, with substitutability, an input like irrigation water would not be essential, because capital (invested in technology) could replace any aspect of it. According to this logic, natural resource degradation as a result of agricultural production can be economically rational if resources can be replaced by capital or technology (Daly and Cobb 1989).

An environmental emphasis usually assumes that the sustainable use of a renewable natural resource means maintaining its stock. For example, in irrigating crops with groundwater, the stock of groundwater in an aquifer is determined by the amount of recharge to the aquifer from rainfall and the return of irrigation water, minus the amount pumped out of the aquifer for irrigation. Therefore, a sustainable level of irrigation water pumping is one that would not lower the level of the aquifer. In comparison, an economic emphasis would define a sustainable level of pumping water from the aquifer in terms of profits, based on the assumption that if the aquifer becomes depleted, water can be imported from somewhere else, or new technologies invented to obtain water previously inaccessible (see section 4.2).

Adherents to an environmental emphasis believe that, rather than the market setting the parameters of resource use (including pollution), scientists should define the limits to which resources can be diverted to human wants without jeopardizing the sustainability of natural systems—then markets could set prices within those limits. This assumes that scientists are capable of making fairly accurate and reliable estimates of sustainable resource use, based on a definition of sustainable agriculture that can be agreed on by society. It also assumes that society (including both private corporations and the public) will be willing to allow scientists to do this, which is usually strongly opposed by the mainstream economic emphasis, as it has been in discussions of how to mitigate anthropogenic climate change.

For the social emphasis on sustainable agrifood systems, natural resources are important but secondary to issues of social justice. It is access and distribution, not the amount or renewability of resources, that matters most because it is only through equity that sustainability can be achieved.

4. HUMAN NATURE

What do we humans value the most—our own material well-being, the natural environment, or the well-being of society as a whole, including future generations? This question is an old one that is still hotly debated today. In this section we will see how different answers affect the sustainability of agrifood systems.

4.1. Basic Values

What are our basic assumptions about our own human nature and welfare, and about the welfare of the rest of the world, including other people living now and in the future? The central assumption by those with an economic emphasis is that we are dominated by materialistic values. They refer to humans as *Homo economicus,* a model species that is "unswervingly rational, completely selfish, and can effortlessly solve even the most difficult optimization problems" (Levitt and List 2008:909). It follows that human happiness is directly correlated with increasing consumption and material wealth. For example, public relations professionals Ted Nordhaus and Michael Shellenberger equate scientists' empirical assumption that there are biophysical limits to physical consumption with a pessimistic belief in limits to human creativity and happiness (Nordhaus and Shellenberger 2007). As mentioned above, environmental economics since the 1980s has given much attention to research on people's willingness to pay for ecosystem and social services that had not previously carried a price tag, so that they can be included in standard economic models (Norgaard 2010).

In terms of social welfare, this idea is captured in the concept of Pareto optimality, a situation in which no one can be made better off without someone else being made worse off—"the most frequently used normative criterion of modern economics" (Cornes and Sandler 1996:22). This implies that rational economic persons seeking to maximize their economic utility, and firms seeking to maximize their profits, together will result in optimizing social welfare if markets get the prices right, without regard to the equity of distribution of resources (Gowdy and Erickson 2005:210).

It follows from these assumptions that the most efficient form of economic organization is based on a free market mediating all transactions. The sum of all of our maximized individual utilities is optimal social utility, assuming that everything of value has a price and so is substitutable. These assumptions result in a "preoccupation with commodity production, opulence and financial success" and a focus "on the characteristics of overall material success [rather] than on the deprivation and development of human lives" (Anand and Sen 2000:2031). Therefore, in terms of the production function (fig. 4.1), the economic emphasis, with its focus on materialistic values, defines rational behavior as investing at the point where profit, or "expected utility," is maximized (Ellis 1993). This is the *economically optimal level of resource use*, where MVP = MFC.

An environmental emphasis does not accept the assumption that natural resources are substitutable, or that the optimal level of investment should be based on monetizing natural resources because they may be indispensible for human life and also because this ignores the fact that humans have evolved an appreciation for the living world, for biological diversity—that is, that humans are biophilic (Wilson 1984)—which means that it is not possible to compensate someone for the loss of biodiversity with money. The economic assumption also ignores the complex nature of ecosystems and our relative lack of knowledge about how they function. Norgaard has argued that the concept of payment for ecosystem services, developed especially since the mid-1990s by ecologists to emphasize the limits to growth and the need for fundamental changes in the economic system, has been hijacked by mainstream environmental economists to avoid dealing with ecological complexity and to support continued growth (Norgaard 2010).

In a social emphasis on sustainable agrifood systems, social relations and concern for others are assumed to be more important than the individualism and materialism of the economic emphasis. These assumptions are receiving support from psychological research into the effects of mental attitudes on behavior—which suggests, for example, that honesty and trust are important, evolutionarily adaptive human traits (Ekman 2003). In a study of consumer goals and environmental attitudes of undergraduate students at the University of Toronto, researchers found that consumerism was negatively associated with agreeableness (compassion, empathy, and concern for others), and environmentalism positively associated with both agreeableness and openness (imagination, creativity, and receptivity to new ideas) (Hirsh and Dolderman 2007). Similarly, research on the effect of money on human behaviors found that exposing people to the idea of money (indirect exposure to images of currency and related images on a computer screen) made them less likely to ask for help when solving a problem, less likely to help others solve problems, and more likely to prefer working alone—in other words, it encouraged less social behaviors (Vohs et al. 2006).

Detachment from material things is often seen as a corollary to compassion or altruism (e.g., Kristeller and Johnson 2005). In an experiment designed to isolate altruistic behavior from self-interested behavior based on delayed reciprocity, researchers found that humans are motivated to contribute assets to those with less, even when the contributors, the active players, have no chance of receiving any material reward (Dawes et al. 2007). The size of income alterations (due to giving or taking) varied depending on the income of the recipient relative to the active participant and did not decline over time. That is, the active players tended to give more to those who had fewer assets (lower "income") than the active players had, and took more away from those with higher "incomes," even though this cost the active players. The researchers interpreted these results as implying that the players' behaviors reflected a concern for equality.

These and other experimental findings about basic human nature and values are important because they challenge the assumption of the economic emphasis that people are essentially selfish.[5] In contrast, if people have the potential for prosocial, altruistic behavior, it opens up a whole new range of solutions, including creating institutional structures that build on and can reinforce this potential, like common property management (chapter 7). This involves a major rethinking of a standard economic tool for determining level of input investment, discounting.

4.2. Discount Rates

An important way in which human nature has been conceptualized and quantified is in the way values are affected by distance in time. An economic emphasis based on *Homo economicus* assumes that human beings want to maximize personal utility in the present, not in the future, because the value—of outputs, the inputs used to make them, and the cost of environmental and social damage—decreases rapidly with distance in time. In other words, the priority is for localized, individual material benefits obtained as soon as possible.

This is captured in the idea of discounting, based on the assumption that people's interests are overwhelmingly short term and that they therefore place less value on a given resource in the future compared to that same resource in the present. For example, if you are hungry, you would probably be willing to pay more for a meal now than for that same meal if you had to wait until tomorrow to eat it. By extension, the rate at which the current value of a future resource is reduced or discounted (discount rate, DR) becomes an implicit indication of how the present generation values, or is forced to value, future generations. For example, water is commonly extracted from nonrenewable groundwater aquifers for irrigation, justified by using a high DR, without considering the future need for irrigation and the fate of future farmers (see Gleick and Palaniappan 2010).

DRs and interest rates are directly related; they are different ways of expressing the same relationship, from different viewpoints. The relationship of present value (PV) and future value (FV) for interest and discount rates is expressed in the following formulae:[6]

- An interest rate greater than zero means that the future value of a resource is judged to be *greater* than its value in the present: $FV = PV(e^{rt})$.

- A discount rate greater than zero means that the present value of a resource is *less* than its value in the future: $PV = FV/(e^{rt})$. In other words, discounting of FV is expressed in the form of lower PV—the higher the DR for a given resource with a given assessment of its monetary value at a time in the future, the lower the present value of that resource will be.

For example, to answer the question "What is the present value of groundwater that would be worth one million dollars in twenty years, if the DR is 10 percent?" we can calculate:

$$PV = FV/(e^{rt}) = \$1{,}000{,}000/(e^{(0.1)(20)}) = \$135{,}335$$

Therefore, with a twenty-year planning period, it would not be "rational" to invest any more than $135,335 to conserve the aquifer for twenty years—that is, by limiting its use in the present.

DRs are also important in determining whether a resource is conserved or destroyed by comparing interest and discount rates between different production functions using different resources. If the DR of one resource, such as groundwater in a poor growing environment, is less than the interest (growth) rate of another resource, such as groundwater in a good environment, then it is rational to use the groundwater in the poor environment as quickly as possible and to invest the profits in crop production in the good environment (Cohen 1995:249–251, Daly and Cobb 1989). In this way it becomes economically rational to extract value from agrifood resources, including groundwater, soil, trees, and processing facilities, to the point that they are destroyed. That is, those resources will not be available to future generations (Daly and Farley 2004: 272).

The logic of discounting in an economic emphasis on sustainability is based on the assumptions that technological innovation will always provide a substitute for any input that becomes scarce, that inputs are substitutable, and that future generations will be better off due to increases in consumption with economic growth.

Environmental and social emphases do not share these assumptions. From the perspective of an environmental emphasis, the assumption that constant growth in productivity will automatically improve the welfare of future people is unjustified given the depletion and pollution of agrifood resources (HCC has been exceeded and we are in the zone). Real productivity is more likely to be negative, and therefore DRs should be zero or negative, which would make the present value of resources the same as or greater than their future value, providing economic incentives to protect them for the future (Daly and Farley 2004:274).

From a social emphasis perspective, discounting is based on the value that people today obtain from "contemplating the welfare of future people," not on an estimate of their actual future welfare or how people in the future might value welfare (Daly and Cobb 1989:154). There is no robust, purely economic rationale for discount rates, and thus they must be explicitly based, at least in part, on ethical and social values (Parks and Gowdy 2013). Therefore, a DR greater than zero means that people today value people in the future less than themselves in the present, and is unethical. If people now value people in the future as much as themselves, then the DR should be zero. In contrast, an economic emphasis views a zero DR as irrational, because it reflects the "pessimistic environmentalist view of pending disaster," and wrongly "assumes that the current generation feels the pain of future generation as much as its own" (Bartelmus 2013:55).

DRs are also affected by the state of their social and biophysical environment. That is why poor people tend to have high DRs—if they don't use all available resources today, they may not survive until tomorrow. Farmers who would otherwise have low DRs may be forced to adopt high DRs because they have no other choice in the short run. I once asked a farmer in Zorse, as we looked at one of his fields running up a steep slope full of eroded channels from the last heavy rain, why he farmed there. He told me that he knew it was not good because the erosion would eventually wash away the field, but he had no other choice, since land was limited and he needed to grow food to feed his family now.

A similar situation may be behind the high average discount rate (28 percent) found among U.S. farmers in an experiment to assess how they valued payments offered by federal environmental conservation programs (Duquette et al. 2012). Tight farm budgets, pessimism about the future of farming, especially intergenerationally, and inadequate information about the shared consequences of environmental and natural resource degradation may all contribute to their high DRs, as they do for valuation of ecosystem services in general (Carpenter et al. 2009).

5. INTERNALIZING EXTERNALITIES

Now that we have seen how the three emphases contrast in terms of their assumptions about natural resources and human nature, we can put these together to analyze a concept that is key to understanding agrifood system sustainability—internalizing externalities. (I discuss internalization further in chapter 6.)

Negative externalities are the costs of a production activity not borne by the person or entity conducting that activity, but rather by other members of society. In other words, the producer gets the full benefit of the production but does not pay the full costs of production, because some costs are externalized onto society in general—locally, nationally, or globally.

Discounting the present value of future costs, as discussed in the previous section, can provide an economic rationalization for imposing negative externalities on people in the future.

Externalities are generated in agrifood systems as the rule rather than the exception. For example, users of irrigation water can decrease the ability of downstream users to irrigate both by excessive extraction of water from a stream, and by reducing the quality of the water leaving their fields (tailwater) that flows back into the stream by polluting it with silt, weed seeds, disease pathogens, or agrichemicals. Another example is when a farmer decides to add lots of manure or ammonium nitrate to his field to increase yield. This will likely increase nitrates leached into ground and surface water, which will have a health cost to others who drink this water, even though they don't receive any of the direct benefits of increased yield (Reddy et al. 2009). It can also have costs for people in the future.

There can also be *positive externalities* of an activity—benefits not received by the person who is carrying out the activity. For example, a farmer who encourages beneficial insects in her field will benefit the farmers whose fields are adjacent to hers, to the extent that the insects she encourages attack pests in her neighbors' fields. Figuring out whether externalities are positive or negative, or both, is not always easy. For example, the large increase in area planted to transgenic *Bt* cotton (engineered using genetic material from the soil microorganism *Bacillus thuringiensis* to produce a pesticide against cotton bollworm, the major cotton pest) in China has been claimed to result in an increase in resistance of the bollworm to the *Bt* toxin (Tabashnik et al. 2012) and an increase in secondary pests (Lu, Wu, Jiang, Xia, et al. 2010), which could negatively impact farmers not planting *Bt* cotton. But the same increase in *Bt* cotton area appears to have resulted in an increase in beneficial insect predators in fields adjacent to those planted with *Bt* cotton (Lu, Wu, Jiang, Guo, et al. 2012) and a reduction in bollworm in other nontransgenic crops (Wu et al. 2008), both of which could benefit farmers not planting *Bt* cotton. Whether the net effect of these externalities is positive or negative is unclear at this point.

A major concern for all emphases in creating more sustainable agrifood systems is how to internalize the cost of negative externalities to the producer, in order to create incentives to alter production to reduce the externalities. The economic emphasis sees markets as the best way to internalize externalities, based on the assumption that when the costs of the externality or the rewards from internalization become large enough they will affect the person or other entity carrying out the activity, who will then be motivated to internalize in order to maintain profits. This could happen directly—for example, if nitrate pollution affects the farmer's own water supply, and she has to pay to treat it—or indirectly—when consumers' concern about nitrate is reflected in their purchasing from producers who have lower negative externalities, although this may mean paying a higher price. According to the economic emphasis, any economically important negative externality of agrifood systems will be internalized via the market so that levels of inputs and outputs will be adjusted accordingly.

In contrast to an economic emphasis that relies on markets to establish a sustainable rate of resource use, an environmental emphasis stresses the necessity for scientific assessment of the total environmental costs of resource use for a given production function (fig. 4.1, above), and internalizing these costs via regulations and laws. This requires adequate investment in monitoring externalities, including developing adequate technologies, and institutional and political structures to implement the laws and regulations. This is especially challenging for dispersed, or nonpoint sources, externalities like water pollution, even in industrial countries with relatively high levels of resources, as is the case for agricultural water pollution in California (Dowd et al. 2008).

A social emphasis requires attention to the unequal distribution of externalities on society, and on equity as the basis for internalizing negative externalities. It sees the economic approach as ignoring the reality that negative externalities are disproportionately relegated to places, or to the future, where people have little if any effect on markets, or on policies that could influence markets, because they are especially poor or powerless. For example, water pollution by industrial agriculture in California's Central Valley has resulted in residents in many low-income communities where farm workers live having to purchase drinking water because the water supply is so polluted by industrial agriculture (Francis and Firestone 2011). In 2010, the U.N. Independent Expert on the right to water and sanitation investigated this situation, and in 2012 the California human right to water bill came into effect, but how effectively it will be implemented remains to be seen.

6. RISK MANAGEMENT

Uncertainty and risk are important influences on how humans understand and respond to situations in the agrifood system, and they are dealt with in very different ways by the three emphases. *Uncertainty* is imperfect knowledge of the probability of events and their consequences and *risk* is the subjective probability of the consequences of uncertain events, especially unfavorable consequences (Hardaker et al. 1997). Some of the important kinds of uncertainty faced by farmers and consumers, especially those in marginal environments, include natural hazards, market fluctuations, social relations, state actions (such as agricultural pricing policy changes), and social instability (such as wars) (Ellis 1993:82–84).

In the industrial world, the risk management process (RMP) is the standard institutional approach to risk for new technologies, such as pesticides or farm machinery, and novel biological entities, such as invasive species or transgenic crop varieties (NRC 1996, 2002). There are four key steps in this process, although they may be organized and labeled in different ways (fig. 4.2): (1) *identification* of a hazard (or potential risk), (2) *risk analysis* of the probability of (a) *exposure* to this hazard and (b) *harm* resulting from this exposure (risk $= P_E \times P_H$), (3) *risk evaluation* (or perception, assessment) of harm, and (4) *risk treatment* (or management, regulation) of risk by reducing exposure and harm (e.g., COP CBD 2000, Dietz, Frey, et al. 2002, Hardaker et al. 1997, NRC 2002:54–55). However, there is much controversy about the way in which these steps are carried out, and about the broader context within which risk management functions (for an example with *Bt* maize in the European Union, see Levidow 2003).

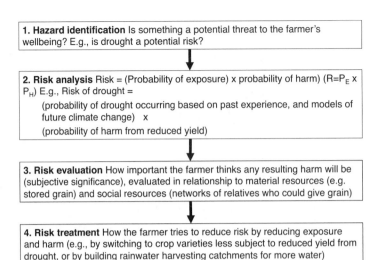

1. **Hazard identification** Is something a potential threat to the farmer's wellbeing? E.g., is drought a potential risk?

2. **Risk analysis** Risk = (Probability of exposure) x probability of harm) ($R = P_E$ x P_H) E.g., Risk of drought =
(probability of drought occurring based on past experience, and models of future climate change) x
(probability of harm from reduced yield)

3. **Risk evaluation** How important the farmer thinks any resulting harm will be (subjective significance), evaluated in relationship to material resources (e.g. stored grain) and social resources (networks of relatives who could give grain)

4. **Risk treatment** How the farmer tries to reduce risk by reducing exposure and harm (e.g., by switching to crop varieties less subject to reduced yield from drought, or by building rainwater harvesting catchments for more water)

FIGURE 4.2. The risk-management process. © 2013 David A. Cleveland.

Lack of knowledge makes *ex-ante* (before the event—i.e., before the technology has been used in context) benefit-cost analysis, including the RMP, difficult. A mainstream economic emphasis favors an *ex-post-trial-and-error* approach in setting environmental hazard alarms (Welsh and Ervin 2006)—that is, conducting evaluations after the technology has been implemented. This approach tends to set the sensitivity of the environmental alarm to the left of the optimal point in figure 4.3, which is the production function graph of figure 4.1, but with a cost curve that has increasing slope. Therefore, the assumption is that technology development should be determined by the market, and that this will eventually correct any errors (and new technology will mitigate any harm) and result in the overall greatest good. With this approach, the burden of proof is on society and scientists to show net costs if they exist, and the focus is on controlling type I error in testing potential for harm. Type I error is concluding that there is a difference between two treatments when there is none (see chapter 3, section 6.2, and fig. 3.5). In other words, a type I error would be concluding that the new technology is more hazardous than the technology it would replace when it is not. The main concern is gaining, and avoiding losing, the benefits of a new technology. We could call this the "stop worrying, go for it, and see what happens" approach.

Environmental and social emphases tend to favor a precautionary approach to risk, and *ex ante* benefit-cost analysis—that is, conducting evaluations before the technology has been implemented—in contrast to the ex-post-trial-and-error approach. A *precautionary* approach tends to set the sensitivity of the environmental alarm to the right of the optimal point (fig. 4.3). This leads to the conclusion that technology development should be determined through the RMP using available knowledge, new research, and social evaluations, and that it will eventually result in low risk and the overall greatest good, perhaps through rejecting the proposed new technology in favor of existing or alternative technologies. The burden of

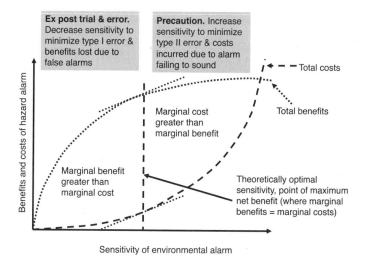

FIGURE 4.3. Contrasting preferences for setting the sensitivity of environmental alarms. Based on (Pacala et al. 2003). © 2013 David A. Cleveland.

proof is on the proponents of a new technology to show net benefit, and the focus is therefore on controlling type II errors in testing the potential for harm. Type II error is concluding that there is no difference, that the new technology is not more hazardous than the technology it would replace, when in reality it is (fig. 3.5). In other words the main concern of this risk-averse approach is in avoiding, or at least not increasing, costs, including negative outcomes. We could call this the "better safe than sorry" approach.

Differences in assumptions about risk among the three emphases in agrifood systems sustainability are important for understanding how farmers deal with one of their biggest sources of risk, uncertainty about future growing conditions. Farmers have to make decisions about when to plant seed and when to begin weeding and harvesting based on their assessment of risk. In the more marginal environments of many Third World farmers, this means that many inputs that could produce higher average (expected) yield are also higher risk because they increase responsiveness—that is, they increase the amplitude of variation from year to year. This is illustrated in figure 4.4, which compares the yields of two varieties of a crop across a series of good and bad years. The variety with the higher mean yield also has the highest variability of yield and therefore the higher risk; it is more responsive to environmental variation than yields of the variety with lower mean yield and lower risk.

Understanding farmers' concepts of risk is important for understanding farmer practices and the potential for farmers to improve their agriculture, and for understanding agricultural development in general. The different emphases in agrifood system sustainability result in differing assumptions about farmers and their attitudes toward risk, leading to different conclusions about the optimal level of inputs to invest in production. Much mainstream economic research on risk assumes that "the farmer is *risk neutral*, and the farmer's

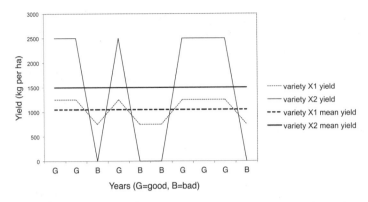

FIGURE 4.4. Variability in growing conditions, yield, and risk through time. © 2013 David A. Cleveland.

objective is to maximize expected profits" (e.g., Zilberman et al. 2007), and these assumptions in turn shape economic policy recommendations (Gowdy and Erickson 2005:209). Figure 4.5. illustrates farmer choices in the face of risk due to the uncertainty of growing conditions from year to year. TVP_G shows the output in good years and TVP_B the output in bad years, while TVP_E shows the average output weighted by the farmer's assessment of the probability of the occurrence of good and bad years, based on her past experience. As a result of this experience, the farmer's subjective assessment in this example is to expect good years 60 percent of the time and bad years 40 percent of the time.

The "rational" level of input for a risk-neutral farmer is at X_2, because this provides the highest average profit over time (f – g), given the expectation of the probability of years with good or bad growing conditions (fig. 4.5). Even though in bad years investing inputs at X_2 will result in a small loss (g – h), this is balanced by the large profit in good years (e – g). If the degree of uncertainty, and therefore of risk, could be significantly reduced so that the farmer could predict annual growing conditions fairly accurately, then it would be rational to invest at X_3 in good years (profit = m – p) and X_1 in bad years (profit = c – d) to gain the largest overall profit. However, given the unpredictability of growing conditions, affected by temperature; rainfall; pest populations; supply and cost of irrigation water, compost, seeds and other inputs; and labor supply, it is rational to invest at X_2.

When farmers don't invest at X_2, mainstream economists view them as irrational. However, environmental and social emphases have different explanations for such behavior. The majority of small-scale farmers have little access to external inputs, credit, savings, insurance, or government support, which means that they cannot survive bad years if they invest at X_2, since they will have a net loss (g – h) and no resources to survive on. Research has found that the majority of these farmers are in fact risk averse and choose to invest at X_1, which means that they forgo the larger profit in good years, but even in bad years they make some profit (c – d). In addition, farmers have motives other than maximizing profits from crop production, so they may not invest at levels above X_1 because they choose to invest their

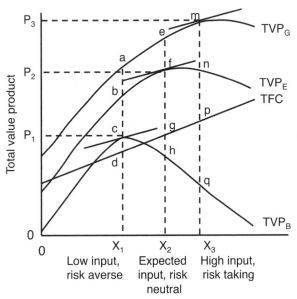

FIGURE 4.5. Farmer selection of input level based on subjective risk. TFC = total factor cost; TVP_E = expected (average) output, farmer expectation; TVP_G = production in good year; TVP_B = production in bad year. Based on (Ellis 1993:86–89).

time and other resources in other activities—for example, in community activities or off-farm labor that may maximize their overall well-being.

Having now contrasted the basic assumptions underlying the three emphases in sustainability (section 2), and discussed how, as a result, each emphasis understands and responds to key variables in agrifood systems (sections 3–6), let's return to the current world food crisis I described in the introduction chapter, and look in more depth at the differences between the mainstream and alternative perspectives.

7. THE FOOD CRISIS FROM DIFFERENT PERSPECTIVES

While the G8 heads of state enjoyed their sumptuous meals at their meeting in 2008, described in section 1, what solution to the food crisis did they offer? They stated: "We emphasize that economic growth must be at the heart of any successful strategy for reducing poverty, achieving the Millennium Development Goals and enabling developing countries to become self-reliant. We shared the view that the private sector is the driver of growth and offers the most effective way to create wealth, jobs and prosperity."[7] In addition to private sector–driven economic growth, the group emphasized increasing food production to achieve food security and highlighted the importance of commercial fertilizers and seeds, including transgenic crop varieties (TGVs).

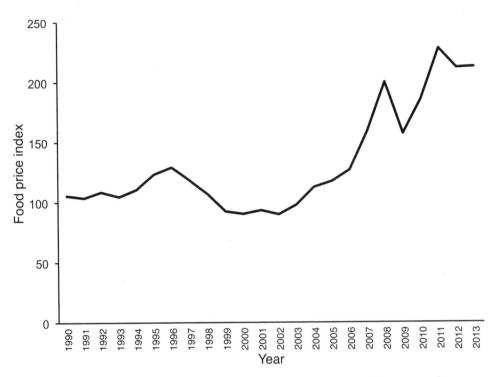

FIGURE 4.6. The FAO Food Price Index. The index consists of the average of five commodity group price indices weighted with the average export shares of each of the groups for 2002–2004: in total, fifty-five commodity quotations considered by FAO commodity specialists as representing the international prices of the food commodities noted are included in the overall index. All indices have been deflated using the World Bank Manufactures Unit Value Index (MUV), rebased from 1990 = 100 to 2002–2004 = 100. Graph created from data in (FAO 2013a).

Several months after the G8 summit, in October 2008, La Vía Campesina, an organization representing Third World farmers, also issued a statement about the world food crisis explicitly laying blame on the "globalized, neoliberal, corporate driven model," in which agriculture is seen as a profit-making venture with resources concentrated in the hands of private industry (LVC 2008:43).[8] In contrast to the G8, La Vía Campesina emphasized its vision of a food system based on food sovereignty, small-scale farmers, local resources, and domestic markets. Clearly there is strong disagreement about the crisis, but there is some agreement about the proximal cause.

Most observers have defined the immediate cause of the food crisis as steeply rising food prices and their effects on consumer access to food. Food prices rose in both the Third and industrialized worlds—and the crisis continues, with prices in 2012 at record high levels (fig. 4.6), a trend that has slowed down slightly due to drops in the prices of sugar and dairy products (FAO 2013c). In fact, food prices in the last half of 2010 and first half of 2011, including prices for cereals, were more than twice their level in 2001–2004. As food prices rise, people's ability to buy food decreases, which can affect their nutritional status. FAO data showed a slight increase in the absolute number of undernourished people in the world in 2010–2012 (868 million)—down 13 percent from 1990–1992 (1,000 million), but 0.001 percent higher than in 2007–2009 (867 million).[9]

Even in the United States, one of the wealthiest countries in the world with a thoroughly modern industrial agrifood system, the prevalence of food insecurity in 2008 reached its highest level since 1995, when the first food security survey was conducted, with 14.6 percent of households "food insecure at least some time during the year, including 5.7 percent with very low food security" (Nord et al. 2009).[10] Where I live, in Santa Barbara County, California, a relatively wealthy county in one of the most productive agricultural regions of the world, the local food bank struggled to keep up with the growing demand for affordable food by the county's poorest. In 2010 the Santa Barbara Food Bank distributed 10.5 million pounds of food to 164,000 people (38 percent of the population)—that is, 2.8 ounces of food per person per day—yet 17 percent of households with children that received food assistance reported that those children had gone hungry at least once during the previous twelve months (FBSBC 2010).

The G8, adhering to the economic emphasis, implied that restrictions on free markets and a lack of food production were important causes of the crisis. That is, they saw the crisis as primarily a *supply-side* problem: food shortages due to inadequate production led to increased prices. This emphasis sees industrial agriculture as a great success and holds up the Green Revolution as a model for the Third World. According to this perspective, the causes of the food crisis include inefficient economic functioning of the globalized industrial agrifood system due to lack of supporting policies and public investment to stimulate private-sector development of new technologies and to provide inputs such as genetically engineered crop varieties and manufactured fertilizers.

Advocates of alternative perspectives, like the international NGO GRAIN, tend to have social and environmental emphases that are often diametrically opposed to those of the G8.[11] In Grain's July 2008 article "Getting out of the Food Crisis" (GRAIN 2008), the organization laid out its assumptions—that industrial agriculture and the Green Revolution have been failures, that local food systems are best, and that traditional knowledge is superior to the knowledge of mainstream "experts." The organization saw the market-dominated system, a key feature of the mainstream agrifood system, as a major cause of the food crisis. In the article, GRAIN refers to the "market sham," claiming that unregulated markets underlie the failure of the dominant industrial agriculture system to make food available to those who most need it, and that those markets encourage land grabbing and speculation (GRAIN 2008, 2009).

In contrast to the mainstream, the alternative perspective sees the crisis as primarily a *distribution and demand-side* problem—the inability of the dominant global agrifood system to make resources for food production and food available to everyone, including the poorest, and the excessive demand on natural resources and on the Earth's capacity to absorb the negative effects of that system. There is adequate food and other agrifood system resources, but the people who need them most are invisible in the marketplace because they cannot trans-

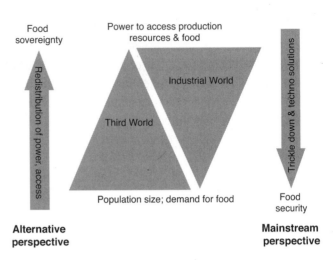

FIGURE 4.7. Alternative and mainstream perspectives about the way the world works for understanding the world food crisis. Food security is movement of food and technology from the industrial world to the Third World, with continued control by the industrial world; food sovereignty is movement of power and access to the Third World so people there can supply their own food and technology. © 2013 David A. Cleveland.

late those demands into money—markets cannot recognize their biological needs. Markets respond instead to those with the most money, and relatively richer people and countries consume food and agrifood system resources at levels much higher than biologically necessary. Hunger and overconsumption are negative social externalities of the mainstream agrifood system.

7.1. The Right to Food

The food enjoyed by G8 leaders at their 2008 summit was obviously expensive, but it was also high on the food chain, with animal products featured in almost every dish. The decision to feature such foods implied a number of mainstream economic assumptions—for example, that the resources (including land, water, energy, and nutrients) needed to produce food will be available over the long term, that those with power *should* be able to use those resources as they wish, and that they do not have to give up anything in order for those on the bottom of the food power pyramid to have food to eat (fig. 4.7). This is a trickle-down approach to the food crisis as a supply-side problem—the solution is to produce more food, not to reduce consumption of the rich, and redistribute resources to the poor, because there is no shortage of resources that new technologies and more efficient markets can't solve. What does the economic emphasis of the mainstream imply for the right to food?

The idea that people have a right to food was first stated in an international declaration on December 10, 1948, when it was briefly mentioned in the Universal Declaration of Human Rights adopted by the UN General Assembly, and it has since been reasserted in a number of international documents. A core difference between the mainstream and alternative perspectives is in their theories of what the "right to food" means, illustrated by the contrasting concepts of food security and food sovereignty (fig. 4.7).

Food security is the way the mainstream tends to think about the right to food, emphasizing the technical and biophysical aspects. For example, in 2000 the UN High Commissioner for Human Rights defined the right to food as including rights of "access to safe and nutritious food . . . adequate food" and freedom from hunger "so as to be able fully to develop and maintain their physical and mental capacities" (UNHCHR 2000). The FAO definition adds "physical and economic access to sufficient, safe and nutritious food that meets their dietary needs for an active and healthy life," and the USDA definition includes the concept of "socially acceptable" ways of acquiring food—that is, "without resorting to emergency food supplies, scavenging, stealing" (USDA ERS 2009).[12]

Food sovereignty is a term that was first coined by La Vía Campesina in 1996. It is a broader, more active definition of the right to food concept, with a social emphasis that maintains that people *have* this right, rather than being passive recipients who have it *given* to them by governments and development professionals. La Vía Campesina's definition emphasizes "new social relations free of oppression and inequality," control of the agrifood system by farmers and their communities, fair trade and social justice, and the notion that food is not a "mere commodity" (LVC 2008:61, 147–148). UN Special Rapporteur on the Human Right to Food Olivier De Schutter defines the right to food in sovereignty terms— an alternative perspective from the United Nations, an organization whose views tend to be mainstream. He emphasizes that it is "not primarily about being fed" but rather about "being guaranteed the right to feed oneself" (De Schutter 2011c).

7.2. Solutions

The mainstream and alternative perspectives have proposed different actions and policies for solving the world food crisis, which can be understood in terms of the differences illustrated in figure 4.7. Proposed solutions could be thought of as hypotheses that can be tested by implementing them and then measuring how effective they were in resolving the crisis. These proposed solutions are implicitly based on each of their respective definitions of the crisis based on assumptions about agrifood systems.

The mainstream does not believe in the need for a redistribution of resources—a reduction in consumption by the relatively rich, or in the industrial world in general, so that the relatively poor, or the Third World in general, can have more to eat. This is because the mainstream perspective does not see resources as limiting—what is limiting is the capital investment and new technology to increase food production. Therefore, the mainstream's proposed solution consists of the following components: loans, a "robust world market and trade system," a "science-based strategy" led by the "international community" of governmental organizations like the United Nations, providing "access to seeds and fertilizers," and helping to "build up local agriculture by promoting risk analysis" for "new technologies" like genetically engineered crop varieties and biofuels.

The Millennium Project, for example, is based on the assumption that economic growth is the solution to poverty (Sachs 2005), and therefore the project seeks to end absolute poverty, based on an established threshold of available calories or income, not relative poverty (Unwin 2007). Ending relative poverty would mean increasing equity, closing the gap between rich and poor by redistributing resources and power. The contrast between the trickle-down and redistribution approaches has been brought out on the global scale at the Conference of Parties (COP) meetings on global climate change—for example, in 2011 in Durban, South Africa, where the richest industrial nations were the most resistant to reducing GHGE or helping poor Third World nations to deal with climate change. Interestingly, U.S. president Barack Obama's past endorsement of "redistribution" for social well-being was perceived as a significant liability in the 2012 U.S. presidential campaign, reflecting the dominance of the mainstream perspective.[13]

By the time of the G8 Summit in France in 2011, the food crisis was overshadowed by the revolutionary movements in the Middle East and North Africa and the earthquake-tsunami-nuclear power generator crisis in Japan. However, their report on health and food continued the mainstream approach, supporting the CGIAR as the organization to lead the way.[14] For example, while the importance of "small" farmers was acknowledged, the main method proposed for supporting them was working with the private sector, which means moving them into modern industrial agriculture (G8 2011), and to market forces that will result in most small farms being absorbed into a few expanding large farms. The mainstream solutions were reiterated by the FAO in its 2011 report on the state of global food insecurity—increased production, international trade, and the role of the public sector in supporting private-sector-dominated solutions (FAO 2011b).

In 2012 the director-general of the FAO and the president of the European Bank for Reconstruction and Development co-authored an op-ed titled "Hungry for investment: The private sector can drive agricultural development in countries that need it most" (Chakrabarti and Graziano da Silva 2012). To solve the world food crisis, they advocated a mainstream economic vision of industrial agriculture where governments provide the policy and physical infrastructure for private investors to "fertilize this land with money." This summarizes the commitment of the dominant global food and agriculture organization to market-based solutions to the world food crisis and sets it squarely within the mainstream economic emphasis, and in vivid contrast with the environmental and social emphases of alternative perspectives.

In contrast to the top-down, market-based approach of the mainstream, GRAIN (2008) emphasized that the capitalist model of agriculture is "unjust" and that a "real shift in power" is needed. It proposed that markets be regulated to benefit the people directly, and since small-scale farmers are "still responsible for most food produced [they] should be the ones setting agricultural policy." The organization stated that "agrarian reform" (land redistribution) was necessary "to empower people to feed themselves and their communities" based on "indigenous knowledge, focused on maintaining healthy, fertile soil, and organized around a broad use of locally available biodiversity."

PART TWO

Moving toward Sustainable Agrifood Systems

A Balancing Act

Part 2 moves from the general discussion of problems and solutions to specific aspects of the long-term food crisis and provides examples of applying the concepts presented in part 1. Chapters 5, 6, and 7 take up the three fundamental changes of the Neolithic revolution—increased management of other species, ecosystems, and people—in more detail, showing how supply-side solutions worked through subsequent revolutions, and how they can be combined with demand-side solutions to create a more sustainable alternative agrifood system. In chapters 8 and 9 I address two of the biggest challenges to creating a more sustainable agrifood system—global climate disruption and economic globalization—and discuss the potential for diet change, food waste reduction, and localization to meet these challenges.

FIVE

Managing Evolution

Plant Breeding and Biotechnology

1. INTRODUCTION

It is an ocean of maize extending in all directions. The maize leaves curl in the dry after-
noon heat. The rains are late, the rains are scarce . . . again. The climate is changing. Teresa
looks out at her field and plans her next move. "Where can I get maize that will produce a
good harvest in a shorter time, that will help me cope with the changing rains, which come
late and leave early?" Teresa Gonzalez is a farmer in Oaxaca, Mexico. She began managing
her family's farm, along with working in the fields and preparing food, when her husband
started extended migrations to work in the United States many years ago. She is assisted on
the farm primarily by one son, Hector, and his wife, and one daughter, as her other four
sons and one daughter are living permanently in the United States. Their family has been
farming in their village in the Tlacolula branch of the Valley of Oaxaca for generations, long
before the European invasion. Their biophysical and social environments have always been
changing, but recently the pace of change has accelerated dramatically, as a result of migration, economic crises, and climate change.

A major criterion for Teresa's choice of new varieties will be the texture and taste of the
tortillas and other foods she and her family make from maize, since this is their staple food
and they grow all of the maize they eat in their own fields. Using her knowledge of the growing environments, cultivation practices, and quality of maize varieties in different locations
in the region, Teresa travels to a community known for its short-cycle maize and obtains
seeds of a variety there that she will plant. She will select seed from her yearly harvest, gradually adapting it to the environments of her farm, just as she saves the seed of the other varieties of maize, beans, squash, and other crops she grows.

In the middle of North America is an even larger ocean of maize, but virtually all of this maize is grown from hybrid varieties, purchased by farmers from multinational corporations like DuPont Pioneer and Monsanto or from smaller companies, and by 2013 almost all—90 percent—of the seed planted was genetically engineered (USDA NASS 2013) and required farmers to pay those seed companies an added "technology fee." Almost all of this maize will be processed into ethanol fuel, animal feed, or industrial products like high-fructose corn syrup—almost none of it will be directly eaten by humans. Farmers here also have to deal with an increased pace of change, and in the summer of 2012 this "corn belt" suffered a major drought, likely a result of anthropogenic global climate warming.

These two ways of growing maize are very different, the ways of using it are very different, and the varieties grown are very different. The genetic structures of *farmer crop varieties* (FVs) and plant breeders' *modern crop varieties* (MVs) reflect differences in the environments in which they are selected, the selection criteria, the methods used, and the knowledge and values of the farmers and scientists that determine their goals for selection. This means that plant breeding and biotechnology, and the crop varieties they create, can be understood only in the cultural, social, and economic contexts in which they operate, in addition to their biophysical contexts. As the eminent plant breeder Norman Simmonds stated, economic criteria for selection are "always implicit in the objectives and in the conduct of selection in plant breeding" and the "nature of a maximum profit/benefit criterion depends on who writes the equation" (Simmonds and Smartt 1999:358). This observation applies to all aspects of the social environment and all phases of plant breeding, and to farmer as well as scientist breeding.

This chapter is about plant breeding, or the human management of other species, from the Neolithic to biotechnology. I focus on the similarities and differences between traditional farmers and modern plant breeders—many traditional farmers embody the alternative social and environmental emphases, while many plant breeders embody the mainstream economic perspective. By analyzing the different perspectives about yield and stability, farmer-breeder collaboration, and biotechnology, we will be able to see how different biophysical and social environments and assumptions affect what farmers and plant breeders think plant breeding is, and what they think it should be. But first, I place plant breeding in broad perspective and discuss its biological basis.

2. PLANT BREEDING IN BROAD PERSPECTIVE

At the simplest level, crop plant breeding (hereafter, "breeding"), whether it involves selective harvesting and environmental management by the earliest foragers on their journey into the Neolithic or twenty-first-century genetic engineers inserting genes from distantly related species into crop plants, involves changing the genetic structure of the plants we eat. Those plants evolved over many millennia in response to selection pressures in their environments, before humans introduced new selection pressures. Beyond the simplest level, however, different forms of breeding differ substantially, as I will discuss below.

An important result of the current dichotomy between traditional farmer and modern scientific plant breeding is that the former is integrated within the household, community, and local landscape, while the latter is much more isolated, with its goals shaped by mainstream modern industrial society. The mainstream assumption of unilineal evolution means that plant breeders and crop improvement policy makers frequently promote MVs as best for traditional farmers, as epitomized by the Green Revolution and the current promotion of MVs, including transgenic crop varieties (TGVs), in the Third World.

2.1. The Separation of Farmer and Scientist Plant Breeding

As discussed in chapter 2, farmers were the first plant breeders, probably achieving most changes indirectly and unintentionally through their cultivation and seed saving (see fig. 2.5). With the beginning of more organized and specialized amateur breeding about two hundred years ago, farmer plant breeding and plant breeding by these specialists began to be separated in the more industrialized, temperate-region countries (Simmonds and Smartt 1999:12).

Modern scientific plant breeding coalesced after 1900 as part of the development of the "modern synthesis" in the theory of evolution, comprising three main contributions. *First* was evolutionary theory and data—most importantly, Darwin's proposal in the *Origin of Species* based on analogy with artificial selection by amateur plant and animal breeders, that species evolve gradually via natural selection acting on individual variation (Darwin 1859). Darwin was strongly influenced by Malthus's idea of geometric increase of organisms leading to increased competition for limited resources, so that individuals with slight advantages would survive in greater numbers to reproduce.

Second was the rediscovery and elaboration of Mendel's research on inheritance after 1900, first published in 1865. This led to the development of Mendelian genetics, which documented discrete hereditary units (later termed chromosomes and genes). Conflict between Mendelians (discontinuous evolution by qualitative leaps) and Darwinians (continuous evolution by small, quantitative changes) was resolved between 1918 and the 1930s by the *third* main component of modern evolutionary science, quantitative population genetics, primarily by Haldane, Fisher, and Wright (Fitzgerald 1990, Provine 1971). In the 1930s and 1940s the field was further defined by geneticists like Dobzhnsky, Simpson, and Mayr, and given its name by Julian Huxley in his book *Evolution, The Modern Synthesis* (Gould 2002:503 ff.).

Farmers and formal plant breeders continued to collaborate in the early years of modern scientific breeding—for example, in making crosses and selections in maize breeding in the United States (Fitzgerald 1990, Kloppenburg 1988), selecting lines of spelt and wheat in farmers' fields in Switzerland (Schneider 2002), and in farmer cooperatives breeding rye and oat in southern Germany (Harwood 2012). But beginning in the 1920s and 1930s, with the increased importance of evolutionary theory in formal breeding in comparison with empirical heuristics, and the growing economic impact of breeding, professional plant breeders and private seed companies came to dominate the development of new varieties in industrialized countries, and the role of farmers decreased. Plant breeders' concepts and methods began to develop independently of farmers' concepts and methods. To the extent

TABLE 5.1. Plant management Functions in Traditional Small-Scale and Industrial Large-Scale Agriculture

Function	Traditional	Industrial
Plant management in general	Nonspecialized activities, integrated within households	Specialized activities, institutionally separated
Production	Family based, many farms	Vertically integrated, few farms
Consumption	Producers eat what they grow	Producers do not eat what they grow
Improvement (plant breeding)	Via farmer selection as part of seed saving	By professional plant breeders
Conservation	*In situ* in fields and storage containers	*Ex situ* in gene banks
Seed multiplication	As part of production	By seed companies, increasingly by a small number of multinational corporations

that farmers are involved by modern plant breeders in their work, this involvement has generally been limited to the stage of evaluating the plant breeders' populations or varieties in their fields (Duvick 2002). Later in this chapter I will discuss the move in recent decades to reunite farmers and scientists in collaborative breeding.

2.2. Similarities and Differences between Breeding by Farmers and Scientists

The continuing separation between informal farmer breeding and formal professional breeding can be understood in terms of the three major components of the Neolithic revolution. *First,* in terms of controlling the evolution of other species, the selection process and products are different in the two forms of breeding. Traditional agriculture, in which farmers are the breeders, is often dominated by genetically diverse FVs, selected for adaptation to environments that are often highly variable and have relatively few external inputs.[1] Industrial agriculture, served by professional breeding, is dominated by MVs, selected for high yield in fields with high levels of inputs, which overall tend to have much less genetic variation than FVs (Frankel et al. 1995:57 ff.).

Second, in terms of ecosystems management, traditional farmers' fields are often more variable in space and time than industrial agriculture fields, and this variability results in greater phenotypic variation for many crop traits, including most yield-related traits in FVs. One important reason for the contrast in growing environments is the ongoing displacement of traditional farmers into relatively more variable, marginal environments by the spread of industrial agriculture into the more uniform, better environments. Environmental variation reduces the proportion of phenotypic variance that is additive genetic variance, thereby reducing a trait's heritability (the extent to which it will be passed from parent to progeny) (Ceccarelli, Grando, and Hamblin 1992) (section 3.3). As a consequence, for many

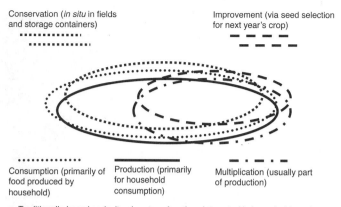

Conservation (*in situ* in fields
and storage containers)

Improvement (via seed selection
for next year's crop)

Consumption (primarily of
food produced by
household)

Production (primarily
for household
consumption)

Multiplication (usually part
of production)

a. Traditionally-based agricultural system: functions integrated in households and
communities.

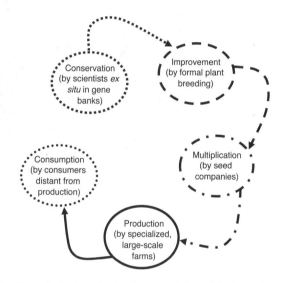

b. Industrial agricultural systems: functions separated, specialized, many institutionalized.

FIGURE 5.1. Plant management functions in traditional small-scale
and industrial large-scale agriculture. © 2013 Daniela Soleri and
David A. Cleveland. Used with permission.

important traits, heritable genetic variation is masked by environmental variation, increasing the difficulty for farmers to make genetic progress in selection.

Third, in terms of culture and society, traditional agriculture is characterized by the integration within the household or community of production, consumption, crop improvement, seed multiplication, and conservation of genetic diversity, whereas in industrial agriculture each of these functions is spatially and structurally separated and specialized (table 5.1, fig. 5.1).

This means that traditional agrifood systems are different than industrial systems in several important ways that affect the goals and results of selection. For example, the inte-

FIGURE 5.2. Farmer crop varieties. Oaxacan farmer Delfina Castellanos is winnowing black maize, a farmer variety she and her family have grown. It is a variety that is relatively low yielding and subject to high losses in storage, yet is prized for its flavor in foods like *tlayudas,* the large crisp tortillas of the central valleys of Oaxaca. Photo © 2013 Daniela Soleri. Used with permission of photographer and subject.

grated nature of traditional agrifood systems means that farmers value FVs not only for agronomic traits, such as drought or pest resistance or photoperiod sensitivity, but also for traits contributing to storage qualities, food preparation, taste, appearance, the potential for developing new varieties, and longer-term conservation of crop genetic diversity (Berthaud et al. 2001) (fig. 5.2).

In contrast to the integration of farmer breeding within plant management in traditional agriculture, formal plant breeding in modern agriculture is restricted to one particular part of the agrifood system (fig. 5.1), and plant breeders think of themselves as applied evolutionists who generate diversity and then select plants to create new varieties with desired characteristics (Allard 1999). The formal plant breeding process can be divided into six steps (Simmonds and Smartt 1999, Stoskopf et al. 1993, Weltzien et al. 2003):

1. Decision about breeding goals and plan

2. Creation of a large amount of genetic diversity through choosing parent germplasm, hybridization (crossing), and recombination in progeny generations, and by direct manipulation (e.g., radiation to create mutations, genetic engineering to insert genes)

3. Selection from that diversity of individual plants and populations that results in a new population with changed allele frequencies and phenotypes (the *phenotype is* the sum of an organism's physical properties)

4. Evaluation of the "best" populations resulting from selection across a wider range of *test environments*

5. Choice of varieties for release in the *target environments* on the basis of their potential to outperform (namely, outyield) the existing varieties

6. Release of a new variety and its subsequent multiplication and dissemination to farmers

Farmers' selection in traditional agriculture differs in many ways from plant breeders' selection because of the high levels of genetic and environmental variance, low heritability, and multiple integrated functions affecting selection goals. However, this does not mean that farmers' understanding of the basic relationships underlying selection is different than those of biologists and plant breeders; they can have quite similar understandings of these basic relationships (Cleveland and Soleri 2007a, Soleri, Cleveland, Smith, et al. 2002) (section 5). Indeed, plant breeders who view plant breeding as a biological process may recognize that plant selection in traditional systems is also applied evolution. "Plant breeding is the current phase of crop evolution and it proceeds by the same mechanisms that are responsible alike for the evolution of wild populations of plants and of cultivated ones in earlier times" (Simmonds and Smartt 1999:27).

For example, farmers appear to understand the roles of genotype and environment in determining some phenotypic differences (e.g., Sperling et al. 1993), and heritability as the basis for selection for short- and long-term change. Farmers can also distinguish between high and low heritability traits and, especially in cross-pollinating crops, consciously select for the former to create populations that express these new traits across a range of environments, while considering it not worthwhile, or even possible, to create similar populations when selecting for low heritability traits (Soleri and Cleveland 2001, Soleri, Cleveland, Smith, et al. 2002). I will discuss farmer knowledge further in relation to farmer-breeder collaboration below.

2.3. Plant Breeding and Crop Genetic Diversity through Time and Space

As we saw in chapter 2, genetic diversity is an essential ingredient of evolution, and this includes applied evolution or plant breeding; without diversity for a given trait, there is no possibility for adaptation to new conditions. The degree of genetic diversity at several spatial scales—crop plants, populations, varieties, and species—has changed significantly through time since the Neolithic, with important implications for adaptive evolution. We can identify five broad periods listed in table 5.2 and illustrated in figure 5.3. In the next two chapters, I discuss diversity at higher levels—in fields, ecosystems, and social systems.

Foragers ate a very large number of plant species—approximately 1,410 species in central Africa, for example (Harlan 1992). With domestication, beginning about twelve

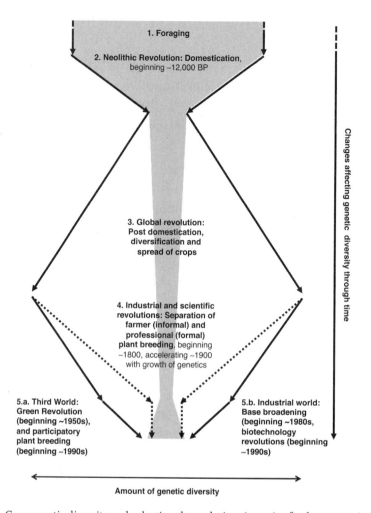

FIGURE 5.3. Crop genetic diversity and selection through time in major food crop species. Shading denotes changes in interspecific diversity (i.e., number of species); arrows denote changes in intraspecific diversity (i.e., number and genetic diversity of varieties); horizontal distance between arrows and width of shading represent relative amount of crop genetic diversity. **1. Foraging.** Large number species exploited (e.g., ~1400 spp. in central Africa). **2. Neolithic Revolution: Domestication,** beginning ~12,000 years ago. Limited number of species and of populations within species domesticated, leading to domestication genetic bottleneck; high rates of genetic change with exploitation of major genes in domestication syndrome **3. Global revolution: Post domestication, diversification and spread of crops.** Cycles of differentiation and hybridization; intraspecific diversification in form of locally selected FVs; great increase in number of varieties within species; slowing rates of genetic change. **4. Industrial and scientific revolutions: Separation of farmer and professional breeding,** beginning ~1800, further separation beginning ~1900 with growth of genetics. **a. Professional breeding.** ~1800; scientific breeding ~1900 leads to increased rate of intraspecific genetic change via Mendelian and quantitative genetics; decrease in number of major species (to ~30); large decrease in number of varieties; MVs and increased agronomic management widely adopted in industrial world. **b. Farmer breeding.** FV diversity decreased through replacement by MVs, decreased area planted to remaining FVs, decreased number of farmers, other socioeconomic and biophysical changes. **5. Green and biotechnology revolutions,** beginning ~1950s, ~1980s repectively. **a. Farmer breeding.** Green Revolution and spread of MVs leads to rapid decrease in FV diversity, limited local increase due to participatory plant breeding and unintentional gene flow. **b. Professional breeding.** Green Revolution MVs developed; TGVs 1990s; MV diversity increased via base broadening (e.g., by crossing with FVs) ~1980s, and genetic engineering; decreased diversity due to consolidation of seed industry, decreased number of varieties. Data from (Harlan 1992, Meyer et al. 2012, Simmonds and Smartt 1999). © 2013 Daniela Soleri and David A. Cleveland. Used with permission.

TABLE 5.2. Levels of Diversity in Agricultural Systems

Level (in time and space)	Small-scale, traditional agrifood systems	Large-scale industrial agrifood systems
1 Plants, genetic and phenotypic diversity	FVs have more diversity (more heterozygous)	MVs have less diversity (more homozygous)
2 Populations within varieties, genetic and phenotypic diversity	More populations, smaller population size; heterogeneous allogamous crops, multilines of autogamous crops	Fewer populations, larger population size; more homogeneous allogamous hybrid varieties, homogeneous autogamous crops
3 Varieties within species	Many, generally more locally adapted	Few, generally more widely adapted
4 Species	Many, often in polycultures	Few, often in monocultures
5 Fields	Smaller, with more environmental variation	Larger, with less environmental variation
6 Ecosystems	More varied, more nondomesticated species, more noncultivated areas	Less varied, fewer nondomesticated species, less noncultivated area
7 Foods	More varied, specific foods linked with specific varieties	Less varied, most foods not linked to specific varieties
8 Sociocultural systems (history, social organization, knowledge, values)	Smaller scale, more varied	Larger scale, less varied

thousand years ago, plant food sources began to be limited to a much smaller number of species than those used by foragers. In addition, the development of new domesticated species from a small sample of wild plants reduced intraspecific diversity but was often followed by high rates of genetic change and by fixation at key gene loci (loss of all but one allele) due to selection of major genes in the domestication syndrome (a recurrent suite of traits present in domesticated species). After domestication, the spread of some crop species to new environments and human groups led to cycles of differentiation and hybridization, followed by artificial and natural selection, resulting in greatly increased intraspecific diversity in the form of many locally adapted FVs within the relatively few species domesticated.

The rise of industrial agriculture and the separation of farmer and professional plant breeding continued the trend of fewer major plant species falling to the approximately thirty species now dominating the world food supply. It also led to a loss of intraspecific genetic diversity, including a decrease in the number of varieties within each species (Harlan 1992). As the agronomic management of growing environments increased in the industrial world, including mechanization and the application of irrigation, manufactured chemical fertilizers, and pesticides, similar growing environments came to cover larger areas, mostly those with better soils and climate. MVs were selected for these similar environments and were adopted across large areas, including large areas with one or very similar varieties.

These MVs came to be called *broadly adapted,* even though this breadth refers to spatial distance, not degree of environmental variation (Ceccarelli 1989). The expansion of relatively similar industrial growing environments increased the economic expediency of formal breeding programs whose products could be used over large areas. Hybrid varieties were created beginning in the 1930s with maize in the United States and had less genetic diversity than the open-pollinated varieties they replaced. Intraspecific diversity in the industrial world decreased not only by replacement of FVs by less genetically diverse MVs, but via the decreased area planted to the remaining FVs and decreased number of farmers maintaining varietal diversity.

Concern about the genetic uniformity of MVs leading to vulnerability to pests and diseases grew slowly beginning in the early 1900s and then accelerated with major epidemics such as the southern corn leaf blight epidemic in 1970, which led to a 15 percent drop in maize yield in the United States (NRC 1972). The understanding that this epidemic was caused by the genetic uniformity of widely planted maize MVs resulted in efforts to increase the genetic diversity of MVs through a process called base broadening. Base broadening involves introducing more diverse genotypes, including FVs and crop wild relatives, as sources of diversity in breeding programs for MVs (Cooper et al. 2001, Ortiz et al. 2007, Smale et al. 2002), which may have resulted in an increase in diversity among the main field crops (van de Wouw et al. 2010).

In the Third World, the Green Revolution was the introduction beginning in the 1960s of industrial agriculture featuring MVs and the inputs they needed for high yields; these varieties were mostly developed and distributed before base broadening was well incorporated into formal breeding programs. Green Revolution varieties were planted in the best environments and replaced FVs, which rapidly declined in number, along with the area planted with them and the number of small farmers. Remaining FV diversity became concentrated in the most marginal areas. More recently, intraspecific diversity has probably increased locally due to participatory plant breeding (Ceccarelli 2012:117) and probably unintentional gene flow from MVs, including transgenic crop varieties (TGVs).

Biotechnology increased genetic diversity of existing MVs by adding new genes using genetic engineering (Gressel 2008), with the first TGVs grown commercially in the 1990s and soon adopted widely and rapidly (James 2011). The biotechnology revolution has also decreased crop diversity as the new TGVs replaced a larger range of MVs and FVs. Furthermore, it has led to major consolidation of the seed industry into less than a handful of global "life sciences industries" (Fuglie et al. 2011, Howard 2009).

There are few data on the crop diversity being maintained by farmers globally, but the FAO attempted to estimate this by soliciting reports from member countries (FAO 2010:36–37). These reports suggested a general pattern of a higher level of diversity "in areas where production is particularly difficult, such as in desert margins or at high altitudes, where the environment is extremely variable and access to resources and markets is restricted." The FAO also noted that scientific research has shown that diversity in terms of numbers of traditional FVs tends to be "maintained on farm even through years of extreme stress" but that this conservation is mostly done by a small number of households in relatively small areas.

3. THE BIOLOGICAL BASIS OF PLANT BREEDING

In order to understand and evaluate different perspectives on the role of plant breeding (including plant biotechnology) in creating a more sustainable agrifood system, we need to begin with the biological basis of plant breeding, which is the same today as it was in the Neolithic. The first farmers interrupted natural cycles of evolution, bending them to their own needs and managing all three components of evolution directly and indirectly to increase food production: adding or creating plant diversity, selecting favored plants based on inherited phenotypes, and speeding up the process over time (chapter 2, fig. 2.3).

3.1. Sources of Diversity

As discussed in chapter 2, variation is a prerequisite for genetic change; without heritable genetic variation for a phenotypic trait, it is not possible to change the genetic basis of that trait through time—that is, to achieve evolutionary change.

Our understanding of genetic variation at the molecular level grew out of the discovery in the 1960s, by Rosalind Franklin, Francis Crick, James Watson, and others, of DNA as the primary information-containing portion of chromosomes. This was a major factor in the development of a powerful model for further research. Information is embodied in DNA as combinations of four nucleotides: cytosine (C), adenine (A), guanine (G), and thymine (T). The basic unit is the *codon,* comprised of a combination of three nucleotides. The four nucleotides can combine into sixty-four possible unique codons, sixty-one of which code for one of twenty amino acids, with each amino acid being coded for by one to six codons (Hedrick 2005:10 ff.).[2]

DNA codons in the chromosomes of the nucleus are transcribed by messenger ribonucleic acid (mRNA), which then moves to a ribosome in the cytoplasm. Transfer RNA (tRNA) brings amino acids to the ribosome, where they are assembled into polypeptide chains, according to the order encoded in the mRNA. Proteins are assembled from one or more polypeptide chains and are essential components of all cells—for example, as enzymes, structural components, and molecule transporters. A *gene* can be defined as a series of codons that code for a protein. The fundamental source of genetic variation is *mutation,* a change in the DNA from one nucleotide to a different one, or by addition or subtraction of nucleotides. This results in a change in codons, which in turn can lead to a change in the protein coded for.

Classical genetics has tended to reify the concept of the "gene" as the basis of all inherited traits (Commoner 2002, Keller 2000). Especially in the last few decades, however, there has been rapidly increasing recognition that the situation is neither so simple nor so deterministic. *Epigenetic* processes modify the expression of the information encoded in the DNA via heritable molecular configurations outside of the DNA.[3] Epigenetics also refers to heritable effects on protein function after translation of the genetic code into polypeptide chains, effects influencing the way transcribed polypeptide chains are folded into functional proteins—for example, as prions (infectious agents created when some proteins are misfolded, which cause other similar proteins to misfold in the same manner). These epigenetically created configurations can be inherited and are additional sources of variation.

Sexual reproduction involving male and female gametes is another key source of variation: the assortment of maternal and paternal chromosomes during *meiosis*, the production of gametes through reduction in chromosome number, and *fertilization*, the fusion of gametes that restores the original chromosome number and creates the next generation of individuals. Sex was probably selected for evolutionarily because it increases the genetic diversity that natural selection acts on, resulting in ongoing adaptation to changing environments (Barrett 2002). In diploid organisms, each set of two homologous chromosomes contains one chromosome each from the mother and the father.[4]

The first step of sexual reproduction (in diploids) is the formation of gametes via *meiosis* in each of the parents, which includes the random assignment of one chromosome from each set of homologous chromosomes to each gamete. Frequently the chromosomes in gametes include new mixtures of DNA different than the original template chromosome due to *crossing over* between two arms of a homologous chromosome pair followed by breakage and *recombination* of parts of different chromosomes. This crossing over and recombination occurs early in meiosis and is another source of variation that appears to play a major role in the evolution of plants (Gaut et al. 2007).[5]

Fertilization is the fusion of two haploid gametes—each containing only half the number of chromosomes as in a full set—one gamete carrying maternal chromosomes and one carrying paternal chromosomes. This *reassortment* of maternal and paternal alleles results in a diploid zygote that is a unique combination of alleles and grows into a unique mature organism that can again reproduce sexually.

Plants that reproduce mostly through cross-pollination between different individuals are called *allogamous*, or outcrossing.[6] Many crop plants are primarily allogamous, including maize, some squash, and many fruit trees. There are two other ways of reproducing that limit the generation of diversity but have other evolutionary advantages. Many other crop plants—including most of the cereals, such as rice, barley, and wheat, and some beans—are predominantly *autogamous*. In other words, they are self-pollinating, which results in genetically homozygous groups of individuals in populations that can maintain adaptive genetic configurations, although they often have 1 to 10 percent cross-fertilization, which adds important variation. Other crop plants, such as potato, banana, and cassava, are primarily reproduced vegetatively by cloning—that is, without sex—so that all plants in a clonal population are genetically uniform. Many fruit tree species are naturally allogamous but are propagated vegetatively to maintain uniformity. The reproductive system of each crop species has a major effect on the next steps in crop improvement by farmers or breeders: choice and selection.

3.2. Choice

In plant breeding there is an important distinction between *choice* of seed lots, populations, or varieties and *selection* of individual plants or lines from within populations or varieties

(Cleveland, Soleri, and Smith 2000). The *choice* of germplasm (seed lots, populations, and varieties) determines the genetic diversity available for selection (by farmers and breeders) and for production (by farmers). Farmers and especially plant breeders make choices between varieties and populations in the initial stages of the selection process when deciding on the germplasm to use for making crosses, and in the final stages when choosing among populations and varieties generated from those crosses for further testing (Hallauer and Miranda 1988:159), or for planting (farmers) or release (plant breeders). In addition, farmers' choices that are not part of crop improvement per se also affect the genetic diversity of their crop repertoires and determine the diversity on which future selection will be based. For example, farmers make choices when procuring grain for eating (which may later be planted as seed) and seed for planting, and when allocating different varieties to different growing environments.

As already mentioned, formal breeders often emphasize adaptation to wide geographic regions but a narrow range of growing conditions, and they tend to focus on developing and promoting the best variety or few varieties. In contrast, small-scale farmers, especially in marginal environments, tend to keep a repertoire of varieties whose different characteristics serve different needs. Therefore, some plant breeders collaborating with farmers support varietal portfolios (Ceccarelli et al. 2003, vom Brocke, Weltzien et al. 2003) available through farmer-to-farmer exchange as an alternative to the development of a small number of varieties for large-scale adoption.

For example, because climate change is reducing the growing seasons in Mali and other countries in the West African Sahel, some policy makers have argued that breeders should focus on improved short-cycle varieties (Dembélé and Staatz 2000:60). However, because in good rainfall years long-cycle varieties generally have higher yields (Adesina 1992) and farmers consider the grain quality of long-cycle varieties to be better (Ingram et al. 2002), farmers do not abandon those varieties. In Mali, a study of farmers' choices among their traditional sorghum varieties in terms of number of varieties maintained, and whether these were short- or long-cycle varieties, found that farmers make these choices in an effort to optimize harvests in the face of variation in the growing environment and in availability of inputs, such as labor and tools. Better rains in 2002 compared with 2001 appeared to be a major factor in the general shift toward more varieties and longer-cycle ones, with 60 percent of farmers adding varieties between 2001 and 2002 (Lacy et al. 2006).

Farmers may also choose more than one variety because of their different quality traits. For example, interviews with 599 Nigerian farmers found that they grow both long- and short-cycle cowpea varieties—short-cycle for food grain, long-cycle for feed during the dry season when fodder sources are scarce (Abdullahi and CGIAR 2003). Some maize farmers in Oaxaca, Mexico, maintain varieties specifically for their colored husks or tassels because of their aesthetic qualities—for example, colored husks used to color the tamales wrapped in them (Soleri, field notes, 1996–1999). Families who make the traditional maize and cacao beverage *tejate* maintain more varieties of maize than families who do not, using them to prepare that drink (Soleri, Cleveland, and Aragón Cuevas 2008a).

3.3. The Determinants of Phenotype and the Results of Selection

In contrast to choice, artificial *selection* by farmers and breeders refers to the identification of individual plants as parents of the next generation; these are the plants from which seeds or other propagules are taken. As discussed in chapter 2, artificial selection can be direct or indirect, and direct artificial selection can be either intentional or unintentional; intentional selection can be for phenotypic traits that are not heritable, for maintaining heritable traits intergenerationally, or for changing varieties over generations (see chapter 2, sector 4.3, and fig. 2.5). Varieties change or new ones are developed when selection is within phenotypically variable plant populations, and the variability is heritable and consistent over generations, resulting in changes in allele frequencies in the population. This is evolution, and it is fundamentally the same whether done by formal scientist breeders or farmers. However, no matter who is selecting, if the phenotypic trait being selected is not heritable, selection will not result in changes in allele frequencies.

The goal of plant breeding is to improve existing crop varieties or create new varieties with more desirable phenotypes. Desirable phenotypes could include taller or shorter plant height, greater vigor, higher yield, drought resistance, larger seed or fruit size, more intense color, greater resistance to storage pests, shorter cooking time, or better taste. As noted above, the *phenotype* is the sum of an organism's physical properties; it is determined by its genotype and the environment in which it develops, and by the interaction between its genotype and its environment:

$P = G + E + G{\times}E$, where
   P = phenotype
   G = genotype
   E = environment
   $G{\times}E$ = the effect of the interaction between the genotype and the environment

The potential for changing the phenotype is described in the elementary biological model on which plant breeding is based; this model is presented in the standard textbooks and is universally accepted among plant breeders (e.g., Falconer and Mackay 1996, Simmonds and Smartt 1999). It is the basis of the ontological comparator I describe below. The model is based on two fundamental empirical observations described in simple equations.

First, determination of variation in population phenotype (V_p) on which choice and selection are based is given by

$V_P = V_G + V_E + V_{G{\times}E}$, where
   V_P = phenotypic variance, for example in yield
   V_G = genotypic variance
   V_E = environmental variance
   $V_{G{\times}E}$ = variance due to the interaction between genotype and environment

While assessment of V_G and V_E can be relatively uncontroversial, the manner in which V_{GxE} is interpreted can vary. One reason is that at this level V_{GxE} is comprised of two components, V_{GxL} (V due to GxE across spatial locations) and V_{GxT} (V due to GxE over time—i.e., over seasons or years). V_{GxL} is relatively constant and predictable, while V_{GxT} is unpredictable, although it can be made more predictable with inputs such as irrigation water to control some of the variation. While V_{GxT} can occur due to the weather or populations of pests or pathogens, it also occurs because of the changing ability of households to acquire needed inputs such as irrigation water or fertilizers. Because most formal plant breeders work in industrial agricultural systems, they usually make the assumption that inputs are used and can be readily acquired, so their focus is on V_{GxL}. This interpretation influences plant breeders' approach to developing and improving crop varieties and their choices of how many and which varieties will be released across agricultural locations (Cooper and Hammer 1996), as I discuss regarding yield stability.

Second, response to selection (*R*) for the traits being measured is the difference between the mean of the entire population from which the parents were selected and the mean of the next generation that is produced by planting those selected parents under the same conditions. That difference is the product of two variables:

$R = h^2 S$, where

>> R = genetic response to selection in one generation, measured as the difference in mean phenotypic value for a given trait between the offspring of the selected parents and the entire parental generation before selection

>> h^2 = heritability in the narrow sense = V_A/V_p[7]

>> S = selection differential, measured as the difference in mean phenotypic value for a given trait between the selected parents and the entire parental population before the parents were selected[8]

Success in selecting for a given trait improves as the proportion of V_p contributed by V_G (especially V_A) increases. Thus, artificial phenotypic selection is a process of identifying the individuals with specific, desirable phenotypic traits within a population that will contribute genetic material to the next generation, but the genetic outcome of that selection will depend on the heritability of these phenotypic traits, which is the result of the genotypes and the environments in which they are expressed and selected.

Phenotypic selection has four primary genetic outcomes ($S \approx 0$; $S > 0$ and $R \approx 0$; $R > 0$ and $mEv \approx 0$; $mEv > 0$) depending on the heritability of the traits involved, the selection intensity, and the number of generations on which selection is carried out (Soleri and Cleveland 2004, Soleri and Cleveland 2009) (fig. 5.4). The extent to which the heritability of phenotypic traits will influence their inclusion in selection criteria will depend on many things, as discussed below in terms of farmer-scientist collaboration. Selection may have outcomes such as $R \approx 0$, which are typically not recognized by formal plant breeders, but which farmers see as useful and desirable. For example, maize farmers in Oaxaca, Mexico, select for

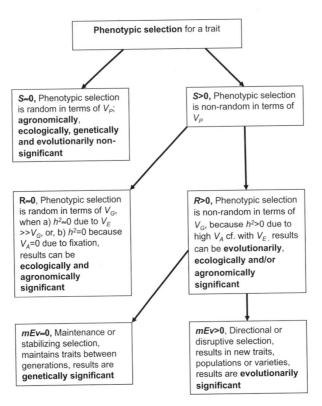

FIGURE 5.4. Phenotypic selection classified according to outcome of selection. See text for key to symbols. © 2013 David A. Cleveland and Daniela Soleri. Used with permission.

large seeds even though there is no response to this selection as documented by field experiments and by farmers' own statements.[9] However, larger seeds have better germination and produce more vigorous seedlings than smaller seeds, important advantages that are acknowledged by some farmers as the reason they select large seeds. Phenotypic selection can also result in genetic change ($R > 0$), but not directional change over generations ($mEv \approx 0$), when farmer selection is focused on maintaining a variety's recognized characteristics, such as seed shape or color, by removing variants in each generation resulting from cross-pollination. Directional cumulative change over time, or microevolution ($mEv > 0$), can occur when farmers consciously select to increase the frequency of a desired phenotype.

The biological relationships represented in the equations described here underlie plant breeders' understanding of even the most complex phenomena they encounter (Cooper and Hammer 1996, DeLacy et al. 1996). For example, two highly respected English-language plant breeding texts state that the relationship between genotype and phenotype is "perhaps the most basic concept of genetics and plant breeding" (Allard 1999:48), and of $R = h^2S$, that "if there were such a thing as a fundamental equation in plant breeding this would be it"

(Simmonds 1979:100). These equations can be thought of as the basic biological model of plant breeding, and they serve as an ontological comparator to assess differences and similarities among farmers, among plant breeders, and between farmers and plant breeders.

The biological basis of plant breeding is key to understanding how approaches to crop improvement differ according to one's perspective on agrifood system sustainability. In addition, comparing formal scientist and informal farmer plant breeding shows how different assumptions result in different practices, and in some ways, these types of crop improvement have come to represent the contrasting perspectives on agrifood systems that I have called "mainstream" and "alternative." In the next three sections, as we continue to look into the sources of differences in approaches to plant breeding in sustainable agriculture, I illustrate how the plant breeding process is the result of the interaction between its biological basis and its sociocultural and economic bases. From a holistic perspective (chapter 4), the knowledge of scientific and farmer breeders is a result of the interaction between their empirically based assumptions about plants, growing environments, and social and economic environments and the value-based assumptions underlying their goals for breeding. I look at how this interaction plays out for three key issues for the future sustainability of agrifood systems: the relationship between yield and yield stability on the kinds of varieties developed, the importance of how the biological basis of breeding is understood for collaboration between farmers and scientists, and the appropriateness of transgenic crop varieties for Third World farmers.

4. YIELD AND YIELD STABILITY

One of the best places to see the differences between mainstream and alternative approaches to plant breeding for agrifood systems is in the relationship between yield and yield stability. Yield is commonly defined as the ratio of edible harvest to the area of land required to produce it, for example, "5 MT per ha," and it is obviously an important criterion for evaluating the success of any agrifood system. A less commonly recognized but equally important criterion is *yield stability*—that is, the variation in yield through time and space, a specific case of $G \times E$. Yield stability is a measure of the interaction between the plant's genotype and the environment it grows in. As mentioned above, the focus of formal plant breeders is on stability through space, while Third World farmers are equally or even more concerned with stability through time, because they are not able to apply the inputs needed to reduce the relatively greater temporal variation in their growing environments.

As discussed in chapter 4, risk reduction is an important factor in farmers' decision making and something that gets a lot of attention from agricultural scientists, too. Low yields and variation in yield are major sources of risk (fig. 4.4), so it is not surprising that yield stability is often just as, or even more, important than yield both for farmers (Asrat et al. 2010, Soleri, Cleveland, Glasgow, et al. 2008) and for formal plant breeders and agricultural economists (Haussmann et al. 2012, Pingali and Rajaram 1999). The problem is that yield and yield stability are often negatively correlated—as average yield increases, stability decreases, and vice versa. Therefore, a major challenge for sustainable agriculture is finding situations where they are positively correlated.

However, plant breeders disagree about how to define the relationship between yield and yield stability and the significance of any negative correlations between them. Therefore, there is no consensus on the extent to which such negative correlations require changes in plant breeding theory and practice, including the choice of selection, test, and target environments. This disagreement is reflected in the different working definitions of yield stability used by plant breeders.

Farmers also have different views of yield and yield stability depending on their levels of production and access to resources, as well as their involvement in the market. This section analyzes these differences both between and among plant breeders and farmers, and relates them to mainstream and alternative perspectives on the future of the agrifood system (Cleveland 2001). Broadly speaking, the mainstream perspective, with its economic emphasis, focuses on yield, and the alternative perspective, with its environmental and social emphases, focuses on yield stability. Although both perspectives seek ways to bring yield and yield stability together, their approaches are very different.

4.1. Defining and Using Yield Stability

The easiest and most intuitively appealing way to illustrate yield stability and its contrasting definitions is with regression slopes (fig. 5.5).[10] This method was introduced to plant breeding by Finlay and Wilkinson (1963) and has been widely used by breeders.[11] The stability of individual varieties is indicated by comparison of their regression slopes over a range of environments, with each environment defined by the mean performance of all varieties in the trial in that environment, so that the population mean has a slope (b) of 1 (A, fig. 5.5) (Eberhart and Russell 1966, Hill, Becker, et al. 1998:Chap. 7).

4.2. Type 1 Stability

In plant breeding, *type 1 stability*, also referred to as static or biological stability, is recognized as the simplest concept of stability. With type 1 stability, the rate of reduction in yield with decreasing environmental mean yield is less for a stable variety (B, fig. 5.5) than for the population mean (A, fig. 5.5). Those with a regression slope (b) of 0 are defined as the most stable, with a slope increasingly positive or negative indicating decreasing stability.

Type 1 stability is the result of homeostasis or buffering, the characteristic of a plant or plant population that confers decreased sensitivity to changes in the environment (Allard and Bradshaw 1964, Borojevic 1990:332–334). While the mechanisms of this phenomenon are not fully understood, it is likely the result of heterozygosity at the individual level (individual or physiological buffering, or developmental homeostasis) or genetic heterogeneity at the population level (population buffering, or genetic homeostasis) (Haussmann et al. 2012).

Use of type 1 stability implies that plant breeding goals are dominated by environmental limits and high levels of variability in the factors affecting yield, such as rainfall, pest, and pathogen levels, many of which vary over time. As such, this is the definition of yield stabil-

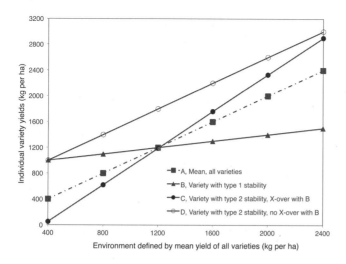

FIGURE 5.5. Regression slopes for individual variety yield on environment mean yield (of all varieties in trial), illustrating different definitions of yield stability. Adapted from (Cleveland 2001). © 2013 David A. Cleveland.

ity that fits most closely with the alternative perspective of agrifood systems with its environmental and social emphases, and its frequent focus on small-scale farmers using limited inputs in more marginal growing environments.

4.3. Type 2 Stability

With *type 2 stability*, also referred to as dynamic or agronomic stability, a stable variety (D, fig. 5.5) is one with a response to environments that is similar to the average of other varieties in the trials under consideration ($b = 1$), but with higher yields in any environment compared with the average of other varieties. Varieties with slopes decreasing below or above the average being increasingly unstable (Eberhart and Russell 1966). In practice, varieties or populations with $b > 1$ and no crossovers are considered superior to those with $b = 1$—for example, in trials comparing wheat purelines with hybrids, hybrids with $b > 1$ were considered superior (Bruns and Peterson 1998).

Use of type 2 stability implies that plant breeding goals are dominated by increasing yield in response to improved growing conditions (Eberhart and Russell 1966, Hildebrand 1990, Romagosa and Fox 1993). This is the mainstream perspective on yield stability. From this perspective, human scientific ingenuity and technology are regarded as the main limits to progress, which means that while a yield plateau is reached for any given input because of diminishing marginal returns, "a succession of new inputs can keep rescuing yield from the plateau" (Evans 1993:29) (chapter 4, section 3.2). Those who raise questions about the environmental sustainability of modern agriculture and MVs as represented by this definition of stability have even been characterized as "antiscientific" and emotional (Borlaug 1999).

As we can see, yield stability, a concept basic to understanding and improving the performance of crop varieties, can be defined and interpreted very differently depending on whether one adopts the perspective of mainstream formal breeding or the alternative perspective of some plant breeders and many smaller-scale farmers. While both perspectives are based in part on empirical evidence, which evidence is included and how it is interpreted is greatly influenced by both empirically based and value-based assumptions. The implications of these differences become even more evident as we follow their practical application in the following sections.

4.4. Yield Spillovers versus Yield Crossovers

For a crop variety created by formal plant breeders, the degree of similarity between the environments in which it was selected and tested and the environment in which it will be grown (target environment) is an important determinant of yield stability (Hill, Becker, et al. 1998: Chap. 7, Simmonds and Smartt 1999:200–207). A fundamental challenge in formal plant breeding is choosing appropriate selection and test environments, so that when crop varieties are grown in their target environments (farmers' fields) they will perform as intended, outyielding or at least equaling in yield the varieties currently being grown there, including FVs with type 1 stability (B, fig. 5.5). When such an advantage observed in the test environment carries over to the target environment, as illustrated by variety D (fig. 5.5), it is called a *spillover;* the change in yields over environments shows *quantitative* G×E interaction. In contrast, a *qualitative* G×E interaction among two varieties is called a *crossover* and is something plant breeders try to avoid. This is illustrated by variety C in fig. 5.5, which has a slope slightly greater than 1 and yield greater than B in good environments, but a yield less than B in marginal environments, and the two varieties crossover, or change ranks.

Plant breeders have debated the relationship between yield stability and selection and test environments for decades (Ceccarelli, Erskine, et al. 1994; Cleveland 2001). In terms of yield stability, the question of whether to choose wide or narrow adaptation as a breeding goal might be stated as: Are high yields and high yield stability compatible breeding goals, and will they result in spillovers? A goal of type 1 stability implies a "no" answer, and a goal of type 2 stability a "yes" answer. The answer can be complicated, and it depends on the component of V_E that is the focus, and on the traits being considered. For example, if the focus is variation through space, and the locations are clearly defined and contrasting, then a relatively narrow adaptation may be appropriate if inputs can be applied to counter temporal variation. However, if the environments vary over time more than over space, then wide adaptation is a more appropriate approach. In addition, different traits can have different responses—for example, in Peru, experiments with potato seedlings in three contrasting locations showed crossover G×E for tuber set, noncrossover G×E for tuber yield, and no G×E for days to flowering (Ortiz and Golmirzaie 2004).

If the assumption is made that there will be *no qualitative G×E* across the range of spatial-temporal target environments, and therefore that high yield in test environments will translate into high yield in target environments (i.e., that there will be a yield *spillover*), then the strategy of selection and testing in optimal environments for production in marginal target environments makes sense. This is based on the fundamental principle that—all else remaining constant—as V_E decreases, h^2 increases—the reason breeders seek to minimize V_E in their selection and testing environments (see section 3.3). This is the approach of those supporting type 2 stability, who also tend to assume that farmers should add the inputs necessary in order to make their growing environments appropriate for varieties with type 2 stability, and if this is not possible then these environments are deemed inappropriate for farming and should be excluded from the target environments. This implies the further assumptions that farmers are risk neutral and seek to maximize profit (Ceccarelli, Erskine, et al. 1994; see also Evans 1993), and that type 1 FVs will (and should) be replaced with type 2 MVs (Mohapatra et al. 2006).

In contrast, if the assumption is made that there *will be qualitative GxE* across the range of target environments, and thus that high yield in optimal environments will not translate into high yield in marginal environments (i.e., there will be a qualitative yield *crossover*), then it makes sense to choose a breeding strategy of selection and testing in marginal environments for production in marginal target environments (Ceccarelli, Grando, and Impiglia 1998). In fact, crossovers in performance between varieties are "common" and reflect different adaptation to different environments (Comadran et al. 2008, Evans 1993:165 ff.). Therefore, from the type 1 stability perspective, plant breeders should adapt their breeding goals to meet the needs of farmers in these marginal environments, who are often risk averse and seek to maximize stability (see chapter 4, section 6).

4.5. Choice of Selection Environment and Environmental Variation

So why do some plant breeders base their strategies on the assumption of yield spillovers across environments while others work from the assumption that qualitative crossovers are common? A major source of the difference may lie in the range of environments being considered (Ceccarelli 1996). For example, different conclusions about adaptation are reached in four articles, all by CIMMYT (International Maize and Wheat Improvement Center, one of the CGIAR centers) scientists, reporting on evaluation of maize genotypes selected in different types of environments across a range of test environments (fig. 5.6) (Cleveland and Soleri 2002b). Two of the articles conclude that selection in optimal environments produces genotypes with higher yields than locally adapted genotypes in marginal target environments (Ceballos et al. 1998, Pandey et al. 1991). The other two conclude that selection should take place in marginal environments that have similar stresses to the target environments (Bänziger et al. 1997, Edmeades et al. 1999). The range of environments used in the former two studies (0.5–3.6, 4.3–6.5 MT per ha) as a proportion of the range in the latter two (0.7–7.8, 1.0–10.4 MT per ha) is 0.32. That is, those reporting spillovers were working across a much narrower range of environmental variation than those who observed crossovers; their experience constrained their knowledge.

Plant breeders who use type 2 stability as a goal may not realize the degree of difference in the range and type of V_E, especially in stress levels, between test environments and target environments (Cleveland 2001). Figure 5.7 illustrates the contrast between the assumptions often made by plant breeders—for example, in plant breeding for the Green Revolution—and the reality that often prevails in farmers' fields in more marginal environments. When I interviewed Dutch plant breeders in the Netherlands about their experience of spillovers and crossovers, they were incredulous when I told them that there were often crossovers

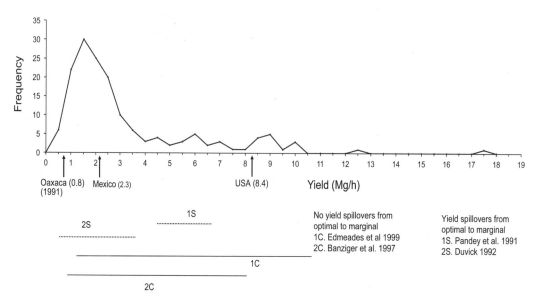

FIGURE 5.6. Yield spillovers and environmental range. Maize germplasm trials from optimal to marginal environments, compared with world maize yields by country, 1999 (n = 158). Adapted from (Cleveland and Soleri 2002b). © 2013 Daniela Soleri and David A. Cleveland. Used with permission.

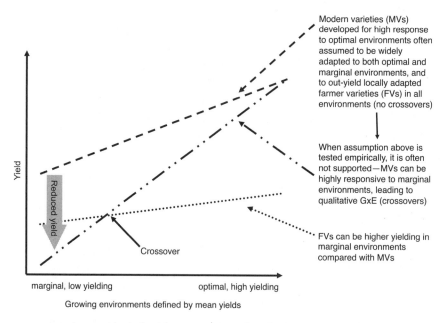

FIGURE 5.7. Assumptions about homogeneity and malleability of landscapes have dominated conventional agriculture but are challenged by empirical data. © 2013 Daniela Soleri. Used with permission.

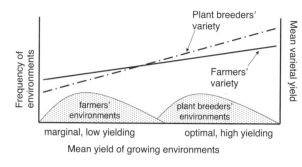

FIGURE 5.8. Visual representation of the hypothesis that experience limits plant breeders' perceptions and theory: range of V_E limiting plant breeders' observations of G x E. Based on (Cleveland and Soleri 2002a) and ideas in (Ceccarelli 1989, Ceccarelli, Erskine et al. 1994) © 2013 Daniela Soleri. Used with permission.

between locations within a single farmer's fields in marginal Third World environments. They could not imagine environments so different and so much more variable than their uniform Dutch fields. Thus, the experience of many conventional formal plant breeders is limited to a narrower range or more optimal environments, and they appear to often generalize their experience to more marginal environments (Soleri, Cleveland, Smith et al. 2002) (fig. 5.8).

The selection and testing environments used by plant breeders may be so different from Third World farmers' fields, that the genes responsible for a quantitative trait such as yield may be different in optimal testing and marginal target (farmers') environments (Atlin and Frey 1990, Atlin et al. 2000, Venuprasad et al. 2007). Some breeders interpret this as meaning that selection and testing should be done in farmers' environments (Ceccarelli, Erskine et al. 1994), and this is a strong argument for decentralized collaborative breeding to develop a larger number of varieties specifically adapted to local environments (Ceccarelli 2012), in contrast to the mainstream breeding goal of a small number of widely adapted varieties.

5. THE POTENTIAL FOR FARMER-SCIENTIST COLLABORATION

We have seen how different perspectives on the world and different social and biophysical environments affect the approach to breeding in terms of yield and stability. One of the keys to positively correlating yield and yield stability to increase sustainability is finding ways for farmers and scientists to better understand one another and work together, to combine the best of both traditional local and modern scientific knowledge.

Recently there has been some rapprochement between formal and informal breeding. This has occurred in the industrial world because alternative farmers—for example, those who practice organic and biodynamic farming—feel that conventional MVs do not meet their needs (Murphy et al. 2005). In the Third World, much of the initiative has been from a small minority of plant breeders who have been questioning the assumptions of conventional breeding (e.g. Ceccarelli and Grando 2002, Gibson, Byamukama, et al. 2008, Weltzien et al. 2003).[12] For these breeders, diversity at the many spatial levels of the agrifood system (table 5.2) is considered a key to alternatives such as organic and low-input agriculture (SOLIBAM 2011). While the focus of plant breeding for industrial agriculture in terms of diversity is narrowly focused at the plant, population, and varietal levels, a more holistic perspective assumes that diversity at higher levels (crop species, growing environments, farmers) can interact synergistically with the lower levels (De Schutter 2009: paragraph 57(b)).

However, many plant breeders believe that farmers have little or no understanding of the basic principles of what determines plant phenotypes, including the interaction between genotypes and environments. This means that these plant breeders believe they need to be in charge of breeding crop varieties for farmers, and farmers are merely the passive recipients of the results. When a plant breeder at CIMMYT reviewed the draft of a paper my colleagues and I had written on collaborative breeding, in which we emphasized farmers' theoretical knowledge (Cleveland, Soleri, and Smith 1999), he vehemently objected that "farmers don't have theory!"

Our research suggests he was wrong. Using the basic biological model as an ontological comparator, we tested the hypothesis that farmers have theoretical knowledge of important aspects of plants and environments, of the ways these interact, and that this knowledge is fundamentally similar to that of plant breeders. We found that farmers and scientists share many important concepts, while also having some different ideas. These findings provide a solid basis for collaboration.

5.1. An Ontological Comparator

An *ontological comparator* is key for understanding differences and similarities among and between farmers and scientists, and also among farmers and among scientists. It can provide the empirical and theoretical basis for collaboration between farmer and formal scientist plant breeders. As described in chapter 3, an ontological comparator is a fundamental model of reality—in this case, assumed to be common to both scientists and farmers. Here I use the basic biological model described in section 3.3 as an ontological comparator. This model is universally accepted by biologists, including plant breeders, but they disagree among themselves about its interpretation at higher levels of generalization—for example, whether selection in optimal or marginal environments leads to genotypes that are better adapted to marginal environments, as described above. Existing research also suggests that this model is the basis for farmers' understanding, although there are also differences among farmers at higher levels of generalization (Soleri and Cleveland 2009, Soleri, Cleveland, Smith, et al. 2002).

The methodological tool for using the ontological comparator to understand differences and similarities in ways that can contribute to collaboration between farmers and scientists are hypothetical *scenarios* (Soleri and Cleveland 2005). These scenarios: (a) are based on this comparator, (b) include components that are both familiar and unfamiliar to farmers, (c) refer to crop-specific reproductive systems and local propagation methods, (d) use lots of visual aids—for example, maize ears from local farmers' fields when talking with farmers about ear length, and (e) address issues and practices central to scientists' and farmers' approaches.

Novel components are key especially for farmers, because they allow farmers to generalize or theorize their understanding to new situations based on their past experiences. An example of a novel component in the scenarios is a field with an *optimal* growing environment that is uniform and with no limits to plant growth, in contrast to a farmer's *typical* field, which is relatively variable and has high levels of biotic and abiotic stress. Asking farmers about crop performance in an optimal field helps them to express their knowledge in terms of the biological model, because in the optimal field the source of any variation for low-heritability quantitative traits among plants will be primarily genetic, not environmental, whereas the opposite is true in farmers' typical fields.

Scenarios present a situation and ask the farmer or scientist to interpret the situation or predict what will happen next (fig. 3.3). The predictions are based on the model of the ontological comparator and on interpretations of this model in specific situations that vary in space and time in terms of variables such as crop genetic diversity, environmental factors, G×E, social context, and farmers' and scientists' preexisting knowledge and epistemology (chapters 3–4). Differences and similarities in specific situations lead to differences and similarities in responses among farmers, among scientists, and between farmers and scientists, which can then be analyzed.

The results can facilitate communication and collaboration. However, when farmers are asked questions in plant breeders' terms, they will have difficulty understanding, and scientists are apt to incorrectly conclude that farmers' knowledge has little in common with scientists' knowledge, and that therefore farmers have to be taught the basics in scientists' terms before they can do effective selection, as appears to have been the case in many projects with a mainstream perspective.

In the following sections I describe our work using an ontological comparator with farmers.

5.2. Farmer Understanding of How Genotype and Environment Determine Phenotype

To investigate farmers' understanding of the relative contributions of variation in genotype and environment to variation in phenotype, one of the two components of our ontological comparator, we tested hypotheses by asking farmers in five countries around the world to respond to several scenarios. These scenarios asked farmers to predict the phenotypes of both high- and low-heritability traits in their major crop, and in both optimal and typical (marginal with high variability in space and time) growing environments (fig. 5.9, table 5.3) (Soleri et al. n.d., Soleri, Cleveland, Smith, et al. 2002).

The first null hypothesis was that farmers would not predict a difference in plant phenotype for a selected trait when seed was planted in typical and in optimal environments.[13] When asked about traits that scientific breeders consider to have relatively low heritability, like maize ear length, farmers predicted that the trait would differ across environments. Therefore, we rejected the null hypothesis for low-heritability traits. However, farmers predicted that high-heritability traits, such as husk, tassel, glume, or seed color, would *not* change across environments, so we accepted the null hypothesis for these traits.

To further test this conclusion, we formulated a second null hypothesis to compare both types of traits simultaneously: that farmers would not predict difference from parental phenotype for either low- or high-heritability traits when seed was planted in a typical

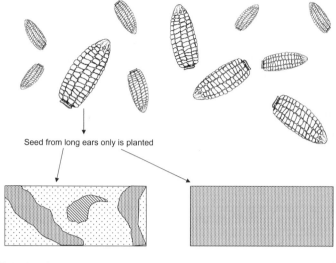

Seed from long ears only is planted

Typical local field, variable and high stress

Hypothetical uniform and optimal field

For each field we asked, *"What length will the ears that grow in this field be? The same as or different from the ears from which the seed was taken?"*

FIGURE 5.9. Sample scenario used to elicit farmers' knowledge of heritability. This scenario, used with farmers in Cuba and Mexico, is about a maize trait with low average heritability (ear length). We asked farmers what the ear length of the next generation would be in a typical relatively variable, stressful field and a hypothetical uniform, optimal field if only seed from long ears was planted. Similar scenarios were used for low and high heritability traits for crops in the five sites in the study (see table 5.3). © 2013 Daniela Soleri and David A. Cleveland. Used with permission.

environment. However, farmers predicted that there would be variation in the low-heritability trait in the variable typical environment, but not in the high-heritability trait, so we did not accept the second null hypothesis.

Figure 5.10 illustrates our overall conclusion: farmers' experiences and the tools and methods available to them mean that the role of genotypic variation (V_G) in determining the phenotypic variation (V_P) for low-heritability traits is obscured by the high variability in their growing environments (V_E). This helps to explain why farmers consciously select for high-heritability traits while often considering it not worthwhile or even possible to make intergenerational changes (*R*) for low-heritability ones, especially in cross-pollinated crops that have high genetic variability (Soleri, Cleveland, Smith et al. 2002).

What about varietal choice? How do farmers' perceptions of how the generally low-heritability trait of yield is affected by environmental variation influence their choice of what crop variety to plant in different spatial environments? To answer this question, we hypothesized that farmers would not perceive qualitative *G×E* interactions across a spatial range of environ-

TABLE 5.3. Understanding Farmers' Perceptions of the Relative Contribution of Genotype and Environment to Phenotype

Location, major crop	Null hypothesis #1: For traits with low and high heritability, farmers predict the same phenotype across typical and optimal environments. That is, farmers do not see a contribution of environment (Env) to phenotype.		Null hypothesis #2: For traits with low and high heritability, farmers predict the same change from parental phenotype across typical and optimal environments. That is, farmers do not see a difference in h² between traits.
	(a) Low heritability trait across typical and optimal Envs	(b) High heritability trait across typical and optimal Envs	Low v. high heritability traits in typical, variable Env
Cuba, maize	Ear length*	Husk color	Ear length v. husk color*
Mexico, maize	Ear length*	Tassel color	Ear length v. tassel color*
Mali, sorghum	Panicle weight*	Glume color*	Panicle weight v. glume color*
Syria, barley	Plant height*	Seed color	Plant height v. seed color*
Nepal, rice	Plant height*	Seed color	Plant height v. seed color*

Source: Based on (Soleri et al. n.d.).
* Hypothesis not accepted, Fisher's exact test, $P<0.05$.

ments (Soleri, Cleveland, Smith et al. 2002). We asked 208 farmers to respond to a scenario in which two varieties of their major crop originating in contrasting growing environments were planted together in two different communities, in two different fields in the same community, and in two different places in one field. A majority of the farmers (57 percent) anticipated qualitative G×E (crossovers) for their major crop between communities, but only a minority (30 percent) anticipated crossovers within a community, and fewer still anticipated crossovers within a single field (18 percent). Perception of qualitative G×E at smaller spatial scales was highest for farmers growing autogamous (self-pollinating) crops, especially those working at a small scale with intimate knowledge of field soil and moisture variations, such as rice farmers in western Nepal. Also, because autogamous crops are more genetically uniform and vary less from generation to generation, it is easier for farmers to see the effect of V_E on V_P.

These results show how farmers understand the basic biological model in terms of their own genotypes and environments, and in terms of their ability to apply this model to novel situations. The results provide the basis for further exploration of farmers' selection—key for collaboration with plant breeders.

5.3. Farmer Selection: Results and Goals

The second major component of the ontological comparator for plant breeding is the results of selection. Just as we have seen for G×E interaction, the genetic structure of crop populations

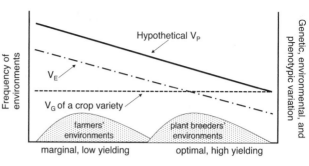

FIGURE 5.10. Visual representation of the hypothesis that experience limits farmers' perceptions and theory: V_E obscuring V_G in farmers' experience of V_P. © 2013 Daniela Soleri. Used with permission.

affects selection and how farmers perceive and manage it. As genetic variation between generations increases from clonally propagated species, such as potato and cassava, to self-pollinated species, such as rice, wheat, and bean, to cross-pollinated species, such as maize and squash, the challenge of making genetic progress in selection, and of maintaining the genetic structure of a variety, increases.

For all three of these types of propagation, research has firmly established that farmers' phenotypic selection can result in longer-term (multigenerational) genetic change or micro-evolution (mEv), intergenerational genetic change or response (R), or only within-generation phenotypic differentiation (S), as described above (section 3.3, and fig. 5.4) (reviewed in Soleri and Cleveland 2009). However, farmers' goals for phenotypic selection (see chapter 2, fig. 2.5) may not result in the outcomes they desire, and regardless of goals, the outcomes of farmer selection can be varied, although many studies of farmer selection that document genetic or agronomic effects do not document farmers' goals (and vice versa).

Perhaps the most common goal of farmer selection has been ignored or dismissed by most plant breeders because it doesn't result in genetic change (R or mEv) (Cleveland and Soleri 2007a). This is because most of the time farmers' primary goal in selecting seed is to obtain good planting material—large, clean, disease-free seeds or other propagules for cross-pollinated (for example, maize; see Louette and Smale 2000, Soleri and Smith 2002), self-pollinated (for example, barley; Ceccarelli, Grando, Tutwiler et al. 2000), and vegetatively propagated crops (for example, potato; see Zimmerer 1996). Farmers' maize seed selection in Oaxaca resulted in high S values for a number of ear and kernel traits, including those explicitly identified by farmers (ear and kernel size) (Soleri, Smith, and Cleveland 2000). However, over three generations these traits did not differ from randomly selected seed from the same lot, resulting in an $R \approx 0$. Reasons given by farmers for selecting large seed included seed quality and purity, but not maintaining or changing (i.e., "improving") a variety. Some farmers said that large seeds resulted in higher germination, larger seedlings, early vigor, and higher yields; however, most farmers attributed their preference for large seed to "custom."

Based on ongoing research, farmers' comments that selection can make seedlings stronger, but not higher yielding, appear correct: days to germination are fewer and seedling vigor is greater in selected versus random seeds from the same maize populations (Soleri n.d.). Thus, while farmers' selection may not affect crop populations genetically, it may make a critical contribution to environmentally important characteristics, such as plant establishment and seedling survival. However, simple mass selection for intragenerational phenotypic differences could result in R or mEv, even if these were not farmers' stated goals. For example, maize farmers in Uganda and Tanzania, like those in Mexico, were reported to select for large clean kernels from large ears, apparently because they believed that these kernels germinated well and produced high-yielding plants (Gibson, Lyimo et al. 2005). Yet this selection also appeared to result in decreased resistance to maize streak virus, since resistant plants had smaller ears and plants with large ears appeared to be nonresistant plants that escaped disease by chance.

However, although it is clear that farmers can understand the principle of phenotypic selection and use it to achieve some goals in different crops, little research has been done on how farmers understand the process of selection per se. To investigate this topic, we presented farmers in five countries with a hypothetical scenario asking them to compare random with intentional selection for ten cycles in a typical field, in populations with phenotypic variation for the low-heritability traits they used as major selection criteria for their main crop (table 5.4, question A) (Soleri and Cleveland 2009:352–353). The null hypothesis was that farmers would not differ from plant breeders—that is, that they would all consider intentional selection to be more effective than random selection for improving or at least maintaining this trait. The majority of farmers (76.2 percent, or 144 out of 189) responded that intentional selection was more effective for increasing yield, although those who disagreed with that idea were sufficient to reject the hypothesis statistically.

So farmers see an advantage to intentional versus random selection, but is their goal S or R or mEv? To discriminate between these possibilities, those farmers who responded to the first question that intentional selection resulted in greater yield were asked to compare random selection for ten cycles followed by one cycle of intentional selection, with eleven consecutive cycles of intentional selection. Of these farmers, only 23.2 percent (20 out of 86) saw eleven years of intentional selection as superior—that is, as providing cumulative multigenerational change (mEv). Therefore, we rejected our null hypothesis of no difference between farmer and plant breeder expectations. For the majority of farmers, the primary selection goal can be inferred to be either (R) or a nongenetic advantage embodied in (S), both of which they believe are fully achieved within one year (table 5.4, question B).

Clearly, farmers see selection as an important process, but not necessarily one that is cumulative and directional—that is, not the way that scientists see it—because farmers have other concerns, such as seed quality and vigor. These results show that we can't assume we know what farmers' motives are and we can't romanticize them, but also that we can't dismiss farmers' knowledge. Working with farmers to investigate how farmers and scientists are similar and different will be valuable to support collaboration.

TABLE 5.4. Farmers' Expectations for Response to Selection for Their Primary Selection Criterion in the Major Crop They Grow

Country, crop, trait (n)	Question A. Farmers responding that response to intentional selection for 10 cycles > random selection for 10 cycles $(IS_{10}>RS_{10})^a$			Question B. For farmers responding $IS_{10}>RS_{10}$ to Question A, those stating that response to intentional selection for 11 cycles > random selection for 10 cycles + intentional for 1 cycle $(IS_{11}>RS_{10}+IS_1)^a$		
	n	%	P^b	n	%	P^b
Mexico, maize, ear length (59)	23	39	* 0.000000	6	26	* 0.000000
Cuba, maize, ear length (29)	27	93	0.245614	12	44	* 0.000002
Syria, barley, plant height (21)	20	95	0.499999	11	55	* 0.000614
Nepal, rice, grain yield (40)	39	98	0.499999	17	44	* 0.000000
Mali, sorghum, grain yield (40)	35	88	0.057662	23	66	* 0.000078
Total (189)	144	76	* 0.000000	69	48	* 0.000000

Source: From (Soleri and Cleveland 2009:353).

[a] RS = random phenotypic selection by farmer, IS = intentional phenotypic selection by farmer, subscript indicates number of years (reproductive cycles) selection carried out.

[b] Results of one-sided Fishers' exact test of the null hypothesis that, similar to plant breeders, farmers would see intentional selection as achieving a greater response than random selection. Calculated using SISA (http://home.clara.net/sisa/).

6. TRANSGENIC CROP VARIETIES

In this section we turn to the most prominent controversy in plant breeding today, a controversy that is also central to the broader debate about the food crisis and the future of the agrifood system—the role of genetically engineered, transgenic crop varieties (TGVs) in creating a more sustainable agrifood system. TGVs are the most economically, socially, and environmentally important component of agrifood biotechnology.[14]

There is some evidence for potential negative environmental effects of TGVs—for example, on the genetic diversity of crops, especially FVs and wild relatives, via transgene flow, and on ecosystem function, via effects on nontarget organisms—but the significance of this evidence is not clear. However, there is much more evidence for potential negative social and economic effects of TGVs as the centerpiece of mainstream industrial agriculture's increasing control over the world's agrifood systems (Cleveland and Soleri 2005, Soleri, Cleveland, and Aragón Cuevas 2006). In this section I describe the mainstream perspective on TGVs, including assumptions about small-scale Third World farmers, and then give some results of our tests of these assumptions.

6.1. The Mainstream Consensus and Its Critics

There is currently a consensus among mainstream development organizations, governments, and corporations that TGVs are the key to increasing production and income and reducing hunger and malnutrition in the Third World; increasing yields and income in the industrial world; and making agriculture more environmentally sustainable. This vision is being implemented via market-based mechanisms with major participation by the private sector. Adherents to the consensus on the necessity of TGVs for the Third World dominate the international agricultural development agenda and often cite the rapid rate of TGV adoption as evidence of their efficacy (James 2010).

The consensus comprises three main groups, with some degree of overlap among them:

- *Agricultural development policy* by many national governments (e.g., G8 2010), led by that of the United States (USDA APHIS BRS 2008); private foundations (e.g., Gates Foundation 2011, Rockefeller Foundation 2007); and international development organizations, including the UN organizations (CGIAR 2006, FAO 2004, UNDP 2001, WHO 2005), some of which are focused on developing TGVs to reduce hunger and poverty in the Third World (e.g., GCGH 2010, Golden Rice Project 2009, James 2009).
- *Development and marketing of TGVs* by private agbiotech corporations (e.g., Monsanto 2009), their affiliated nonprofit organizations (e.g., Syngenta FSA 2009), and some national governments, namely that of China.
- *Research on TGVs and small-scale Third World farmers* by public and private universities, research organizations (ABSPII 2010, UC 2011), and mainstream "development professionals," who "have increasingly agreed to something like a standard narrative of biotechnology . . . an optimistic but cautious consensus" (Herring 2007:7).

The consensus is part of the mainstream economic emphasis regarding agrifood sustainability for small-scale Third World farmers, which it assumes needs to be accomplished by following the example of agriculture in the most industrialized countries (Hazell et al. 2007). This is spelled out clearly in the goals of the Alliance for a Green Revolution in Africa (AGRA) (Toenniessen et al. 2008), one of the largest and most prominent consensus projects, which promotes increased use of purchased inputs like fertilizers and the privatization and commercialization of agriculture along with TGVs. The consensus promotes an image of concern about and support for small-scale farmers—for example, on the websites of its members, such as Monsanto, the U.S. Agency for International Development (USAID), and the Consultative Group on International Agricultural Research (CGIAR). However, the underlying goal is that these farmers should abandon farming and move to cities "on an unprecedented scale and with unprecedented speed" (Hazell et al. 2007:2–3), eliminating small-scale, traditional agriculture in a process referred to as "creative destruction" (ICABR 2008) after the expression of economist Joseph Schumpeter (Schumpeter 1975 [1942]:82–85).

Critics have examined the development and marketing of TGVs in terms of the sociopolitical forces shaping agricultural development policy for small-scale Third World farmers

(Brooks 2011, Glover 2010a), including the potential influence on research by consensus proponents on policy for and development and marketing of TGVs to farmers. For example, Glover showed that Monsanto funded economic research that "helped to generate a substantial body of economic studies which claim that transgenic crops (primarily insect-resistant cotton) have produced a range of significant benefits" for farmers (Glover 2010a:83–84). Glover (2010b, 2010c) and Stone (2012) have critiqued this research in detail, showing that it is both unscientific and dominated by biases in favor of consensus assumptions about the effects of TGVs. Other studies have found that adoption of TGVs may be due to a passing fad (Stone 2007) or to lack of freedom to choose (Witt et al. 2006), and that purported benefits like higher yields and reduced pesticide use may be reversed after several years because of the emergence of secondary pests (Wang, Just et al. 2006).

6.2. Thinking Critically About Consensus Assumptions

While most of the criticism of the consensus in relationship to farmers has focused on problems with research methods and results, much less attention has been paid to testing the assumptions about small-scale Third World farmers and agriculture on which it is based. These include empirical assumptions about *what agriculture is* (how agriculture works, how farmers make decisions) without adequately testing them, making the consensus *unscientific,* and value-based assumptions about *the way agriculture should be* (the way farmers should participate in evaluating TGVs, what the future of agriculture should be) without any meaningful input from those affected, making the consensus *undemocratic.*

Because the consensus assumes that Third World agriculture either is or should be like industrial agriculture, and therefore that TGVs are appropriate because they are scale neutral, they also assume that TGVs will have the same risks and benefits in the Third World as they do in the industrial world (Brooks 2011). For example, the U.S. government explicitly does not evaluate the risk of TGVs outside of the United States even though it exports large quantities of TGV seeds and grain (Cleveland and Soleri 2005). However, as discussed above, industrial and Third World agriculture are very different. For example, in the United States close to 100 percent of maize is planted from seed purchased new every year, whereas in Oaxaca, Mexico, over 90 percent of farmers save their own maize seed or obtain it from other farmers and often have small fields, facilitating cross-pollination (fig. 5.11). Therefore, one of the main risks of TGVs—transgene flow to nontransgenic varieties via cross-pollination—would be difficult if not impossible to remedy in places like Oaxaca, as compared with the United States. Another important difference is that the United States is not the center of domestication or diversity for most crops, unlike some other regions where TGVs are being introduced and promoted, such as TGV maize in Mexico (Soleri, Cleveland, and Aragón Cuevas 2006).

Studies by mainstream consensus economists of *Bt* cotton, maize, and rice have concluded that farmers readily adopt TGVs because they increase yield and income, reduce pesticide applications, or improve farmer health (Gouse et al. 2006, Huang, Hu, Pray et al. 2003, Huang, Hu, Rozelle et al. 2005, Morse et al. 2006, Qaim and Zilberman 2003). Those

FIGURE 5.11. Maize fields in the Central Valleys of Oaxaca, Mexico. These long, narrow fields, some only a few meters wide, result from both land reform and generations of dividing among family members. One result is a high level of pollen and gene flow among populations growing in adjacent and nearby fields. Photo © David A. Cleveland.

economists assume that farmers are "risk neutral" and that their objective is to "maximize expected profits" (Zilberman et al. 2007). The implication is that if farmers do not adopt TGVs, it is due to market failure ("imperfect markets") resulting from inadequate or inaccurate knowledge, legal restrictions, price distortions, or farmers' unfounded, nonscientific perceptions.

My colleagues and I carried out research with small-scale maize farmers in six communities in Cuba, Guatemala, and Mexico to test the assumptions of the consensus that farmers are risk-neutral profit maximizers. We assessed attitude toward risk by presenting farmers with a hypothetical choice between two maize varieties whose yields vary in response to variability in growing environments (defined in terms of rainfall, the key limiting factor for maize production in the six communities studied) (fig. 5.12 A) (Soleri, Cleveland, Aragón Cuevas et al. 2005). These two maize varieties have qualitative crossover G×E interaction as presented (fig. 5.12 B), so farmers' choice was between higher average yield or higher yield stability (fig. 5.12 C). Overall, a majority of farmers preferred the stable variety (X), although there was variation between communities and countries. Only farmers in the one more modern Guatemalan community, which grew mostly commercial crops, favored the variety with higher average yields (Z). In a follow up scenario, variety Z had characteristics similar to TGVs—higher yield coupled with higher yield variability over time as yields declined with the evolution of resistance in pest populations controlled by the hypothetical TGV, so that farmers had to purchase commercial seed after six or so years, while variety X had lower, more stable yield, and they could save the seed (Soleri, Cleveland, and Aragón Cuevas 2008).

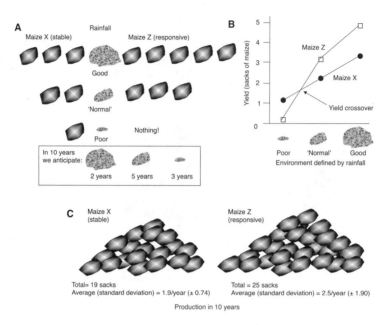

FIGURE 5.12. Risk scenario presented to maize farmers: yield versus yield stability in response to environmental variation. Rocks of different sizes represent different annual rainfall, and sacks of maize grain represent yield response of a variety to that rainfall. (A) Rainfall and yield for two varieties as presented to farmers. The scenario presented represents the following, which was not presented to farmers: (B) Crossover of yields for the two varieties. (C) Total and average yields for the two varieties. From (Soleri, Cleveland, Glaslow et al. 2008). © 2008 Daniela Soleri and David A. Cleveland. Used with permission.

In this second scenario the large majority of farmers chose variety X, again indicating that they were risk averse, contrary to the consensus assumption that farmers are risk neutral. Farmers preferred to save their own seed rather than to have to buy new seed even every six years, in part because of the unreliability of markets and government extension, and in part because of uncertainty about seed quality from such sources.

We also asked farmers about what they thought of transgenesis per se when this was described to them neutrally, including providing a positive example of a maize TGV with the potential to decrease pest damage, similar to *Bt* maize. The majority of farmers did not object to transgenesis per se, which shows that they are capable of making distinctions between some aspects of transgenic technology and its hypothetical consequences in TGVs, and also that they are not closed to new ideas and methods for solving their problems (Soleri, Cleveland, Aragón Cuevas et al. 2005).

To test the consensus assumption that farmers are profit maximizers, we presented them with a scenario that included four varieties in a ranking exercise: farmers' own variety (FV), a conventional modern variety (MV) they were familiar with, and those same varieties as backgrounds for a transgene—a transgenic farmers' variety (TGFV) and a transgenic modern variety (TGMV) (Soleri, Cleveland, Glasgow et al. 2008). Farmers were asked to rank these first as maize seed for sowing in their own fields, and then again as maize grain for their family to eat. The FV and MV represented two seed systems (informal versus formal, respectively) and different agronomic, storage, and culinary characteristics with which farmers were already familiar. Farmers had no previous experience with TGVs. Providing these four choices allowed us to distinguish farmers' preferences for varieties or genetic backgrounds (FV versus MV) from their preference for a genetic technology (TGV versus non-TGV), a distinction either overlooked or confounded in most research with farmers. The results were that for both sowing and especially for eating, the great majority of farmers preferred FVs and nontransgenic technology (Soleri, Cleveland, Aragón Cuevas et al. 2005, Soleri, Cleveland, Glasgow et al. 2008). Since farm families often rely on their own crop production for a significant proportion of their food, traits related to storage, food preparation, taste, color, texture, and specific uses are important when choosing crop varieties. That is, farmers are not profit maximizers, but rather seek to maximize a range of attributes simultaneously.

The majority of farmers who rejected hypothetical varieties with traits similar to TGVs had many of the characteristics typical of farmers most in need of support from agricultural research and development investments, as well as those most likely to be conserving globally important genetic resources *in situ* and associated cultural and linguistic diversity. These research findings do not support mainstream assumptions about farmers. While the consensus believes that critics of TGVs are starving Third World farmers of science (Paarlberg 2008), it may be more accurate to say that they are being starved of a sound scientific process for evaluating TGVs in comparison with other approaches to achieving goals for agriculture that are democratically agreed on.

SIX

Managing Agricultural Ecosystems

The Critical Role of Diversity

1. INTRODUCTION

In rural Durango, in north-central Mexico, a colleague and I were being shown gardens in a rural area near the capital city, escorted by a small group of urban women, wives of agronomists working for the government (Cleveland 1986a). These gardens were being promoted by a federal agency, overseen by the women from the city, who took us to several project gardens consisting of neat rows of onions, lettuce, carrots, beets, zucchini, strawberries, and cucumbers inside the walled yard of the local health clinic (fig. 6.1). The seeds that had been planted were mostly of U.S. varieties and the use of pesticides was encouraged. Project personnel complained that local women did not use the beets and some of the other vegetables being promoted.

Walking away from one of the demonstration gardens, we looked over an adobe wall around a nearby house and saw a mass of trees, vines, and herbaceous plants, tumbled together in no apparent order. The woman whose house this was participated in the health clinic garden project and had just given us a monotone, rote description of the project garden at the clinic. When we asked her about her home garden, her face brightened, and in animated tones she began to tell us the names and uses of the many plants. This was a thriving traditional mixed garden, of the sort that we were then able to pick out in most of the other houses in the villages where we were taken to see project gardens (fig. 6.2). We became much more interested in learning about these gardens than the neat little project gardens, much to the dismay of the project promoters.

The traditional gardens contained pomegranate, lemon, peach, apple, and fig trees, along with maize, beans, squash, pepper, chayote, tomatoes, and many herbs for both

FIGURE 6.1. A project garden in Durango, Mexico. Photo © David A. Cleveland.

FIGURE 6.2. A traditional household garden in Durango, Mexico. Photo © David A. Cleveland.

medicinal and culinary use, as well as flowers. Indian fig cacti, which produce large, edible fruits (*tunas*) and large thick leaves (*nopales*) were often grown as fences; mulberry trees were frequent in the yards, providing both shade and fruit. Households with enough space in their gardens were often growing small patches of maize with pumpkin, squash, and other vegetables interplanted. Talking with the women, my colleague and I discovered that the seed and other planting materials they used came mostly from the gardeners' own harvests or those of neighbors. They said that most of the vegetables they were growing in the project gardens were not their choice, and that they had never been told how to use them.

The contrast between these two types of gardens in Mexico reflects the contrast between the mainstream and alternative approaches to agricultural ecosystems management at the household level. As we have seen, increased ecosystem management was needed in the Neolithic to ensure successful yields from newly domesticated plant populations (chapter 2). However, compared with industrial agriculture, which focuses on increased yields through controlling and replacing natural ecosystems and reduces diversity to a few varieties of a few crop species, traditional and small-scale agriculture often maintains more ecosystem processes, more wild and weedy plants in and near fields, and a larger number of different crop species and varieties. The replacement of traditional with industrial agriculture often decreases total diversity at many levels. That is, it is a reduction in number of kinds of a given component, from alleles to cultural groups, as well as a reduction in types of interactions among components, and it can also decrease yield stability (table 5.2 classifies the levels of diversity in agrifood systems). Decreased diversity in turn often increases the need for ecosystem management.

Diversity has been a lightning rod for debates about the future of the planet and agrifood systems. It has gained wide popular recognition since the Convention on Biological Diversity, which was presented at the 1992 Rio Earth Summit and went into effect in 1993 (CBD 2012). The summit was "inspired by the world community's growing commitment to sustainable development," a commitment that continues to inspire—for example, in the 2010 Strategic Plan for Biodiversity, which was based on the assumption that "biological diversity underpins ecosystem functioning and the provision of ecosystem services essential for human well-being," including "food security."

We know that diversity plays a critical, yet often complex and controversial, role in the yield and stability of agroecosystems, key components of most definitions of sustainable agrifood systems. In this chapter I focus on diversity at the field and ecosystem levels, and its relation to yield and stability, building on my discussion of these variables at the plant, population, and species level started in chapter 5. This includes an in-depth look at one of the most important strategies for managing diversity in agroecosystems—polyculture, or growing more than one crop in a field. Polyculture is common in most traditional farming—as in the household gardens in Durango—but is the exception in modern industrial agriculture, where centralization and the increase in scale at all levels has driven those systems toward extreme levels of uniformity, including vast farms made up of fields of monocultures, often comprised not only of one species, but of only one variety of one species.

2. THE DIVERSITY-STABILITY DISCUSSION IN ECOLOGY

Ecology and agroecology developed together as research disciplines in a similar way as the fields of evolutionary genetics and plant breeding, sharing problems and discoveries in mutually stimulating ways. One of the core concepts in both ecology and agroecology is diversity and its relationship to productivity, yield, and stability.

2.1. Diversity Begets Stability, or Does It?

The idea that diversity begets stability became widespread in ecology in the third quarter of the twentieth century (Elton 1958, MacArthur 1972) and was "one of the most influential beliefs in ecology" from the 1960s until the mid-1970s, reaching the status, in some cases, of a "core principle" (McNaughton 1988:204; see also Pimm 1991:6–9). This idea has been described as a secularization of the Western religious tradition of the great chain of being (Sagoff 1993), an idea that seems to have resonated with the popular sentiment of the time in the Western world.

Those who see an intrinsic value in nature, characteristic of the environmental emphasis in sustainability (chapter 3), also tend to see a natural balance there maintained by diversity, and want to preserve this balance and diversity against disruptive human forces. In the United States, John Muir is often cited as the founder of the preservationist faction of the environmental movement, and he was part of a nineteenth-century natural science tradition that saw species and landscapes as links in "God's creation," melding the aesthetics and science of nature (Norton 1991:32). Muir wrote: "The universe would be incomplete without . . . the smallest transmicroscopic creature that dwells beyond our conceitful eyes and knowledge" (quoted in Norton 1991:33, 20). Environmentalist Aldo Leopold similarly conflated human values and natural processes: "Science has given us many doubts, but it has given us one certainty: the trend of evolution is to elaborate and diversify the biota. . . . A thing is right when it tends to preserve the integrity, stability, and beauty of the biotic community. It is wrong when it tends otherwise" (Leopold 1970 [1949]:253, 262).

However, when the theory and data began to be scrutinized (Goodman 1975, May 1974), the diversity-stability paradigm failed to hold up (Norton 1987:73–77). Ecologist Robert May's book on stability and complexity (and diversity) (1974) was very influential; it used mathematical models to show that increased complexity leads to increased community instability (in the sense of increased fluctuation).

2.2. A New Sophistication?

May pointed out that his models applied to random complexity, and that real-world natural ecosystems might be mathematically atypical as a result of evolution. This observation stimulated much subsequent research, the results of which have supported a causal relationship between diversity and stability (McNaughton 1988:204, Mougi and Kondoh 2012, Pimm 1991:9–11), and the idea in ecology that diversity is positively related to stability has received renewed research attention and resulting support.

Norton suggested that the original diversity-stability formulation was too simplistic, and that to conclude that the relationship is negative in the face of persisting and powerful indications to the contrary was premature. Pimm emphasized the complexity of the relationship between stability and diversity and the importance of clear definitions in any discussion; he identified five definitions of stability, and three each of complexity (diversity) and levels of organization as used by ecologists, making a total of forty-five "possible questions about the relationships between community complexity and stability" (Pimm 1991:15). Ecologists realized that the formulation had to become more sophisticated. For example, compartmentation and connectedness appear to be common in natural ecosystems and provide mechanisms that simultaneously increase stability and diversity.

However, many ecologists have moved beyond simplistic calls for the conservation of all biodiversity in natural ecosystems on the grounds that all biodiversity is inherently good and are beginning to ask what role biodiversity of different kinds, at different levels, plays in specific ecosystems (Angermeier and Karr 1994, Baskin 1994, Cardinale et al. 2012), a question made more critical as we enter the Anthropocene epoch, which is likely to bring mass extinctions and the necessity for triage among ecosystems and species in terms of conservation efforts (Naeem et al. 2012).

Like other resources, too much diversity, as well as too little, may have adverse effects, with optimal levels differing in each situation. Effects in different parts of the system may vary. For example, ecological diversity measured as number of species can be positively related with ecosystem productivity and sustainability (persistence) but negatively related with survival of individual species (Tilman 1996, Tilman et al. 1996). Modeling of communities that contain antagonistic and mutualistic interactions suggests that diverse interaction types are key for ecosystems to be self-sustaining (i.e., stable) (Mougi and Kondoh 2012), and the current state of ecological knowledge about the relationships among diversity, stability, and productivity is that diversity begets both stability and productivity (Cardinale et al. 2012).

At the same time, some believe that the success of academics in linking biodiversity and sustainability in national and international policy has resulted in unlinking them from their scientific base (Redford and Sanderson 1992), where they have become elements of the new dogma of "sustainable economic growth." This was clearly evident at the Rio+20 conference in June 2012, which was dominated by private corporations and industrial nations, with the main result being the endorsement of "green growth" (Barbier 2012). For example, the World Bank sees no conflict between increasing economic growth and protecting the environment—including biodiversity—based on increasing production and market efficiencies (2012:36 ff., 118). However, from alternative perspectives, Rio+20's green growth solution is unscientific and unsustainable (Confino 2012).

3. PRODUCTION, YIELD, STABILITY, AND DIVERSITY IN AGROECOSYSTEMS

More intensive ecosystems management in the agrifood system has meant increasing human expenditures of labor, knowledge, and resources to achieve greater human control over natural flows of water, nutrients (such as nitrogen, phosphorus, and potassium), and micronutrients (such as magnesium and sulfur) in order to make them more available to crops (see chapter 1). It has also involved a massive diversion of energy from natural cycles directly, as crop plants replace natural vegetation and their solar energy capture, and indirectly, as fossil fuels are used to power the modern industrial agrifood system.

Data from traditional and modern agroecological systems suggests that diversity is negatively correlated with short-term yield (harvest per unit of land area). FAO global data on crop yield show a pattern of highest yields in the most industrialized countries with relatively lower agrobiodiversity (FAO 2013b); for example, yields of genetically uniform maize MVs in the large industrial fields of the United States are about ten times greater than maize yields on the small farms of Oaxaca, Mexico, where farmers are growing much more diverse FVs (Cleveland and Soleri 2002b).

These broad hypothesized relationships assume that most other things are equal, which of course they are not. As we saw in chapter 5, there is often a negative correlation between genetic diversity and yield in good environments, but in poor environments there is a positive correlation. That is, when comparing genetically uniform MVs and highly diverse FVs, MVs have higher yields in good environments, while FVs have higher yields in marginal environments. In other words, there is a crossover between environments. However, how the temporal boundaries are defined is important, as discussed in chapter 3; will the short-term yield advantage of less diverse modern agroecosystems disappear over the long term?

There is evidence that over the long term diversity provides more stability and higher yields, but this does not apply to diversity in general. Just as ecologists have found that the effect of diversity on stability and productivity in natural systems depends on the specific types and combinations of diversity, as discussed above, agroecologists are finding that this is true of the relationship between diversity, yield, and stability in agroecosystems. For example, while some combinations of plant diversity can suppress pests and diseases, other combinations can encourage them (Ratnadass et al. 2012). In the example from Yunnan described below, some rice varietal mixtures suppressed fungal disease and some did not, and in Ethiopia, some traditional combinations of wheat and barley *(hanfets)* had significantly higher yield stability than pure crops and some did not (Woldeamlak et al. 2008). The effects of diversity also vary at other levels, as discussed in section 3.4 for the introduction of modern rice in southeast Asia (Loevinsohn et al. 1993).

In the following sections I briefly describe the changing relationships among diversity, yield stability, and yield for traditional, modern, and sustainable agriculture, especially as they relate to ecosystems management. This discussion provides the framework for analyzing the current problems with modern industrial agriculture and the theoretical potential for a combination of traditional and modern elements to create a sustainable agrifood system.

3.1. Traditional Agroecosystems

In traditional small-scale agriculture, yield (harvested food per unit of land) is often lower than in modern agriculture, but high levels of diversity at genetic, ecological, and sociocultural

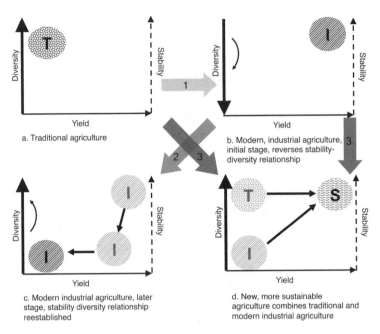

FIGURE 6.3. Model of possible diversity, stability, and yield relationships in the development of sustainable agriculture. T = traditional, I = industrial, S = sustainable combination. See text for discussion. © David A. Cleveland 2013.

system levels can support high levels of yield stability (fig. 6.3.a). A positive correlation between traditional farmers and biological diversity has been accepted internationally in a number of different fora—including Agenda 21 of the UN Conference on Environment and Development, Article 8(j) of the Convention on Biological Diversity (CBD 2012), and the FAO's Global Plan of Action for crop genetic resources, which calls for more emphasis on *in situ* conservation based on evidence that "the rich diversity that exists today offers ample testimony of what has already been achieved" through farmer management and development of crop genetic resources (FAO 1996:para. 26).

Advocates of traditional farming cite its diversity as an important component of sustainable agriculture because the presumed relationship between this diversity and greater yield stability reduces the nutritional and financial risk for farm households, and because traditional agriculture—which is biologically diverse and uses low levels of inputs—has less negative environmental impact (Cleveland, Soleri, and Smith 1994). There is indeed some evidence to support the hypothesis that the greater diversity of crops and fields in some traditional farming systems may have lower yield variance—and, under some conditions, even larger yields—compared with less diverse modern agriculture (Swift and Anderson 1994). As discussed in chapter 5, an important reason for the nonadoption of MVs may be the extent to which their average yield under farmer conditions is lower than that of FVs, or if average yield is higher it also has higher variance (Ceccarelli, Erskine et al. 1994, Cleveland 2001, Kelley et al. 1996). The contribution of crop genetic diversity to stability at supra-population levels—fields, regions, human communities—might be called "system" buffering, and is analogous to individual and population buffering (Souza et al. 1993:216).

At the ecosystem level (table 5.2), there is also evidence supporting a causal connection between diversity and stability. For example, a recent review of theory and empirical data regarding the conversion of tropical forest to agricultural land indicates that small-scale agroecological production provides conservation and production benefits that modern industrial production cannot provide (Perfecto and Vandermeer 2010). This type of small-scale agriculture covers much of the tropics, creating a matrix of fields interspersed with islands of remaining natural areas, which conserves natural biodiversity by providing interconnected migratory pathways—preventing local, and ultimately general, extinctions (Perfecto et al. 2009).

3.2. Modern Industrial Agroecosystems

The modernization of agriculture in pursuit of higher yield and overall production (total usable harvest) has been accompanied by a simultaneous decrease in diversity across a range of scales, from plant genetic makeup to social systems, with the adoption of monocropping of uniform MVs, the move to larger, more uniform fields, and the economic and spatial concentration of agrifood systems (table 5.2). This is the change from figure 6.3a to 6.3b indicated by arrow 1. Modern industrial agriculture makes use of economies of scale to increase output-input ratios like usable harvest for a given amount of land (yield) or human labor, although harvest per unit of energy and nutrients (like nitrogen) is reduced (i.e., there are trade-offs in output-input efficiencies for different inputs).

Loss of diversity can be mitigated, at least over the *short term,* by increasing inputs to compensate for the decrease in stability that often results from reduced diversity; in other words, the positive correlation between diversity and stability is reversed (fig. 6.3b). For example, increasing levels and types of pest and pathogen pressures that accompany a switch to monocropping of genetically uniform MVs can be mitigated by increasing pesticide and fungicide application. Advocates of mainstream agriculture also assume that agroecosystem control increases stability over the *long term*—that is, that the system remains as depicted in figure 6.3b. For example, Connor et al. state that "with increasing farm and field size, bigger, more powerful machinery has contributed to improved quality, uniformity, and timeliness of tillage and harvest with major improvements in stability as a result" (2011 [1992]:19).

However, loss of diversity and increasing ecosystem control by increasing levels of inputs often lead to the positive correlation between diversity and stability being reestablished, resulting in increasing *instability* and decreased yield over the longer term (arrow 2 and fig. 6.3c). There are several reasons for this. *First,* as we saw in chapter 5 at the plant and variety level, increasing responsiveness of MVs to high inputs in the form of increasing yields often means high regression slopes across the range of environmental variation, because plant resources have been redirected from stability to yield by plant breeders. This high responsiveness to environmental variation is also true at an ecosystem level, as a result of resources being redirected from maintaining ecosystem stability—for example, hedgerows and polycultures—to increasing yields and total production.

Second, yields across spatial and structural levels are increasingly correlated due to the linking of local farms with regional, national, and global variation, via input networks (irrigation, fertilizers, energy) and output (marketing) networks. Therefore, local farms are influenced by variation in input and output networks, which can result in decreased yield stability—for example, with increased reliance on irrigation (e.g., Mehra 1981). This is why Merrey states that if irrigation water could be guaranteed, irrigation would increase both yield and stability (Merrey 1987). But he points out that since there are always factors that make water supply unreliable and unstable, irrigation actually increases yield instability. This can be seen in India, where fluctuations in availability of diesel fuel make pumping irrigation water unreliable, and therefore yields unstable (Pandey 1989).

Third, the hierarchical social organization created to implement the high levels of ecosystem control increases the probability of major and sudden loss of stability (Merrey 1987, Pandey 1989), as occurred with large-scale premodern systems (Adams 1978, Diamond 2005) and with the centralization and integration through markets and government management that is characteristic of modern agriculture (Anderson and Hazell 1989a:3–4). For example, a significant rise in instability of cereal yields in China since the mid-1960s (measured as the coefficient of variation) may be primarily due to increasing interregional correlation of yields, caused in turn by the rising influence of both central government policy and increasing market involvement (Stone and Zhong 1989).

Fourth, in addition to decreased diversity over the short run, industrial agriculture also leads to decreasing diversity at larger ecosystem scales due to degradation of natural ecosystems—for example, depletion and pollution of water resources, soil salinization, and waterlogging under irrigation (chapter 7).

Thus, a major challenge for increasing agrifood system sustainability is to find the unique combination of agroecological and social variables that support a positive correlation among diversity, stability, and yield.

3.3. Traditional + Modern = Sustainable

Just as in natural ecosystems, where evolution has resulted in a nonrandom high concentration of diversity and stability, a combination of selection by the environment and humans appears to have often resulted in agricultural systems where relatively high levels of diversity and stability (and sometimes yield) have a relatively high probability of coexisting. Later in this chapter I give a detailed example of rice in Yunnan, China, showing how this can work.

Therefore, an important approach to increasing sustainability of agrifood systems is strengthening existing and creating new sociocultural, ecological, and genetic spaces where high levels of diversity, stability, and yield can coexist (fig. 6.3d). In the past, these spaces were located by farmers through experimentation and trial and error over generations, but today such relatively slow change is not practical given the other rapid changes occurring globally, and it would be imprudent to not take advantage of modern science and technology to increase the levels at which diversity, stability, and yield can coexist through a combination of traditional and modern industrial agriculture (fig. 6.3, arrows 3), sometimes referred to as agroecology. Agroecology has grown since the 1970s into an eclectic field of research and application that integrates not only modern and traditional technologies and knowledge, but also social science and natural science, with an emphasis on food sovereignty (chapter 4) and social and biological diversity (Altieri et al. 2012, Gliessman 2013).

This approach, of course, means evaluating modern innovations using quite different criteria than those of the mainstream economic emphasis that dominates modern agrifood systems. In the following sections I illustrate this idea with several examples.

3.4. Synchrony and the Green Revolution in Rice in Southeast Asia

The effects of diversity can change over time with changes in the amounts and kinds of diversity—if additional diversity is "good," there will always be a limit where adding even more diversity will be "bad" (e.g., too many different crop varieties or cultural values), as with the effects of ecological diversity mentioned above. In other words, like other "inputs" to production, diversity is subject to the law of diminishing returns (see chapter 4, section 3.1). In addition, the effects of different kinds of diversity can be different and conflicting in different parts of the agrifood system.

These concepts provide an explanation of Loevinsohn et al.'s research results on the level of synchrony among rice farmers' crop schedules in the Philippines (Loevinsohn et al. 1993). Farmers had switched from photoperiod-sensitive (onset of flowering is triggered once a particular minimum daylength is reached during the annual seasonal cycle), more genetically diverse FVs, which had to be planted at the same time by all farmers for them to produce a harvest, to photoperiod-insensitive (onset of flowering not affected by daylength), less genetically diverse MVs of rice, which allowed farmers to plant at any time of the year. The replacement of FVs by MVs resulted in increasing asynchrony between different sequential tasks in rice production, measured in terms of the sum of variances for intervals between different tasks. This increasing asynchrony in the rice cropping cycle between farms led to increasing asynchrony in irrigation, which decreased irrigation efficiency since irrigations were more spread out in space and more water was lost in conveyance between fields being irrigated (table 6.1). It also increased ecological diversity (rice fields at many different stages of the crop cycle at any given time of year), leading to increased losses to pests and pathogens, since there were rice plants growing year round to support their populations, which increased for eight rice-eating insects and one viral rice disease (Loevinsohn et al. 1993:427, 437).

However, asynchrony of cropping cycles also had positive effects—it decreased social diversity by spreading out labor requirements over the year, thereby contributing to increased yields, and it evened out the supply of rice in the market, thereby increasing price stability. An informal survey of farmers suggested that they believed asynchrony would also decrease vulnerability to natural calamities, because families who suffered crop losses could depend on other families who were less affected (Loevinsohn et al. 1993:430–431). Thus, a decrease in crop genetic

TABLE 6.1. Changes in Diversity at Different Levels in Philippine Rice Farming Systems

Level	Before Green Revolution	After Green Revolution
Rice variety	Farmer crop varieties (FVs) are photoperiod insensitive and more genetically diverse	*Decreased* crop genetic diversity when modern crop varieties (MVs) replace FVs; MVs are photoperiod insensitive, less genetically diverse, and higher yielding
Rice fields at point in time	Low diversity due to synchronous crop cycles (all fields at same stage of crop cycle at any point in year)	*Increased* diversity with increasingly asynchronous crop cycles (fields at many stages of crop cycle at any point in year)
Rice fields through the year	High intra-annual diversity through time due to cropping synchrony; limits pest and pathogen populations and increases vulnerability to natural disasters	*Decreased* intra-annual diversity through time due to cropping asynchrony; supports *increased* pest and pathogen populations year round and *decreases* vulnerability to natural disasters
Social system	High intra-annual synchrony in diversity of labor demand through time due to synchronous cropping systems; leads to labor bottlenecks	*Decreased* intra-annual diversity of labor demand through time smoothes out labor demand and helps to *decrease* labor bottlenecks

Source: Based on (Loevinsohn et al. 1993).

diversity with the introduction of photoperiod-insensitive MVs led to decreased crop genetic diversity, increased ecological diversity in space, decreased ecological diversity in time, and increased social diversity, which in turn had *opposing* effects on stability (and sustainability) (table 6.1).

Loevinsohn et al. concluded that the goal of the project should be to balance *minimizing ecological asynchrony* to reduce pest problems, increase irrigation efficiency, and increase yields, and *maximizing social asynchrony* to spread labor demands and risk from natural disasters (1993:432, 436–437). Clearly the role of diversity at different levels and the effect of changes on this diversity is complex, and can be difficult to predict.

3.5. Soil Diversity and Crop Yield

In recent decades, soil scientists have discovered the wealth of biological diversity in soils, much of it important in crop production. How agroecosystems are managed affects soil biological diversity, and this diversity in turn affects the growth and yield of crop plants. Recent research supports the conclusion that biological diversity is positively related to stability of soil agroecosystems when that stability is measured in terms of variation in yield. It also indicates lower yield correlated with more diversity, at least over the relatively short term of the experiments (fig. 6.3.a).

A series of experiments in Switzerland has compared the effects of different farming practices on soil diversity (Birkhofer et al. 2008). Mäder et al. analyzed data for long-term

(1978–1999) plot experiments for two organic (biodynamic and bioorganic) and two conventional (only mineral fertilizer and mineral fertilizer plus farmyard manure) treatments (Mäder et al. 2002a). They implicitly defined sustainable agriculture as having good yields and minimum environmental impact, and they hypothesized that organic and biodynamic practices are more sustainable than conventional practices, which cause severe environmental damage. Both conventional systems were modified to adhere to integrated farming (rotation with pasture) in 1985; crop rotation, varieties, and tillage were identical in all systems.

Mäder and his colleagues found that nutrient input (N, P, K) over twenty-one years in organic systems was 34 to 51 percent lower than in the conventional systems, but the mean crop yield (edible harvest per area) was only 20 percent lower, indicating greater productive efficiency per unit of those nutrients. Energy input was also lower in organic compared to conventional systems, measured either per unit of productive land (36 to 53 percent lower) or crop dry matter (20 to 56 percent lower). The biggest differences were in soils, with organic systems having higher levels of biological activity and chemical and physical characteristics better for agriculture (all differences were significant except for bulk density, organic C, and K). There was also a significant positive correlation between aggregate stability and microbial biomass, and between aggregate stability and earthworm biomass, suggesting that soil biota contribute to important soil physical characteristics.[1] Overall, the organic systems also showed efficient resource utilization and enhanced floral and faunal diversity, features typical of mature ecosystems.

Efficiency was measured as metabolic quotient (production of carbon dioxide per unit of microbial biomass; the higher the quotient, the lower the metabolic efficiency) and was negatively correlated with microorganism biodiversity. In other words, as soil microbial communities became more genetically diverse, they became more energy efficient—they used less energy for respiration and more for growth, building up a higher microbial biomass.[2] This resulted in a higher rate of mineralization, making nutrients more rapidly available to growing plants. Organic agriculture that uses higher levels of organic matter can also process this organic matter faster, thereby compensating to some extent for the lower nutrient availability compared with conventional fertilizers. Mäder et al. concluded that "organically manured, legume-based crop rotations utilizing organic fertilizers from the farm itself are a realistic alternative to conventional farming systems. Healthy ecosystems are characterized by high species diversity."

Supporters of mainstream conventional agriculture rejected Mäder et al.'s claim that organic agriculture is more sustainable on the grounds that it had lower yields and would therefore not be able to sustain the growing human population. In response, Mäder and his colleagues noted that in their experiments, organic agriculture was more efficient in terms of both energy and nutrients on both a land and crop basis, and that while organic might require more land to produce the same amount of food, that land is maintained sustainably, whereas conventional agriculture has "degraded soils irreversibly in large areas of the world" (Mäder et al. 2002b, Mäder et al. 2002c). This returns us to the problem of determining temporal boundaries in definitions of sustainable agroecosystems—the trade-off between greater production in the short versus long term. The mainstream economic perspective implicitly assumes that maximizing short-term production will not undermine long-term production or stability because technology will solve any problems that arise.

The influence of different emphases (economic versus environmental or social) in discussion of agrifood systems can be seen in another recent exchange about ecosystem management and agriculture. In 2012 a meta-analysis of yield in organic and conventional agriculture was published (Seufert et al. 2012) that found yields in organic systems to be anywhere from 5 to 34 percent lower compared to conventional systems. From the mainstream perspective with an economic emphasis, findings such as these are taken as evidence that "organic farming cannot produce the amount of food that is demanded in today's world," in the words of former U.S. secretary of agriculture John Block (2012). By contrast, from the environmental emphasis, the other attributes and benefits of organic systems, such as various ecosystem services, are evidence that yields alone should not be the basis of decisions to support organic systems and invest in research for them (Gomiero et al. 2011).

3.6. The Question of Scale

Another important variable that influences the relationships among diversity, stability, and yield is physical scale, or the size of the farm. There are data to support competing hypotheses—that small scale is inherently higher yielding versus that large scale is inherently higher yielding. As discussed in chapter 3 regarding comparisons of farmer and scientist knowledge, when investigating the effect of scale it is important to avoid defining "traditional" or "modern" as monolithic and essentialized categories, or confounding these with agriculture at particular scales. Rather, *we need to objectively analyze the different sociocultural, historical, and environmental factors that influence the sustainability of the various components of the agrifood system in each case, and to clearly define the assumptions involved.*

Overall, the evidence supports the conclusion that yield across a range of farm sizes is influenced by a large number of other factors that interact with size, as well as by *how yield is defined.* Choosing which of these other factors to include and which definition of yield to use depends on the assumptions underlying the chosen definition of sustainability (chapters 3, 4).

For example, yield is commonly used to mean edible harvest per unit of land, but it is also used as a generic output–input ratio or as a measure of "factor productivity"—for example, edible harvest per unit of irrigation water, nitrogen fertilizer, or labor. Yield is also conflated with "production," which is an output–input ratio with gross output as the numerator (that is, it does not control for size of area cultivated) and an implied denominator, such as year, farm, farm family, region, or nation. For example, the 2011 average maize yield in the United States was 147 bushels per acre; maize production for that same year was 12.4 trillion bushels for the United States. Therefore, it is important when comparing yields in small- and large-scale farming to ensure that the same definitions and measurements are being used and to account for other factors.

The ecological anthropologist Robert Netting documented details of small-scale agriculture that often has *higher* yields per unit of land and other inputs than large-scale agriculture

(Netting 1993). This is especially true in more intensive systems, such as household gardens, as compared with more extensive systems, such as fields. It has been accepted by some researchers that scale is inversely related to yield, and that this is "one of the oldest puzzles of development economics" (Barrett et al. 2010:88) because it runs counter to the predictions of mainstream economic theory of economies of scale. Even the mainstream economic perspective sometimes acknowledges this (Hazell et al. 2007, World Bank 2007:90–93). However, this higher yield often requires greater time investment by farmers, so that labor productivity—that is, yield defined as harvest per unit of time—often decreases, a situation that Boserup viewed as a major reason that farmers avoid intensification (chapter 1).

However, it is also the case that when comparing more diverse traditional small-scale agriculture today with less diverse larger-scale industrial agriculture, the former often has *lower* yield per unit area, suggesting a trade-off between diversity and yield as discussed above. One explanation for this is that industrial agriculture has marginalized small-scale traditional agriculture (Wiggins 2009) in several ways: (1) *spatially,* onto the poorest growing environments; (2) *economically,* by enacting discriminatory laws and regulations, with industrial agriculture receiving a disproportionate share of subsidies and greater research support and externalizing a large proportion of its negative environmental and social effects; (3) *socioculturally,* by modern society and culture, which do not value the skills, knowledge, or values of traditional agriculture; and (4) *demographically,* by high fertility and migration rates in traditional agriculture resulting from the first three aspects of marginalization. All of these factors can contribute to creating a "poverty trap," a situation in which farm families can never produce enough to have the resources to improve their system but are instead caught in a downward spiral of degrading biophysical and social resources, as described in the neo-Malthusian scenario for the Kusasi in chapter 1.

4. POLYCULTURE

Polyculture, introduced in chapter 2, is a method for increasing diversity that can also contribute to increasing yield and yield stability. In this section, I will explain how it can be quantitatively evaluated, and in the following section, I provide a detailed example of how polyculture is significantly increasing sustainability in Yunnan Province, China.

Traditional small-scale farmers often plant polycultures—and the smaller the area, the higher the number of crops tends to be, with an average of forty-five species and up to fifty named intraspecific varieties or named types in some tropical home gardens (Galluzzi et al. 2010). Just as modern crop nutrition research began by identifying the nutrients that crops need and then adding single nutrients, modern agronomy has focused on single crop variety monocultures—one crop variety grown in larger and larger fields with increasing levels of inputs to obtain higher and higher yields. While polycultures are sometimes used in mainstream modern industrial agriculture—for example, planting annual crops between young fruit trees in an orchard—the only criterion for evaluating them is total yield and net income (Connor et al. 2011 [1992]:58–66). This situation is judged rational and desirable, based on the mainstream economic assumption that the purpose of agriculture is to maximize short-term profit. However, this assumption overlooks the reality of Third World and industrial world small-scale farmers, who are growing food for their families and communities, lack the resources for high-input monoculture farming, and are interested in longer-term profitability and ecosystem stability. Many of these farmers grow polycultures instead of monocultures, and the land equivalent ratio is a way of quantitatively evaluating their choice.

4.1. The Land Equivalent Ratio

The land equivalent ratio (LER) is a quantitative measure for testing hypotheses about yields in polycultures compared with monocultures. The LER is the relative land area required under monoculture for each crop (or variety) to produce the same amounts of these crops as in polyculture (mixed cropping, intercropping, alley cropping—see chapter 2, section 5.2 for definitions). I use the term "crop" in this context to refer to both variety and species, unless otherwise indicated.

$$\text{LER} = \text{yield in polyculture} / \text{yield in monoculture, and}$$
$$\text{LER}_T = \text{LER}_1 + \text{LER}_2 \ldots + \text{LER}_N = Y_{P1}/Y_{M1} + Y_{P2}/Y_{M2} \ldots + Y_{PN}/Y_{MN}, \text{ where}$$
$$\text{LER}_T = \text{total LER}$$
$$\text{LER}_1 = \text{partial LER for crop 1 etc.}$$
$$Y_{P1} = \text{yield of crop 1 etc. in polyculture}$$
$$Y_{M1} = \text{yield of crop 1 etc. in monoculture}$$

The LER provides a measure of the yield difference obtained by growing two or more crops together in polyculture compared with growing the same crops in separate monocultures. It assumes that other variables influencing yield, such as water availability, are controlled (equally for both poly- and monocultures). It also assumes that the goal is simultaneous maximization of production for *all* crops in the polyculture, not just maximum yield for the highest-yielding crop (chapter 4), as is assumed in an economic emphasis (Smale 2002). Instead, the logic here is that it is rational for farmers to grow the crops they want in combination because this strategy can (1) provide the maximum yield for all crops grown, (2) spread their production and income risks, and (3) meet the goal of producing a diversity of crops that provide desired foods and required nutrients. An example of this last point is the traditional intercropping of cereal grains and legumes—for example, maize and *Phaseolus* beans throughout North and South America—which provides food with complementary amino acid content and culinary attributes.

An LER of 1 is the break-even point, indicating that there is no difference in yield between the polyculture and the collection of separately grown monocultures. Any value greater than 1 indicates a yield advantage for the polyculture, a result called "overyielding." For example, an LER of 1.4 means that the area planted to monocultures would have to be 40 percent larger than the area required to obtain the same yields of these crops in a polyculture.

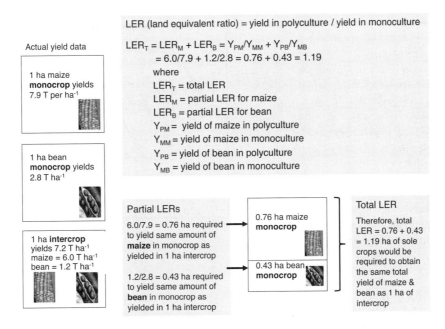

Actual yield data

1 ha maize **monocrop** yields 7.9 T per ha^{-1}

1 ha bean **monocrop** yields 2.8 T ha^{-1}

1 ha **intercrop** yields 7.2 T ha^{-1}
maize = 6.0 T ha^{-1}
bean = 1.2 T ha^{-1}

Partial LERs

6.0/7.9 = 0.76 ha required to yield same amount of **maize** in monocrop as yielded in 1 ha intercrop

1.2/2.8 = 0.43 ha required to yield same amount of **bean** in monocrop as yielded in 1 ha intercrop

0.76 ha maize **monocrop**

0.43 ha bean **monocrop**

Total LER

Therefore, total LER = 0.76 + 0.43 = 1.19 ha of sole crops would be required to obtain the same total yield of maize & bean as 1 ha of intercrop

FIGURE 6.4. The land equivalent ratio. General layout based on (Vandermeer 1992:21), using data on maize and common bean intercropping field trials in South Africa from (Tsubo et al. 2003:Table 3).

Figure 6.4 illustrates the calculation of the LER using data from the low-density treatment in a plant density trial of an intercropping experiment with maize and common bean (*P. vulgaris*) in South Africa (Tsubo et al. 2003:Table 3). Recall that the logic of polyculture is not simple maximum yield, which in this example would be achieved by growing just maize, but achieving the highest yield for the *combination of crops desired*.

Figure 6.5 illustrates the calculation in figure 6.4 graphically. If the two crops do not compete with or otherwise affect each other, and if both are planted at half the density that they would be planted in monocropping, then LER = 1. This would be equivalent to planting each crop as a monocrop on half of the polycropped area.

4.2. How Does Polyculture Work?

There are two main ways in which polyculture may work:

Competitive production is greater exploitation of niche space because each of the crop components of a polyculture exploits a different part of the space. This includes partitioning light resources via different canopy structures or partitioning soil resources via different root structures and nutrient-uptake mechanisms. While competitive production has been accepted as a common mechanism for superior performance of polycultures, Vandermeer points out that there is very little scientific research documenting it (Vandermeer 2011:77–80).

There is more evidence for *facilitation,* defined as one or more crops positively altering the environment of another crop. This occurs, for example, when one crop makes nutrients

FIGURE 6.5. Graphic illustration of the meaning of LER, showing partial LERs and total LER for the example in Figure 6.4, using the same symbols. General layout based on (Vandermeer 1992:32), data from (Tsubo et al. 2003); see Figure 6.4. In area above heavy dotted line, intercrop yields > monocrop yields and LER > 1; in area below heavy dotted line, intercrop yields < monocrop yields and LER < 1. Note that the units on the *x* and *y* axes are normalized on the yield of each crop in monoculture, which = 1.

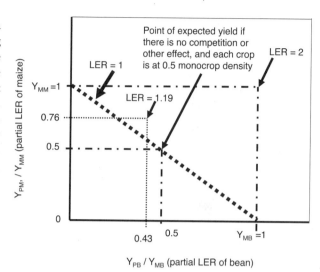

available to another crop, as with N from N-fixing legumes, although this may occur primarily the following season; when one crop enables water uptake by another crop, as with pigeon peas, which make available water from deeper in the soil; or when one crop deters, dilutes, or physically blocks pests of another crop (Letourneau et al. 2011). The traditional practice of interplanting maize and squash has been shown experimentally to produce better results than monocropping, but only under some conditions: low squash density, no moisture stress, hand cultivation, and with certain weeds (Fujiyoshi et al. 2007).

Genotype-by-environment interaction (*G×E*, discussed in chapter 5) is an important variable in both of these processes—it appears as though only certain combinations of genotypes (species, varieties, populations) and environments (e.g., moisture, nutrient availability, non-crop organisms such as weed, insects, pathogens) provide an advantage in terms of LER. In the example mentioned above, Ethiopian farmers have planted *hanfets*—mixtures of wheat and barley—for millennia, but research has shown that while the LER for *hanfets* was about 1.5, the yield stability varied significantly depending on which varieties of wheat and barley were planted (Woldeamlak et al. 2008).

4.3. Advantages and Disadvantages of Polyculture

There are several reasons why polycultures are not used more widely in industrial agriculture compared with uniform monocultures (Wolfe 2000; see also Connor et al. 2011 [1992]:64–66). *First*, polycultures require more management and labor for tasks like irrigation, cultivation, and harvesting. *Second*, polyculture does not use enough technology to be attractive to modern agriculture, implying that there is not enough profit potential for the private corporations that increasingly dominate the agrifood system. *Third*, polycultures are often assumed to be unpredictable in terms of quality. *Finally*, there is simply a lack of research data, especially at field or regional scales, although this appears to be changing.

The characteristics of polyculture that can make it a disadvantage in industrial agriculture can be advantageous in small-scale Third World agriculture. For example, harvesting mixtures like grains and legumes together provide greater nutrient quality as animal feed (Wolfe 2000). Labor, such as the hand harvesting in Yunnan rice polycultures described in the next section, is often less expensive and more reliable than fossil-fueled machinery. Still, in some places small-scale traditional farmers also find the increase in management and harvesting work required by polycultures to be undesirable (Soleri, field notes).

5. POLYCULTURE, DIVERSITY, STABILITY, AND YIELD IN YUNNAN, CHINA

Innovative research on polycultures is being done by Youyong Zhu's group based at Yunnan Agricultural University in Kunming, Yunnan Province, China. The experiments of Zhu et al. are based on the theory that intercropping reduces the evolution of resistance (because increasing crop diversity decreases selection pressure), dilutes the concentration of pathogens, leads to cross-immunization among different crops, and creates physical barriers (for example, to pathogen spore movement), all of which can reduce disease. This work is important for understanding the relationship among diversity, yield stability, and yield, especially in terms of its relevance to small-scale farming and the potential for collaboration between traditional and modern scientific knowledge.

5.1. Intercropping Traditional and Modern Rice Varieties

Chinese researchers were the first in the world to develop successful hybrid varieties in a self-pollinated species—rice—and hybrid rice has been credited with large yield increases in China and other Asian countries (Li et al. 2009b). However, despite their higher yields, these MVs have not replaced rice FVs for several reasons. In Yunnan, for example, increasing severity of rice panicle blast, caused by the fungal pathogen *Magnaporthe grisea,* in monocultures of hybrid rice MVs has resulted in reduced yields and caused farmers to increase application of fungicides.

Compared with rice FVs in Yunnan, the MVs planted there are shorter in stature, less susceptible to *M. grisea,* and have higher yield, but they have a lower market value. In contrast, FVs (glutinous or sticky varieties) are tall, more susceptible to rice blast, and have lower yield, but they are more desirable for eating and command a higher price in the market. Therefore, farmers want to grow both FVs and MVs for their different traits. So the problem Zhu's team defined was how to produce both FVs and MVs of rice while reducing fungicide applications. They were inspired by seeing some farmers intercropping these two varieties.

The researchers' general hypothesis was that crop heterogeneity (intraspecific varietal mixtures) decreases vulnerability to disease compared with monoculture, in the case of rice and *M. grisea.* Their specific hypotheses were that (a) farmers' existing practice of sowing one row of a disease-susceptible glutinous rice FV between four rows of more disease-resistant but less desirable hybrid rice MVs would reduce disease spread overall; and (b) spatial scale would not affect the results of this practice—that is, larger-scale experiments would confirm farmers' experiences and results obtained in small experimental plots regarding varietal diversity and disease spread.

The experimental design consisted of:

- treatment plots with intercropping rice FVs and MVs according to the pattern developed by local farmers (alternating four rows of MVs with one row of FVs); for MVs this pattern and spacing were the same as in monoculture, but for FVs this density was much reduced
- control plots with monocultures of each rice variety
- identical farmer management for all treatments
- all rice fields in five contiguous townships in Yunnan Province (812 ha) by 1998
- all rice fields in ten contiguous townships in Yunnan Province (3,342 ha) by 1999

The results of the researchers' experiments were impressive: panicle blast severity (mean percentage of panicle branches that were necrotic, or dying, due to infection by *M. grisea*) was significantly reduced for mixtures, especially for FVs, compared with monocultures for all combinations of FVs and MVs over time (seasons) and space (counties) (fig. 6.6). This was accompanied by no loss of yields in MVs, an increase in yield for FVs, and an increase in total yield per ha of 13 to 28 percent (average LER = 1.18) (Zhu et al. 2000:Tables 1, 2). The disease-reduction effect actually increased with the scale of the experiment. Farmers also gained financially (about USD $187.50 per ha, especially from sales of high-value glutinous rice, not counting savings from reduced fungicide purchase and application) (Zhu, Wang et al. 2003:160).

Zhu and his colleagues' field experiments showed that a polyculture of intercropped tall sticky rice FVs and shorter rice MVs ("varietal mixtures") reduced plant disease significantly and increased overall yields (Zhu, Wang et al. 2003, Zhu et al. 2000, Zhu, Fang et al. 2005). They concluded that mixtures can help maintain the yield gains of the Green Revolution while minimizing environmental costs.

The researchers also speculated that the results were due to several factors operating via facilitation. The intercropping pattern increased the distance the pathogen spores had to travel between plants, especially for the FVs, and thus diluted the concentration, reducing the incidence of contact with the more susceptible FVs. The change in canopy microclimate due to the variation in plant height, measured at one experimental site, may have resulted in conditions less conducive to pathogen infestation for FVs, such as increased light penetration and reduced humidity. Further research by the team found that intercropping significantly reduced lodging (falling over) of FV plants (Revilla-Molina et al. 2009). However, dilution and microclimate change were not likely explanations for the MVs' reduced disease incidence because they were planted at the same spacing and density as they are in monoculture. The taller FVs could, however, have physically blocked *M. grisea* spore dispersal among MVs.

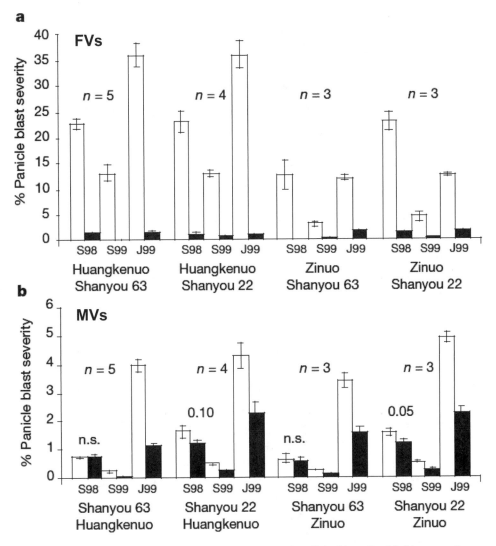

FIGURE 6.6. Effect of intercropping rice varieties on severity of rice blast. Panicle blast severity (average percentage of panicle branches that were necrotic due to infection by *Magnaporthe grisea*) of rice varieties planted in monocultures and mixtures. A. The susceptible glutinous varieties Huangkenuo and Zinuo. B. The resistant hybrid varieties Shanyou22 and Shanyou63. S98 = Shiping County, 1998; S99 = Shiping County, 1999; J99 = Jianshui County, 1999; open bar = blast severity for a variety grown in monoculture control plots; black bar = blast severity of the same variety when grown in mixed-culture plots in the same fields. Error bars are one standard error of the mean; n = number of plot means that contribute to individual bars for each of the four combinations of susceptible and resistant variety. All differences between pairs of monoculture and mixture bars are significant at P, 0.01 based on a one-tailed t-test, unless indicated by 0.05 (significant at P, 0.05), 0.10 (significant at P, 0.10), or n.s. (not significant at P = 0.10). Reprinted from Zhu et al. 2000 by permission from Macmillan Publishers Ltd: Nature.

Induced immunity to the pathogen, through facilitation via $G \times E$, was likely another factor accounting for the results in these field experiments. MVs and FVs are susceptible to a different range of *M. grisea* races, and so increased pathogen genetic diversity or richness (more races of *M. grisea*) and their more even distribution in numbers, facilitated by greater rice diversity, may have resulted in induced resistance (immunity) to rice blast in the MVs.[3] This immunity would have resulted from an increase in the probability that plants of one rice variety would be infected by a race of *M. grisea* less virulent to that rice variety, but that stimulated an immunity in the rice that then reduced the severity of disease when that rice variety was infected by a pathogen race more virulent in that variety.

Experimental work demonstrated that intercropping FVs and MVs was a powerful method of controlling epidemic diseases when resistance is race specific, but *not* for controlling disease when resistance is race nonspecific (Zhu, Wang et al. 2003). In other words, *not all ways of increasing diversity have the same effect.* As the researchers stated: "The effect of varietal diversification will vary among diseases and agro-ecosystems. Further, one can not expect all variety mixtures to provide functional diversity to a given plant pathogen population, nor can one predict the time for which they may remain effective. Indeed, we have identified variety combinations that provide little or no blast control in Yunnan Province" (Zhu, Wang et al. 2003:160, Zhu et al. 2000:721).

Therefore, it is important to consider how generalizable these impressive results are. The authors conclude that "the current world population of over six billion does not allow us to return to agricultural production practices of the past" and that "we need to maintain the benefits of modern agriculture while addressing its drawbacks" (Zhu, Wang et al. 2003:160, Zhu, Fang et al. 2000:721). The success of this strategy assumes that there will be ongoing development of improved, disease-resistant varieties, that farmers will be willing to cooperate with diversification programs (there will not be free riders who plant only high-market-value components of mixtures, hoping others' practices will reduce pathogen activity), and that diversification results can be generalized to other crop species at this scale (they cite some experimental evidence that supports this). The success of their experiment with short and tall rice varieties was critically dependent on hand harvesting the plants of very different heights, which was made possible by the fact that they were working with small-scale farmers. For other crops, like wheat, the advantages of polyculture can be gained from mixtures of varieties that are similar in height and maturity, making machine harvesting feasible.

An additional benefit of the rice polycultures is their great potential for *in situ* conservation of rice FVs that are otherwise disappearing, in part due to their greater susceptibility to disease and lodging (Zhu, Wang et al. 2003), although *M. grisea* continues to evolve and new resistant genotypes are being found among existing FVs—for example, in Yunnan (Li, Li et al. 2012). The number of FVs and the total acreage devoted to FVs has increased dramatically since the start of the rice polyculture research program in Yunnan.

This example of rice polyculture clearly illustrates the theoretical possibilities of selectively combining the best of traditional and modern agriculture for a more sustainable agriculture (Fig. 6.3d, above), and it supports the suggestion that a major challenge for increas-

FIGURE 6.7. Intercropping maize and soybean, Zhongtao, Yunnan Province, China. Photo © David A. Cleveland.

ing sustainability is to find those unique combinations of crop plants and social and biophysical environments where diversity and stability are positively correlated with yield and production. The success of the experiments in Yunnan have attracted much attention and have led to further experiments with intercropping rice varieties—for example, in Indonesia and the Philippines (Leung et al. 2003).

5.2. Intercropping with Other Crops

Zhu's group at Yunnan Agriculture University has also extended experiments to other crops with notable success, implementing plot and in-field experiments with farmers on 15,302 ha of farmland in Yunnan Province (fig. 6.7). Results of the plot experiments of four intercrops (tobacco with maize, sugarcane with maize, potato with maize, and wheat with broad bean) showed impressive yield increases compared with monocrops in the same season, with LERs ranging from 1.31 to 1.84 (Li et al. 2009a).

In the summer of 2009 I accompanied a Chinese national evaluation team to the project sites in northern Yunnan Province and had the opportunity to talk with the scientists and government officials on the team, all of whom seemed genuinely impressed with the research. One question I had when I first read about this research was how the scientists convinced thousands of small-scale farmers to adopt the research protocol in their fields. Both farmers and scientists seemed puzzled by this question—the scientists' answer was that of course farmers would cooperate, as it was in their best interest!

This kind of cooperation would not be possible in Western industrial nations, with their cultural emphasis on individuality and private property. I also realize that this was not the farmers' own response and that the reality may well be far more complicated, and not entirely so positive. What kind of positive social mechanism, not tied to strong central governments like that in China, would be needed to harness this level of cooperation in managing agrifood systems, cooperation urgently needed to solve the world food crisis? This is the topic of the next chapter.

As discussed in section 3, biological and socioeconomic uniformity has increased greatly with the growth of large-scale industrial agriculture, which tends to increase risk resulting from greater instability of yield over the long term. Many of the trends in the modernization of agriculture mark a trade-off between increases in production and decreases in stability via the loss of diversity (as with FVs and MVs). On the other hand, the diversity and stability characteristic of traditional small-scale agriculture are often positively correlated, but both are also sometimes negatively correlated with yield and production.

Mainstream economically rational behavior (investing at the level where profit is maximized) may not be at all rational for a farmer dealing with high levels of uncertainty, especially if she has limited resources. But things are complicated, and empirical research shows that relationships among diversity, stability and yield can change between spatial, temporal, and social levels, and between components of the system. Farmers and agricultural scientists can often manage risk by reducing uncertainty due to instability through collaboration for the wise deployment and management of diversity, as in the Yunnan projects.

SEVEN

Managing People

The Common Property Option

1. INTRODUCTION

The Zorse chief's wife was renowned as the best brewer of sorghum beer in the village. So after a long, hot day of biking between house compounds in the middle of the dry season, my research assistants and I headed to the chief's house for a calabash of her delicious *dam*. When we arrived, we found the shaded area under the *sok* filled with the male village elders talking animatedly, with lots of onlookers on the fringes. It appeared that the elders were interviewing someone from outside the village—a man was at the center of the group, speaking in Twi (the language of the Asante in southern Ghana) through an interpreter.

I discovered that this person from the south was claiming to be the son of a Zorse man who had migrated to work in the cacao-growing region in Asante land and never returned home. The son had never been in Zorse, but he now wanted to return to his ancestral home and farm there. The elders were asking him questions to evaluate his lineage claims. Eventually they were satisfied, and they allocated him a piece of land to farm—land that was not being used because the family that had been farming it had itself lost many members to migration. There was no question of selling or buying the land. Its allocation was a matter for the community to decide, based on the inherited rights of all community members to land. What I was witnessing was common property management of a common-pool resource, although at that time I had no idea there were lofty academic words for it.

European colonialists in Africa and elsewhere spent decades uprooting traditional community resource management institutions based on the assumption that private ownership was the only efficient way of managing resources productively. Those beliefs are still common in the mainstream today. We live in an age when socialism and public sector management in general are widely perceived as discredited, especially since the fall of the Soviet Union in 1990–1991, and privatization of resources has been elevated to the level of orthodoxy by most industrial nation governments and international agricultural and economic development organizations, such as the World Bank, which vigorously promote it (Klein 2007, Sheppard and Leitner 2010). It may come as a surprise to some, then, that over many generations communities around the world have developed successful ways of managing agrifood resources based on community control and *not* private ownership or central government control. Common property management is a widespread and often very effective method that groups have developed for managing human behavior in ways that optimize the benefit of natural resources, agriculture, and food for the group.

This common property option gained greater attention when the late Elinor Ostrom won the Nobel Prize for economics in 2009 for her pioneering work on how natural resources can be managed cooperatively in common (Ostrom 2010). But in the face of pervasive mainstream economic assumptions, there is still much work to be done to better understand the common property option—the focus of this chapter.

2. CHARACTERISTICS OF AGRIFOOD RESOURCES

Managing resources for food assumed whole new dimensions with the adoption of agriculture in the Neolithic and the need to manage the growing environments of the newly domesticated crops and to care for those crops directly by saving and storing seeds. This led in turn to new ways of organizing societies to manage agrifood resources, which is determined in part by their characteristics. Ostrom has clearly identified and defined key qualities of these and other human-managed resources, and the options for managing them (1992), and I follow her definitions.[1]

2.1. Classification of Agrifood Resources

Agricultural (and natural) resources can be defined according to their characteristics when used by farmers (table 7.1). *Costs of exclusion* are the costs (in terms of time and other resources) of keeping others from using the resource via physical or social means. *Subtractability* is the extent to which use of the resource by one person reduces the ability of others to use it, either through decreasing its availability, its value to other users, or both. This decrease can occur because of competition for the resource, and because of negative externalities (discussed in chapter 4) generated by its use.

Agricultural resources can also be classified according to the type of management regime with which they are most amenable, as affected by their characteristics under use (table 7.1). Resources such as air (nitrogen, oxygen, carbon dioxide) and solar radiation, which all plants need to grow, have high costs of exclusion and low subtractability, so are classified as public resources and usually not managed at all—that is, they are treated as open-access resources. Resources such as domestic animals, seeds and plants, and specific fields and gardens generally have low costs of exclusion and high subtractability, and are rel-

TABLE 7.1. Classification of Agricultural Resources According to Characteristics under Use, and Possible Management Systems

Resource type	Characteristics under use		Examples	Management systems possible
	Cost of exclusion	Subtractability (externalities, rivalry)		
Public	High	Low, nonexistent	Air (O_2, N_2, CO_2), solar radiation, abundant water in mountains	Open access (sometimes government)
Private	Low	High	Domestic animals, tools, seeds, crop plants, fields, gardens, compost	Private (sometimes common property or government)
Common pool	Moderate	Moderate/high	Land, crop varieties, knowledge (sometimes), irrigation water (often)	Common property, government, open access, private

atively easy for one person or group to control and use, so are classified as private resources and can be managed as private property.[2] Resources such as land, water, crop varieties (and crop genetic resources in general), and agricultural and food knowledge are often intermediate between public and private resources—they have moderate costs of exclusion and moderate to high subtractability. Therefore, they are classified as *common-pool resources* (CPRs) and their intermediate characteristics mean that they have the broadest range of management options—most often as common property or government property, but also as private property or open access (see Oakerson 1992, Ostrom 1992:295–296).[3] Variation in the way that CPRs are managed can directly affect their environmental sustainability. In practice, the different forms of management may often be intertwined—for example, in the case of water in the Mediterranean region (Ruf and Valony 2007).

2.2. Variation in Space and Time

The classification of a particular resource can vary because its costs of exclusion and its subtractability under use can vary in space and time. For example, when I worked on an irrigation project in Pakistan, I noticed how different the water resources for irrigation were in the valleys and at higher elevations. In the foothills of the Hindu Kush in northern Pakistan, water flows in innumerable rivulets down the slopes, and it is so abundant and dispersed that it would be classified as a public resource. When a farmer diverts one rivulet into his field, as many do, it does not detract from others doing the same with other rivulets, and it would be extremely difficult to exclude others from this widespread resource. However, as they descend the slopes, those rivulets coalesce, finally merging into major rivers like the

Swat, which then flow into the broad, arid valleys of the Khyber Pakhtunkhwa (formerly North-West Frontier) Province. As this happens, the characteristics of water for irrigation are transformed—the costs of exclusion decrease and the costs of subtractability increase— so that we would now classify that water as a CPR. Water in the Swat valley—for example, in the Swabi Irrigation District, where I worked—is managed by the government and by water users' associations (WUAs), although large, private landowners also exert much influence. I will discuss irrigation water as a CPR further in section 5. In places where rainfall is highly seasonal, the nature of water resources vary over time. In northeast Ghana during the growing season, water is a public resource, falling as rain on peoples' fields. During the dry season, that rainwater becomes concentrated underground in riverine aquifers that can be tapped by shallow hand-dug wells, typically located within walled gardens that are watered from those wells (fig. 0.1). The water then takes on many of the characteristics of a private resource and is managed privately. Because the water is restricted to underground flow in a narrow aquifer under the dry streambed, it is easy to exclude access to the wells and dry-season gardens, especially to roaming goats and cattle, by erecting thorn-topped mud walls, and the water has high subtractability because it is limited in quantity.

The characteristics of agricultural resources also vary over longer time periods, often as the result of increasing pressure on the resource or of new technologies that increase the ease of exploitation. In many cases these changes increase subtractability and decrease the costs of exclusion, so that formerly public resources take on the characteristics of CPRs. With further pressure, they may even become private resources. This has been the case in many areas where relatively abundant water for irrigation gave it the characteristics of a public resource. With increased pressure by an increasing number of farmers (as well as urban dwellers and industries), water has become a CPR managed under different regimes, which tend to become dominated by private management as the pressure increases and water takes on more of the characteristics of a private resource.

Political and institutional change also strongly influences the ways in which CPRs are managed, as is well documented for the case of water. The World Bank and other major multinational organizations dominated by the industrial world have been pushing for privatization of agrifood system and other resources especially forcefully since the beginning of the 1980s. For example, in Mexico, neoliberal policy changes have resulted in the privatization of formerly state-managed irrigation water, ostensibly based on the assumption that this would result in greater efficiency and sustainability. However, evidence suggests that while in some cases privatization encouraged the participation of farmers, in general it did not increase efficiency or sustainability. Instead, the main effect of, and motivation for, privatization has been increasing the opportunity for private profit (Wilder and Lankao 2006). The push to export fresh produce from Mexico to the United States has also resulted in the export of nonrenewable virtual water and increasing inequalities in access to water, for example, in northwest Mexico (Zlolniski 2011). The movement toward privatization has led to major campaigns in Mexico and elsewhere to maintain water under community control and common property management (CPM) (e.g., Barlow 2008).

Another example of change over time is the case of crop genetic resources, including crop varieties, that many experts have classified as common-pool resources because it is difficult, although not always impossible, to exclude others from gaining access to them, and because using them does not always, but sometimes can, reduce the ability of others to use them. Traditionally, these resources have been managed as private property, public property, and common property (Cleveland and Murray 1997). However, with the introduction in industrial nations of intellectual property protection, including industrial patents, the characteristics of crop genetic resources have moved dramatically toward those of private resources—not only in industrial countries, but in the Third World as well, under pressure from the industrial countries and the international organizations they support, such as the World Trade Organization (Andersen 2008). I discuss genetic resources as a CPR later in this chapter (section 6).

3. COMMON-POOL RESOURCE MANAGEMENT

As discussed above, the intermediate nature of CPRs means that they often can be, and have been, managed under a wider range of management regimes than public or private resources. In this and the following section we will explore answers to the question "What variables promote sustainable management of common-pool agricultural resources?" Many agricultural resources are CPRs that can be managed in all four of the ways described above (open access, private, government, common property), and these management regimes have varying effects on sustainability, in terms of the emphases on conserving ecosystem function and supporting equitable social systems (see chapter 4) (table 7.2). The theory of CPR management was first extensively developed in terms of ocean fisheries in the 1970s (Clark 1990).

A key characteristic that differentiates management systems is the way in which they internalize negative externalities. In chapter 4 I discussed how externalities in agrifood systems have negative effects on other members of society, both directly and indirectly. In this section I will analyze the three management options other than CPM for agrifood CPRs, focusing on the ways in which each option internalizes negative externalities. I illustrate this by expanding the graphic representation of the production function, introduced in chapter 4 (fig. 4.1).

3.1. Unmanaged Common-Pool Resources: Open Access

When a CPR is used without any restrictions on access, there are no mechanisms to internalize negative externalities. Each farmer will seek to invest that CPR in production to maximize her own profit, but this will eventually push the aggregate production beyond the private economic optimum for the resource as a whole, into the area of *private economic overinvestment* (to the right of I_2, fig. 7.1, refer also to fig. 4.1). That is, investment continues even though the *marginal value product* (MVP) is less than *marginal factor costs* (MFC), and overall profit (the distance between TVP, or total value product, and TFC-P, or private total factor cost) is decreasing, because each new investment will still return some additional

TABLE 7.2. Management Systems for Common-Pool Resources (CPRs)

CPR management system	Characteristics	Means of internalizing externalities	Examples of managing irrigation water
Open access	Individual decisions taken independently without collaboration; rights to resources undefined	None	Each farmer takes the water she needs without consideration of other farmers or limits on availability; can lead to "tragedy of the anticommons," depletion or loss of the resource
Private	Individual (including corporate) rights to and management of resources	Markets	Water owned by individual farmers, or by a private entity who sells it for a profit to farmers who can afford it
Government	Government rights to and management of resources	Pigovian taxes, regulations, subsidies	Government regulates the allocation of water among farmers
Common property	Group rights to and management of resources, based on communication, collaboration, and on sanctions established and enforced by group	Functioning community structures; benefit of contributing to public good greater than cost; may include extension of personal utility in time and space	Farmers' organization creates and enforces rules for allocation of water and penalties for rule breakers

profit for the farmer making that additional investment, even though it will reduce the profit of every farmer not making additional investments. In other words, farmers are motivated to invest beyond point I_2, where MVP = MFC, because other farmers are doing this; according to this logic, if they refrain from investing, their profit will be reduced by other farmers continuing to increase investment, and they will be worse off than if they did not invest.

The same incentives to invest continue even after the TVP passes the point of short-term maximum output (c, fig. 7.1), where the MVP becomes zero and starts to become increasingly negative. As this occurs, the production function moves into the region of *biological overinvestment*, where the *total* production is decreasing and the CPR is very likely being degraded. Individual farmers will continue to increase their investment even though their profits—the distance between TVP and TFC–P divided by the number of farmers investing—continue decreasing at a more rapid rate, because if they do not invest they will have even lower returns, since others are increasing their investment and diminishing the overall profit that is available. This will continue to the bionomic equilibrium, the point where TVP

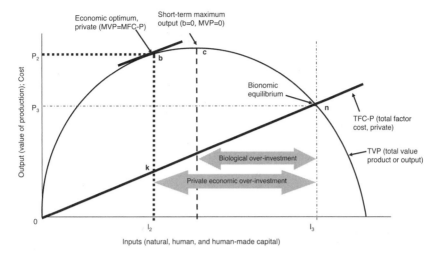

FIGURE 7.1. Effects of increasing inputs of agrifood resources on production. See text for definition of acronyms. Based on (Clark 1990, Ellis 1993).

has decreased to equal TFC–P, and farmers *on average* would be getting back no more than they invested (n, fig. 7.1), as in the well-known fisheries example (Clark 1990). Hardin's classic (but tragically misnamed) "tragedy of the commons" (Hardin 1968), is a case of open access—the *non*management of CPRs—that results in degradation of resources and a reduction in human carrying capacity (HCC), because there are no mechanisms to internalize externalities.

The prisoner's dilemma game is often used to illustrate the psychological-social process governing contribution to the public good, internalizing externalities, in open-access nonmanagement of CPRs (fig. 7.2). This game defines a situation in which "rational" behavior (maximization of personal utility or profit) results in an outcome that is suboptimal for both the individual and society (Cornes and Sandler 1996:310–311). The basic assumptions of this game are that the cost to a player for contributing to the common good exceeds the benefit to that player, and that players do not communicate with each other and therefore do not have the chance to cooperate. The maximum benefit to both players (4, 4) would result if both contribute; however, if player A tried to achieve the maximum benefit by contributing, assuming player B will do the same, but player B does not contribute, then player A will be a net loser, (–2) while player B, who contributed nothing, will gain (+6). In this situation, player B's strategy earns the title "free rider." The most likely outcome, however, is that neither player contributes to the public good (0, 0), because neither player trusts the other to contribute. This is referred to as a *Nash equilibrium,* a situation in which neither player has any regrets, given the other player's choices.

3.2. Management of Common-Pool Resources as Private Property

So what are the alternatives to nonmanagement of CPRs? Let's begin by considering the management of CPRs as private property. We can add to figure 7.1 to illustrate how the dif-

		B's strategy	
		Do not contribute	Contribute
A's strategy	Do not contribute	A: +0, -0 B: +0, -0 **Net: A = 0; B = 0**	A: +6 B: +6, -8 Net: A = 6; B = -2
	Contribute	A: +6, -8 B: +6 Net: A = -2; B= 6	A: +6, +6, -8 B: +6, +6, -8 Net: A = 4; B = 4

FIGURE 7.2. Game theory interpretation of open-access management of CPRs as "prisoner's dilemma," a tragedy of the anticommons. Possible behaviors: contribute 1 or 0 units of public good. For every unit of public good contributed, contributor spends (loses) 8, and every player gains 6. Cell with bold print is the most likely outcome.

ferent management options internalize negative externalities (fig. 7.3). As discussed previously, the economically rational level of input for a CPR managed as private property without internalizing externalities is I_2, the point of maximum private profit (the distance b – k, where MVP = MFC–P, fig. 7.3). Internalizing externalities results in increasing the costs of production to the TFC–S curve, which means that the economic optimum level of investment shifts to the left to I_1, because the profit at this point (a – f, where MVP = MFC-S) is now greater than the profit at I_2 (b – g).

Economics textbooks often make the normative assumption that private ownership is the best way to solve externality problems—for example, through contracting (e.g., Silberberg 1995:297 ff.). Consolidating management under a single private owner has been suggested as a possible way of internalizing externalities generated by multiple private owners, but the model is complicated, based on many assumptions, and has contradictory results (Simpson 2003). In contrast, voluntary action by multiple private owners has also been suggested as superior to government management in industrial capitalist societies for internalizing externalities, in part because of the assumption that local, private owners are better able to perceive and respond to ecological realities (Gottfried et al. 1996).

However, internalization of externalities via the market seems to be the exception for CPRs, and market failures, situations in which not all the costs of production are captured by the market and therefore not internalized, are common (Daly and Farley 2004). This means that relying on the market to internalize externalities through individual profit maximization leads to an open-access situation, subordinating the long-term health of society and of the environment to the short-term economic gain of individuals (Daly and Farley 2004). As we saw in chapter 4, this is unsustainable even economically because the economy depends on society and the environment.

Again, irrigation water provides a good example because it is critical for food production and is often a CPR. Irrigation water is frequently in limited supply, so if some farmers decide to use proportionately more water, the other farmers will have lower yields because they have less water to irrigate with. The result is that the overall social benefit of this scarce resource will not be maximized in terms of food available to the whole community. Because externali-

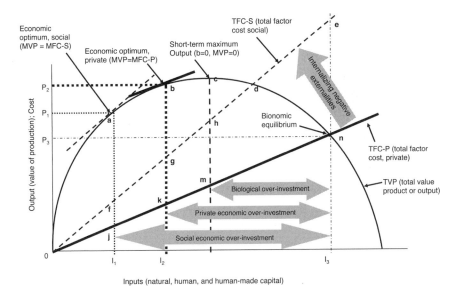

FIGURE 7.3. Options for internalizing externalities in the management of common property resources (CPRs). Based on (Clark 1990, Ellis 1993).

ties are not included in the resource user's benefit-cost calculations, her optimal level of investment (inputs) (I_2, fig. 7.3) will be higher than if externalities were internalized (I_1, fig. 7.3), and the difference between private and social costs at that point (g – k) represents the cost that is externalized to society at this level of private investment. If the externalities were internalized, the offending farmer would be motivated to reduce inputs to maximize her profit, resulting in a reduction in production, from P_2 to P_1. Therefore, investment beyond I_1 is social overinvestment, because it decreases net social benefits.

3.3. Management of Common-Pool Resources as Government Property

Hardin's alternative to private ownership as a solution to his "tragedy of the commons" scenario was government management of CPRs—"mutual coercion, mutually agreed upon by the majority of the people affected" (Hardin 1968:1247). Governments internalize externalities via laws, regulations, and Pigovian taxes, which are meant to increase the cost of production to the point at which the individual or group reduces externalities to avoid having to pay for them. Some economists consider these mechanisms to be market based since they rely on markets to respond to government action and do not directly limit production of the externality (Bartelmus 2013:3, 20). The challenge of using these mechanisms is how to determine the cost to be externalized so that internalization will result in a level of investment that is sustainable. Unlike private management, where the "invisible hand" of the market is assumed to optimally determine the price that will internalize externalities, governments must directly or indirectly set this price, which would ideally be based on scientific research and public discussion of values, but is too often based on politics and heavily influenced by short-term political goals and private-sector lobbying (UCS 2012).

Politicians are notorious for having extremely high discount rates (DRs), so scientific findings are often ignored or deliberately changed to meet political criteria. For example, there have been several major food contamination outbreaks in the United States in recent years, many of which are the result of private agrifood system companies externalizing costs by not investing in adequate preventive measures. Indeed, the Union of Concerned Scientists' analysis of the food safety regulatory situation concluded that the level of political and corporate interference in the interpretation and use of scientific evidence in policy making was most intense when scientific evidence was most important (UCS 2010).

The current international political maneuvering around the scientific consensus on anthropogenic climate change, much of it due to the agrifood system, and the refusal of politicians and private businesses to adequately address the predicted huge negative externalities of business as usual, is another example of high DRs among politicians, which I will discuss in chapter 8.

4. MANAGEMENT OF COMMON-POOL RESOURCES AS COMMON PROPERTY

In most farming societies, many key CPRs have historically not been owned individually but instead been managed as common property, as in the example of Zorse land-use rights at the beginning of this chapter. Over time, many traditional commons in the industrial world have been privatized, or "enclosed." For example, in Great Britain the "most active phases of enclosure began in the 1500s and extended into the 19th century" (Runge and Defrancesco 2006:1714). As European colonialists imposed their ideas on the peoples and societies they subjugated in the Third World, a favorite theme was the need to privatize ownership of land and other resources. Cadastral surveys were the first step in separating land from communities, turning it into a private commodity that could be sold. The Green Revolution, with its higher yields that required higher inputs, stimulated land privatization, as detailed by Goody for northern Ghana (Goody 1980). More recently, the dramatic increase in world food prices after 2008 has led to land grabbing in the Third World, facilitated by the promotion of privatization as social progress and the lack of understanding and respect for traditional common property management (Anseeuw et al. 2012:50–52).

CPM as an alternative to private or government management of CPRs has been largely overlooked by scientists, governments, and policy makers from the industrial world until recent decades. An important reason that CPR was ignored in the mid-twentieth century was the influence of the idea of the "tragedy of the commons" I referred to above, popularized in Hardin's article of the same name (1968). By July 2012, Hardin's influential article had been cited over 4,728 times, according to the foremost academic literature database, Web of Knowledge, and 18,452 times, according to Google Scholar.

In light of our greater understanding of CPRs and options for their management, a tragedy of open access, or the *anticommons*, seem more accurate terms for what Hardin described. Hardin assumed that the users of the rangeland in his example were not able to communicate effectively and reach a communal management strategy. Thus, Hardin concluded that resource destruction, and the population pressure that drives it, can be halted only by "mutual coercion mutually agreed on"—that is, by government authority. In fact, research since Hardin's essay was published has shown that many communities around the world have created institutions of common property management that effectively eliminate the tragedy of open access to CPRs and internalize the externalities of overinvestment and production often more effectively than private or government management (Becker and Ostrom 1995, Dietz et al. 2002a, Netting 1993, Ostrom 1992).

Elinor Ostrom has played a pivotal role in establishing the field of research on CPM of CPRs, carrying out most of her work in India, Nepal, and Kenya, and, as already mentioned, winning the 2009 Nobel Prize in economics for her work. She, along with her colleagues, did more than any other researcher to counter the misconception promulgated by the misnamed "tragedy of the commons." The Nobel Prize press release highlighted the challenge her work has presented to mainstream economics: "Elinor Ostrom has challenged the conventional wisdom that common property is poorly managed and should be either regulated by central authorities or privatized. Based on numerous studies of user-managed fish stocks, pastures, woods, lakes, and groundwater basins, Ostrom concludes that the outcomes are, more often than not, better than predicted by standard theories. She observes that resource users frequently develop sophisticated mechanisms for decision making and rule enforcement to handle conflicts of interest, and she characterizes the rules that promote successful outcomes."[4]

4.1. Key Factors Associated with Successful Management of Common-Pool Resources

The theory that CPRs tend to be managed by common property institutions when key criteria are met is supported by data from CPM of irrigation water and other CPRs. The most important feature of successful CPM organizations may be their ability to unite individual benefits in a common group benefit that facilitates internalizing the limits of sustainable human impact on resources, and many traditional social institutions and cultural values have done this. The result limits the optimization of individual short-term gain in favor of shared community gain, which results in greater individual gain, especially over the long run.

Based on extensive research, Ostrom and others have found that successful CPM regimes display a number of common characteristics (Bardhan and Dayton-Jones 2002, Ostrom 1992, Tang 1992):

- The resource is important, moderately to highly subtractable, and used, but not (yet) drastically overused.
- Access to the resource is restricted, with moderate costs of exclusion.
- Benefits of investing in the public value of the resource are greater than costs, creating incentives for users to invest—in terms of game theory, there is a fully privileged pay-off matrix.

- Evolved norms are translated into rules, with graduated penalties for transgressors.
- A small spatial scale facilitates face-to-face communication, or good communication among larger groups—for example, via the Internet.
- Transparency provides users with accurate knowledge of boundaries and ecosystems and indicators of resource conditions.
- Users have a shared image of how the resource works and a common understanding of the problems and alternative solutions.
- There is perception of mutual trust among the users.
- Users are interested in social and environmental sustainability (i.e., they have low DRs).
- There is a fairly stable economic, social, and political environment.
- Local organizations are nested in a hierarchy of organizations in which they are protected from externalities such as government interference.
- Governments recognize and respect local CPM organizations, or at least don't interfere to the extent to destroy their effectiveness.

4.2. Internalizing Externalities Via Uniting Personal and Social Utility

The institutional incentives and sanctions of successful CPM can provide the payoff structure to successfully internalize externalities. In terms of game theory, CPM management could be considered a fully privileged game, meaning that each player is motivated to contribute to or to privilege the other player given her expectation of what the other player will do. In this game the relative benefits and costs are reversed compared with the prisoner's dilemma game; the cost to a player for contributing to the common good is *less* than the per-player benefit. This defines a situation in which "rational" behavior results in a Nash equilibrium when both players contribute (fig. 7.4, lower right cell, Net: A = 10; B = 10) (Cornes and Sandler 1996:310), but also with an optimal outcome for society.

The outcome of this game benefits the CPR and society, because both players are better off by contributing to the public good—that is, internalizing the negative externalities of their production, because each will be a net winner. This simplifies CPM by assuming that people will contribute because they are aware that they will be better off by contributing, and because they assume the other players will also contribute. In reality, the benefits are often not immediate, and people may feel they cannot afford to contribute because of the short-term costs, leading to the problem of free riders who don't contribute but receive the benefit of others' contributions. Therefore, a key to successful CPM is the development of norms translated into rules for behavior with graduated penalties for breaking the rules. The result is that "communication, trust, the anticipation of future interactions, and the ability to build agreements and rules sometimes controls behavior well enough to prevent tragedy" (Dietz et al. 2002a:5).

		B's strategy	
		Do not contribute	Contribute
A's strategy	Do not contribute	A: +0, -0 B: +0, -0 Net: A = 0; B = 0	A: +8, -0 B: +8, -6 Net: A = 8; B = 2
	Contribute	A: +8, -6 B: +8, -0 Net: A = 2; B = 8	**A: +8, +8, -6** **B: +8, +8, -6** **Net: A = 10; B = 10**

FIGURE 7.4. Game theory interpretation of common property management of CPRs, a "fully privileged" payoff matrix. Possible behaviors: contribute 1 or 0 units of public good. For every unit of public good contributed, costs to contributor = 6, benefits to each player = 8. Cell with bold print is the most likely outcome. Based on (Cornes and Sandler 1996).

The game theory approach to CPM assumes that individuals are motivated to invest in the public good based on self-interest, and this has also been the assumption in much of the CPM literature (e.g., Wade 1988). But what if humans voluntarily acted altruistically, as discussed in chapter 4? It would mean that the payoff matrix in the fully privileged game could result from subjective mental changes as well (Ostrom 2010:660). This is altruism in the sense that the individual receives no material gain, but rather a psychological payoff—contributing to the public good makes the individual feel better. In other words, CPM may also facilitate the internalization of externalities by the expansion of "self-interest" to include future generations (reducing DRs), and in space, to include other individuals and households and even communities, in cases where the management system is larger than a single community.

Altruistic motivations have been documented—for example, among members of a common property irrigation organization in Australia. Based on interviews with 235 farmers, Marshall found that farmers were motivated to cooperate more by social factors like "perceptions of community benefits, and trust" than they were by private and materialistic factors like "distributive fairness and business security," which have been the focus of government programs attempting to motivate cooperation (Marshall 2004).

4.3. Challenges for Common Property Management in a Global Commons

Analysis of the components of successful cases of CPM does not translate directly into the ability of local communities or outside development agencies to create successful CPM structures in situations where they are unsuccessful, have disappeared, or never existed. While there are many examples of existing CPM regimes, the practical application of CPM theory in the creation of community-based natural resource management is rife with problems, as detailed for Malawi and Botswana (Blaikie 2006).

Some of the challenges to the successful functioning of existing or new CPM institutions, especially their extension to the global commons, include:

- It is difficult to "know" what the objective reality relevant to CPM is, including through scientific research.

- There are often many different values, even in relatively homogeneous communities, yet there needs to be agreement on key issues.

- CPRs are often interlinked but require different kinds of management.

- There are accelerating rates of environmental and social change that affect the definitions of CPRs and their management.

- Globalization and rising demand for resources mean that interest in CPRs is being increasingly extended in space to additional communities.

- There is often a lack of experience in scaling up from local systems to global systems.

- The possibility for repeat experiments is decreasing—there is only one planet, and ecological and social functioning is increasingly stressed.

Probably the biggest challenge is using the examples of CPM systems that have been successful on the local level, affecting a very small proportion of the Earth's people and resources, to create management systems that affect the majority of people and resources. Global climate change is one prominent example of this, with a massive failure by the global community, dominated by nation-states and private corporations, to reach any meaningful agreement to deal with what many see as the greatest challenge to sustainability for humans (see chapter 8). As Ostrom and her colleagues have so eloquently stated: "Building from the lessons of past successes will require forms of communication, information, and trust that are broad and deep beyond precedent, but not beyond possibility" (Ostrom et al. 1999).

Even while those with a social and environmental emphasis are working to develop new management institutions of unprecedented scale, the push for privatization of agricultural resources, such as land, is being fueled by the food and climate crises. There has been a dramatic increase in the buying up of land in sub-Saharan Africa by governments and private investors to grow food for export (Anseeuw et al. 2012, Cotula, Vermeulen, Mathieu et al. 2011), such as the Brazilian project to grow soy for Japan on vast tracks of land in Mozambique currently farmed by many small-scale food farmers (LVC 2012). The growth of carbon markets, including the growing of biofuels, has also led to land privatization, referred to as "green-grabbing" (Leach et al. 2012, Tienhaara 2012). From a mainstream economic perspective, land acquisition is not only a rational choice for investors, but a boon to Third World nations who give up their land, as it will help them to modernize agriculture (Connolly et al. 2012, Cotula, Vermeulen, Leonard et al. 2009, Robertson and Pinstrup-Andersen 2010). However, this may often not be the case, as with biofuels development in Ghana, which has resulted in increased poverty and the alienation of local peoples from access to resources (Schoneveld et al. 2011).

From an alternative perspective, the problem is not only depriving local farmers of their land, but the capital-intensive, high-input, export-oriented agriculture that is the goal for this land, which critics see as having a net negative effect on Third World agriculture (De

Schutter 2011b). De Schutter has pointed out that Western ideas of private land ownership will play into land grabbing and suggested that communal forms of ownership should be formalized to protect local communities (De Schutter 2011a) (section 4.0).

5. MANAGING IRRIGATION WATER AS A COMMON-POOL RESOURCE

Agriculture accounts for almost two-thirds of freshwater withdrawals globally, or almost two trillion m³ per year (FAO 2012a). Irrigated crops yield nearly twice as much as unirrigated crops; only 16 percent of cropland is irrigated globally, but this cropland produces about 36 percent of the global harvest. Irrigation is most important in Asia, with 80 percent of food production there from irrigated lands.[5] Yet water resources will become less reliable, and in many places scarcer, as a result of climate change, and competition with other uses will increase. At the same time, up to 50 percent of water extracted from rivers, lakes, and groundwater aquifers to be used for irrigation is not used by crop plants; instead, it is lost to seepage from canals, drainage below the root zone, and excess evapotranspiration. CPM presents an option for managing both the social and biophysical challenges of irrigation for future agrifood system sustainability.

5.1. The Challenge

The agronomic objective of irrigation is to maintain the amount of water in the root zone of plants in the field at 50 to 100 percent of the water-holding capacity of the soil, which maximizes yield. Five related problems that affect the ability to achieve this, and also contribute to unequal benefits, are common to irrigation systems (Mabry and Cleveland 1996):

- inefficient, undemocratic, and inequitable management structures;
- a varying supply of water in rainfall runoff, water courses, and stocks in reservoirs and groundwater;
- inefficient conveyance of water;
- inefficient application of water; and
- insufficient drainage and leaching, leading to waterlogging and salinity.

Conveyance and application efficiencies are often only 50 percent or less, especially in traditional agriculture, and these are the problems most under the control of individual farmers and local communities (figs. 7.5, 7.6). Increased application efficiency would mean that actual evapotranspiration is about equal to maximum potential evapotranspiration, since this is optimal for plant growth. Yield decreases when soil water is less than 50 percent of field capacity due to the stress of inadequate water. When soil water is greater than field capacity, water is lost to drainage below the root zone, or if drainage is poor, the soil becomes waterlogged and plants are stressed due to inadequate oxygen for root respiration as water replaces air in the soil.

FIGURE 7.5. Unlined earthen canals with earthen gates in the Peshawar basin, Pakistan. Photo © David A. Cleveland.

FIGURE 7.6. Lined irrigation canals and concrete gates in the Peshawar basin, Pakistan. Canals lined with bricks and concrete increase conveyance efficiency, and gates of concrete make cheating more difficult. Photo © David A. Cleveland.

If these problems can be avoided, then both yields and yield stability can be higher than without irrigation, and benefits can be more equally distributed. It is a common misperception that irrigation itself usually does this, while in fact research suggests that irrigation often increases yield but decreases yield stability (Pandey 1989). Mehra points this out in her analysis of Indian data (Mehra 1981).

But yields and production can be maximized in different ways. From a mainstream economic perspective, the goal of irrigation is maximum short-term production. From an alternative social and environmental perspective, the goal is maximum long-term stable production that provides equitable benefits to the community. There is increasing evidence that CPM is an effective strategy for managing societies' irrigation water, which can internalize negative externalities to achieve the goal of long-term stable production and equity.

Ancient, centralized irrigation societies often appear to have overshot local or regional carrying capacities. But the archeological evidence can be difficult to interpret in terms of the interacting effects of social organization, environment, secular climatic change, and regional political conflict. Still, centralized systems tend to let demand determine production (Daly and Cobb 1989), allowing allocation to determine the scale of activity (flow determines stocks), an economic perspective on sustainability.

Most of the data on irrigation management supports the hypothesis that water is managed most "efficiently" when irrigators at the local level have a major degree of control. Irrigation water and other CPRs tend to be managed by water users' associations (WUAs) (Hunt 1989, Tang 1992), as is true in Pakistan (Mabry and Cleveland 1996, Merrey 1987). The most important feature of successful WUAs may be their ability to internalize sustainable scales (Daly and Cobb 1989) that directly limit the optimization of individual short-term gain in favor of long-term community gain. Some WUAs, for example, set limits on production based on an analysis of the local carrying capacity, and then let limits, not markets alone, determine prices and allocation (i.e., stocks determine flows, part of an environmental perspective on sustainability).

5.2. Peru

Trawick's study of the irrigation system in the community of Huaynacotas in the Peruvian Andes exemplifies the relationship among local social institutions, physical geography, and the larger polity (Trawick 2001). Trawick claims that successful CPM of irrigation water is common in the Andes, and in the past—before Spanish invaders stole indigenous peoples' land and water, or destroyed CPM institutions—such management systems were probably even more common.

While there is much variation in irrigation systems in the Andes and elsewhere, Trawick's goals were to outline some general principles from an outside perspective that are not necessarily recognized by local irrigators, but that are derived from the explicit rules irrigators do recognize. His goal was based on the hypothesis that traditional irrigation system CPM is a "highly effective means of managing a scarce and fluctuating resource," which contradicts the "tragedy of the commons" misconception popularized by Hardin.

Huaynacotas was able to conserve its CPM system by ceding control over its better lowland fields to the European invaders and retreating to the more marginal higher-altitude fields. The main purpose of the irrigation system in Huaynacotas is to extend the short high-altitude growing season. The irrigation system provides water to fields by gravity flow from two higher-altitude springs through two halves of the irrigation network to maize fields before the start of the rainy season, and in years of drought, it provides additional water during the growing season. People take turns watering, since the supply is inadequate to allow for simultaneous irrigation.

The success of the system rests on following a set of explicit rules that are enforced by officials elected by the community for limited terms.

- Land sectors in the village are given water consecutively based on altitude and microclimate.
- Plots within sectors are given water contiguously and serially.
- Adjusting to drought is done by eliminating fields at the upper (higher-altitude, more agronomically marginal) end first.
- All fields are terraced and formed into *atus* (earth-bermed basins), to assure equal distribution.
- No departures from procedures are allowed except for agreed-on reasons, such as sandy soil.

From these rules, Trawick deduced a set of principles:

- Autonomy: The community has and controls its own flows of water.
- Contiguity: Water is distributed to fields in a fixed, contiguous order based only on their location along successive canals, which facilitates transparency and increases conveyance efficiency by minimizing water flow in dry canals.
- Uniformity among water rights: Everyone receives water with the same frequency and uniformity in technique, meaning that everyone irrigates in the same way.
- Proportionality (equity) among rights: No one can use more water than the extent of their land entitles them to, nor can they legally obtain water more often than everyone else.
- Proportionality among duties: People's contributions to maintenance is proportional to the amount of irrigated land that they have. Proportionality establishes the basic moral principle that defines everyone's rights in the face of an unequal distribution of land.
- Transparency: Everyone knows the rules and has the ability to confirm, with their own eyes (because of contiguous irrigation), whether or not those rules are generally being obeyed, to detect and denounce any violations that occur.
- Regularity: Things are always done in the same way under conditions of scarcity; no exceptions are allowed, and any expansion of irrigation is normally prohibited.
- Graduated sanctions for rule violations: Sanctions are imposed by the community and graded according to the gravity of the offense.

According to Trawick, these principles of the Huaynacotas irrigation system increase its sustainability by greatly reducing the level of conflict compared with that of communities with less severe water scarcity but much less equitable CPM. In terms of environmental sustainability, the principles provide the incentive to conserve natural resources; in other words, they are a mechanism for translating peoples' recognition of the need to limit human impact into social mechanisms, such as CPM, to avoid exceeding the HCC that would harm everyone (chapters 1, 3). In terms of game theory, the principles result in a system where the benefit to an individual farmer is greater than the cost of following the rules. This is made possible because the rules create a system in which, unlike open access, the effects of cheating by taking a longer turn at irrigating increases the time between irrigations in a manner obvious and significant to other farmers on the same canal who can easily detect the cheating. In addition, the rules also result in a longer interval until the next irrigation for the free rider.

6. MANAGING AGRICULTURAL KNOWLEDGE AND CROP GENETIC DIVERSITY AS COMMON-POOL RESOURCES

The other example of CPR management I look at in this chapter is agricultural knowledge and crop genetic diversity. In chapter 3 I discussed farmer and scientist knowledge and the similarities and differences between them, especially in terms of the potential for collaboration between farmers and scientists. In contrast to water for irrigation, which is a physical resource that can be spatially localized, knowledge is a mental or intellectual resource that can move rapidly via many forms of communication (Cleveland and Murray 1997, Dove 1996a). Rights in these kinds of resources are termed *intellectual property rights* (IPRs).

6.1. Crop Genetic Resources

Genetic diversity is the basis for the evolution of crops to adapt to new environments and new human requirements, as discussed in chapter 5. Crop genetic resources (CGRs), like other biodiversity resources, are a mixture of the physical and the intellectual. For example, seeds, like music or computer software files, are physical things, but the information they contain can be duplicated indefinitely, as can knowledge about how to use them. To protect farmers' access, sociologist Jack Kloppenberg (2010) suggests modeling intellectual property protection of crop genetic resources on the organizational and legal structures developed for open-source software.

CGRs have been important in traditional societies for millennia, but the ability to manipulate and profit from them, and hence battles over rights to them, have increased dramatically with the increasing power of biological science—especially biotechnology and genetic engineering—and economic globalism (Cleveland and Murray 1997). In the modern period, increasing conflict over control of CGRs began with the rise of scientific plant breeding at the beginning of the twentieth century, the subsequent growth of seed companies, and the increasing privatization of CGRs through plant variety protection. Privatization increased greatly with the extension of IPRs in the form of industrial patents on life forms, beginning in 1980 with the U.S. Supreme Court decision allowing the patent of a genetically engineered *Pseudomonas* bacterium in Diamond v. Chakrabarty (Blakeney 2011), and it continues today with the growing domination of plant breeding by private multinational corporations focused on genetically engineered, transgenic crop varieties (TGVs).

Much of the debate over farmers' IPRs in CGRs centers on whether farmers have the right to control outside access, and what compensation they are entitled to when CGRs are used by outsiders. The CGR example illustrates clearly the concept that rights are not "inherent" but are socially negotiated within the existing power structure. To what extent are individuals' and communities' rights to these resources independent of their management of those resources versus contingent on the effectiveness of their management in (a) conserving those resources per se, and (b) optimizing benefit to the larger, even global, society (Cleveland and Murray 1997)? To the extent that rights are contingent, when new research changes our understanding of the effectiveness of management, then rights are under pressure to change. This is the difficulty with advocating farmers' rights based on the justification that they are conservators of CGRs.

Even if local farmers' rights are recognized, the extent to which they will benefit farmers is often problematic. The issue of local people's rights to biodiversity resources in general has been a topic of research for a number of social scientists. Anthropologist Michael Dove suggests that compensation will likely fail to reach the indigenous peoples who maintain biodiversity because their rights are often not recognized by national governments, and, if it does reach them, it's likely to change them and undermine the basis for any biodiversity conservation in the process (Dove 1996a:44). He sees IPRs as just another form of outside intervention in indigenous communities that facilitates the destruction of those communities' biological resources. Still, Dove recognizes that poor indigenous communities are sometimes able to intensify linkages to world markets on their own terms and for their own ends. However, because economic and political goals are so intertwined, it will be extremely difficult to change the economic position of indigenous peoples by establishing their rights to biological resources via IPRs unless there is also a shift to establishing their political rights.

Andersen contends that access to and use of CGRs has become increasingly restricted—both by efforts based on the Convention on Biological Diversity to protect farmers' rights, and by trade-based efforts to standardize national and international regulations to the benefit of the industrial nations, resulting in an "anticommons tragedy" (2008:44, 351–352). The challenge of crafting equitable policy is that the two goals—access by the mainstream global agrifood system and respect for farmers' rights to benefits—are governed by fundamentally different assumptions about how the world should work: private property rights and maximization of private profits versus common property rights and maximization of shared benefits. The International Treaty on CGR has the potential to change this, but a basic problem is the influence (via "structural power") of private interests' attempts to control profit, either directly or indirectly (via industrial nation governments), which often overwhelms any bottom-up efforts by or for farmers (Andersen 2008:351). But what do traditional farmers themselves feel about IPRs in their CGRs?

6.2. Zuni Traditional Crop Varieties and Intellectual Property Rights

The Zuni people of what is now the western part of New Mexico have been in a battle with European invaders, including the U.S. government, for generations (Cleveland, Bowannie, et al. 1995). A major contested resource has been their land and its natural resources, including crop varieties that Zuni farmers have been maintaining, planting, and eating for many centuries. The recent interest by the dominant society in traditional crops and foods, along with the advent of modern IPRs, has complicated the situation.

In the early 1990s, as part of the Zuni Folk Varieties Project, Daniela Soleri interviewed Zuni tribal members about their attitudes toward IPRs in their traditional crop varieties (FVs), as part of a project we were working on with the Zuni tribe. We were asked to do this because tribal members were upset by the use of Zuni religious symbols by non-Zunis in jewelry and, increasingly, in food products. Soleri interviewed several groups, including the Zuni Cultural Advisory Team, whose role is not in policy making, but rather in developing recommendations as a group that reflect Zuni cultural values (Soleri, Cleveland, Eriacho et al. 1994). The Advisory Team stated that "Zuni seeds should not be sold or given to outsiders for profit, resale, breeding, or trade marking because of their significance to the Zuni people. This statement applies to all long-time food crop varieties of the Zuni people including corn, beans, squash, melons, gourds, chilies, and peaches." They said that seeds should not be used as a commodity for profit, in part because according to Zuni oral narrative and history, the Zuni people's corn had once before disappeared, and this was followed by a warning that the next time the Zuni people would not have a second chance to recover the core of their livelihood and identity. The Advisory Team added that once you let something as important as these seeds go you don't know how they will be used, because once they are out of your control, there are no guarantees of appropriate use.

The Advisory Team's statements reflect an ideal that they believe in but will not always reflect the changing world that the Zuni people live in and what is actually occurring in the Zuni community. Therefore, while emphasizing the ideal, the team also responded in a way that reflected actual choices people must make. The team felt that when outsiders are given access to Zuni crops, policy agreements should be demanded that protect and compensate the Zuni people in culturally appropriate ways. They did not believe that Zuni crop FVs should be crossed (interbred) with other varieties, and they felt that Zunis should be compensated for any such crosses made by plant breeders in the past. Regarding food products, the team suggested that fresh produce, such as sweet corn, and processed foods, such as cornmeal or parched corn, be sold to non-Zunis, but that whole seed be sold only within the community. This would ensure that no viable seed would be sold to outsiders. On the issue of trademarking, the Advisory Team believed that cultural resources such as Zuni crops, foods, and the Zuni name should be protected for use by tribal members only, as a matter of sovereignty.

The Advisory Team's statements indicate how complicated developing policies for protecting IPRs in CGRs is. A substantial additional challenge is finding agreement among the Zuni as to who fairly and honestly represents the Zuni people and tribe. After a long tradition of freely sharing seeds, Zuni and other traditional farmers have observed the increasing control and manipulation of seeds by private corporations for profit. As they are left out of the benefits and their culture is used and misrepresented for others' profits, these farmers and communities are increasingly reluctant to share their seeds. Many indigenous and other local groups do not have any policies for addressing these situations; however, this does not mean that they are not concerned, but that they need honest information and a meaningful voice in the discussion of these issues. When these are lacking, policies are decided by those with the most power and influence, who have very different interests (Soleri, Cleveland, Eriacho et al. 1994:37).

As resources with moderate to high subtractability and moderate exclusion costs, CGRs and the knowledge and cultural significance associated with them are CPRs that until now have been managed as such within relatively localized areas and among groups of people interacting directly with one another. With increasing globalization and dominance of the quest for profit, the previous arrangements are no longer satisfactory, especially for local communities. If we agree that the mainstream approach with its economic emphasis does not offer a solution, then the challenge we are all faced with is to do as Ostrom indicated and develop CPM mechanisms enforcing trust, transparency, and cooperation at a previously unprecedented scale.

EIGHT

The Big Solutions

Climate Change, Resource Cycles, and Diet

1. INTRODUCTION

Imagine the world twenty to thirty years in the future. How old will you be? How old will your children be? Will it be a world where humans have cooperated to reverse global warming? Will we have decided to pursue prosperity decoupled from increasing consumption? Will we have learned how to share resources so that everyone has the necessities of life and the opportunity for happiness? Or will it be a world of extreme weather where the former homes of many millions of people are under the ocean, where human society and the environment have changed irrevocably and for the worse in terms of human happiness? What kind of a world will all of those living today create for those living in the future?

Anthropogenic climate change is emerging as the biggest threat to planet Earth as we have known it, to the Earth that enabled the emergence of our species and has supported its continuing evolutionary success—for better and worse. Because our activities, including feeding ourselves, generate huge amounts of greenhouse gases (GHGs), our planet is warming more rapidly than at any time in the last million years. The results are already being felt in increasing land and ocean temperatures, ocean acidification, melting glaciers, increasing frequency and intensity of storms, and increasing floods and droughts. These changes are making it more and more difficult to produce, process, and distribute food. Even if we were able to completely eliminate anthropogenic greenhouse gas emissions (GHGE) overnight, the inertia of the effect on the climate of what we have already done will continue to warm our planet for many years into the future. The outlook is not rosy.

Not only will climate change have mostly negative effects on the agrifood system, but the agrifood system is one of the main contributors to climate change through emission of the three main anthropogenic GHGs—for example, from land clearing and energy use that transforms trees, soil organic matter, oil, and coal into carbon dioxide (CO_2); nitrogen fertilizer and manures that emits nitrous oxide (N_2O); and ruminant animals, manure, landfills and paddy rice that emit methane (CH_4). Therefore, we urgently need to turn the agrifood system from a cause and casualty of climate change into a solution, by reducing its contribution to GHGE. To do so we must get a better understanding of the full impact of many of our food production practices using methods, such as life-cycle analysis, which I describe below.

Since continued economic growth is assumed by the mainstream to be a requirement for sustainable agrifood systems, the only option for mitigating anthropogenic climate change becomes increasing the efficiency of GHGE—that is, reducing the amount of emissions per unit of growth, measured for example as gross domestic product (GDP) per person. For the agrifood system, this means *increasing the efficiency of production, processing, and distribution*—that is, the GHGE per unit of food produced and consumed. There are many ways to do this, reducing and making more efficient our influence on the natural cycles of carbon, nitrogen, and other resources—for example, by changing practices such as field cultivation to store more carbon in the soil and to reduce CO_2 emissions, by increasing the efficiency of nitrogen fertilizer use to reduce N_2O emissions, or by creating new breeds of animals with reduced CH_4 emissions. We can also increase the energy efficiency of cultivating, irrigating, harvesting, packaging, cooling, processing, and cooking. Changes like these are important and need to be pursued, yet many times they require more research and new skills and technologies, and they may take years or even decades to have significant effects.

However, in order to stabilize Earth's temperature at no more than about 2° C above preindustrial levels, *increasing efficiencies alone won't work, but on the contrary will lead to continued warming unless net emissions are reduced to zero or even negative, which will probably require reducing economic growth to zero or negative*. The IPCC (International Panel on Climate Change) has suggested a target for maximum concentration of atmospheric CO_2 of 450 ppm (IPCC 2007a), while others, most prominently the climate scientist James Hansen, believe that 350 ppm will be necessary to avoid catastrophic change (Hansen, Sato, Kharecha et al. 2008).[1] The 350 ppm target is especially challenging since it would require a reduction in the current concentration, which in 2013 is almost 400 ppm. The NGO 350.org was named for Hansen's target, and many other organizations have adopted it as a goal (see section 5.2 for further discussion).[2] This means, as we have seen (in chapters 1 and 3), that continued growth in human impact (HI) will eventually lead to extreme environmental and social stress, or even collapse.

History is not encouraging. Just considering population and consumption (measured as per capita income) growth: from 1990 (the Kyoto protocol base year) to 2007, world population increased 1.3 percent per year and income rose 1.4 percent per year, while C efficiency (emissions of C per unit GDP) increased only 0.7 percent per year, resulting in an almost 40 percent increase in GHGE over the period (Jackson 2009b:78–79). But this does not consider equity; in order for everyone in the world to achieve the average European income by 2050 and to reach even the conservative IPCC target of 450 ppm CO_2 atmospheric concentration, emission efficiency would have to increase *over 130 times* (Jackson 2009b:81)! Yet even such a spectacular increase in efficiency, with no historical precedent, would not eliminate absolute growth in GHG concentrations, and would only postpone the ultimate collapse of society as we know it.

But what about reducing GHGE and concentrations by *decreasing both production and consumption by changing diets and reducing food waste, especially in the industrialized world?* I focus on these agrifood system changes in this chapter because they could dramatically and rapidly reduce GHGE, because they can be accomplished with the resources, technologies, infrastructure, and skills we have right now, and because rather than costing money to implement, they could produce short- and long-term net savings. These changes could also generate a lot of positive externalities, such as reducing water pollution, restoring and conserving biodiversity and increasing sequestered carbon on land no longer needed for agriculture, reducing burdens on landfills, and improving human nutrition and health.

However, these changes would require major changes in the way most people, especially in the industrial world, think about food, and in the way food producers, processors, and marketers do business, in order to reduce waste and support consumers' behavioral changes in diet and food waste. This approach is a much greater challenge to the mainstream economic system than increasing production efficiency, because its focus is on reducing consumption in excess of that needed to optimize human health and happiness. This approach will likely be successful only as part of a larger revolution in the way the world thinks about and moves toward sustainability, recognizing the limits to human carrying capacity (HCC) and rejecting the double oxymoron of "sustainable green growth" (World Bank 2012a). That is, it will require figuring out how to have *prosperity without economic growth* (discussed in more detail in chapters 3 and 4) (Jackson 2011, Victor and Jackson 2012).

A critical requirement for making this strategy work is information that will motivate behavioral change. I have long thought that it would be great if food labels stated not only ingredients and price, but also where the ingredients were grown, who grew them, how they were grown, and how they were processed and transported—or at least some informative shorthand addressing these processes. To deal with climate change we will also need to know the effects of production, packaging, processing, transport, and consumption of the food we eat on GHGE directly and indirectly (via effects on water and other resources), as well as their effects on the short- and long-term health and happiness of people throughout the agrifood system. Research on these topics is beginning to be done, but the results are rarely made available to those affected—consumers, farmers, farm workers, processors, retailers, and distributors—especially in ways that can help them make decisions. We need transparent, reliably documented links between food choices and their climate and other impacts.

Having all this information on food labels may not be practical, and many foods do not even have labels. Still, clear, scientifically vetted, publically available information of this sort provided by public-interest government entities or independent nonprofit organizations is essential if there is to be an engaged discussion of how best to change our agrifood system to reach our goals for sustainability. This is not impossible or unprecedented; for example, laws were passed in the United States requiring the listing of food ingredients in 1966, food nutritional content in 1990, and trans fatty acids content in 2003 (US FDA 2012). Recent evidence of the potential power of such information can be seen in the effort to require food ingredients from transgenic crop varieties (TGVs) to be labeled in the United States—for example, in California in 2012—and the many millions of dollars spent by mainstream seed and food corporations to fight this initiative. Also needed will be a fundamental rethinking of the culture of consumerism (Jackson 2011).

2. THE EARTH'S CHANGING CLIMATE

Life on Earth has adapted to the geochemical composition and dynamics of the environment as part of the process of its evolution, but it has also affected the geochemical environment. Indeed, human effects on the environment since the Neolithic, and especially since the industrial revolution, have been so great that this recent period has been labeled the Anthropocene geological epoch by some Earth scientists (see chapter 2). Given enough time, humans and other species might be able to successfully adapt to these dramatic changes by evolving biologically, but the speed of environmental change is currently much greater than the speed of evolution. Extinction rather than adaptation has been the most likely outcome in similar cases in the past, and anthropogenic changes appear to be already driving current high extinction rates of other species (Barnosky et al. 2011). The only choice for humans seems to be to evolve socioculturally to reduce our impact on the environment—to become fast learners (chapter 1).

2.1. Biogeochemical Evolution of the Earth's Atmosphere

The Earth's atmosphere evolved over a long period into one that today has the characteristics needed to support life as we know it—for example, the high levels of oxygen (~21 percent by volume) needed for respiration; the stratospheric ozone layer that provides protection from ultraviolet radiation; and the atmosphere's component gases like CO_2 that fuel photosynthesis and help to warm the surface and prevent freezing.

Photosynthesis evolved early in the Earth's history, about 3.5 billion years ago in cyanobacteria. Oxygen-producing (oxygenic) photosynthesis evolved about 2.5 billion years ago, first in bacteria, and later in plants (Hohmann-Marriott and Blankenship 2011), and was a key event in the evolution of life and the atmosphere (Catling and Claire 2005, Johnston et al. 2009, Rothschild 2008). Oxygenic photosynthesis removes CO_2 from the air and uses energy from the sun to create carbohydrates, which store solar energy in their chemical bonds and in the process produce oxygen (O_2). Initially this O_2 was captured in terrestrial rocks through oxidation, but over many millennia cumulative release of O_2 from photosynthesis led to saturation of those deposits, and O_2 started to accumulate in the atmosphere, making the evolution of animals—including humans—possible. At the same time, C concentrations in the atmosphere were reduced as a result of increasing O_2, and C in plants was

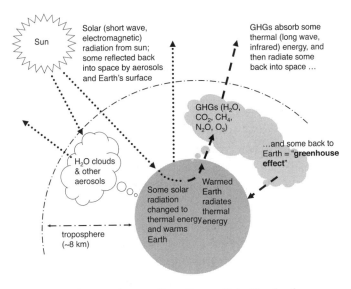

FIGURE 8.1. The greenhouse effect. © 2013 D.A. Cleveland.

taken out of circulation as those plants died and were turned into peat, coal, and oil deposits on land, and into calcium carbonates and other compounds in the oceans, mainly during the Carboniferous period, 300–350 million years ago (Raupach and Canadell 2010).

A dynamic equilibrium evolved in which the amount of CO_2 in the atmosphere remained fairly stable, with plants removing it through photosynthesis and microorganisms, plants, and animals emitting it during the process of respiration, oxidizing the carbohydrates to release energy. Because CO_2 in the atmosphere is known to have a warming effect on the Earth, it is called a greenhouse gas (GHG). Fluctuations in CO_2 levels over millions of years are positively correlated with changes in temperatures on the Earth, although temperature is also influenced by other variables, including other GHGs, variations in solar radiation, and the cooling effect of the albedo (reflectivity) from aerosols, clouds, and the Earth's surface (IPCC 2007b).

2.2. Global Warming

Global warming is the result of the "greenhouse effect," the absorption by GHGs of thermal (i.e., long-wave, infrared) radiation from the Earth that has been warmed by solar (i.e., short-wave, electromagnetic) radiation (fig. 8.1), followed by the radiation by GHGs of some of this thermal energy back to the Earth, resulting in increasing average surface temperatures.[3]

The evidence for geologically recent and rapid global climate change due to human activity is overwhelming (IPCC 2007b, Riebeek 2010). During the million years since the ice ages, as global temperatures fluctuated, it took about five thousand years for temperatures to rise a total of 4 to 7°C (Riebeek 2010). Over the last hundred years (1906–2005), the average global surface temperature on land has increased 0.74°C ±0.18°C, when estimated by a linear trend. The rate of warming over the last fifty years is almost double that over the last hundred years (0.13°C± 0.03°C vs. 0.07°C ± 0.02°C per decade) (IPCC 2007a:237), and about ten times faster than the average rate during temperature rises over the last million years, and the predicted rate of warming in the next century of 2 to 6°C is about twenty times faster (Riebeek 2010).

Human activities both heat and cool the Earth, with their effects estimated in terms of radiative forcing—energy in units of watts per square meter (W m^{-2}) of the Earth's surface (IPCC 2007b). The largest heating effect is that of the GHGs, including CO_2, CH_4, and N_2O, with a heating effect of +2.30 (+2.07 to +2.53) W m^{-2}. Ozone created in the troposphere (the lowest layer of the Earth's atmosphere), water vapor, and halocarbons (such as refrigerants) are the other major GHGs (see fig. 8.1).[4] The IPCC estimated that the average net anthropogenic radiative forcing by 2005 was 1.6 W m^{-2}, with a total temperature increase from 1850–1899 to 2001–2005 of 0.76 °C (IPCC 2007a:3–5).

Some human activities also result in cooling, including the production of aerosols, which in the atmosphere reflect shortwave radiation, and changing land covers that increase surface albedo. However, the net effect of human activity on climate is overwhelmingly to increase temperatures. In section 4 I will discuss the contribution of the agrifood system to global warming, but first, let's look briefly at the effects of global warming on the agrifood system, and the possibilities for adaptation.

3. THE EFFECTS OF CLIMATE CHANGE ON THE AGRIFOOD SYSTEM AND THE POTENTIAL FOR ADAPTATION

The two main consequences of global warming that will affect the agrifood system are increase in temperature and change in precipitation patterns. Much of this is predicted to occur in areas that already have low per capita food production, such as Africa, South Asia, and the Middle East, making the potential for worsening the food crises there even greater. A related effect of GHGE is the effect of increased CO_2 concentrations on photosynthesis and plant physiology and composition.

3.1. Water Availability, Temperature Change, and Increasing Carbon Dioxide Concentrations

Crop production will be affected by all the major components of climate change—both increases and decreases in regional water availability, warmer temperatures, and increasing CO_2 concentrations. The results will also depend on interactions among different plant processes that are not well understood, and this is an active area of research (Long and Ort 2010, Wang, Heckathorn et al. 2012).

The supply and distribution in space and time of precipitation and freshwater for irrigation is changing as a result of global warming. The majority of agricultural production (60 percent) and land (80 percent) is rainfed, but the 40 percent of production that depends on irrigation currently accounts for about 70 percent of human use of fresh surface- and groundwater globally (WWAP 2012). Global warming will bring an increase in precipitation to areas with already-high precipitation, and a decrease to areas with low precipitation, and potentially both in some areas, leading to more droughts and floods, and increasing frequency of extremely hot temperatures, all of which will reduce yields (Hansen Sato, and Ruedy 2012, IPCC 2007a). There is evidence from data on changes in ocean salinity that the rate of this intensification of the global water cycle in response to increases in average ocean surface temperature are much greater than previously estimated (Durack et al. 2012). For example, Durack et al. predict a 16 to 24 percent intensification, which includes increasing frequency and violence of storms, with a 2 to 3°C increase in ocean temperature.

There will also be a shift in many areas toward a lower proportion of precipitation as snow, causing problems in seasonal storage and distribution of water available for irrigation. One reason will be an increase in runoff in the winter due to a higher proportion of precipitation falling as rain instead of snow and earlier snow melt, which will exceed storage capacity of reservoirs, followed by a decrease in runoff in the spring due to less snowmelt, leading to depleted reservoirs in the summer and water shortages (Barnett et al. 2005). Production problems in areas with increasing water shortages will be exacerbated by rising temperatures, which will shift the areas where crops can be grown, with some net losers and some winners, but likely an overall loss in productivity.

Increasing CO_2 concentrations could increase the efficiency of photosynthesis, and some claim that this will be a major offset to the decrease in yields due to changes in precipitation and temperature. It could mean an increase in water-use efficiency because plants' stomata would not have to open as long (to obtain the CO_2 needed for photosynthesis, but which also allows more water to evaporate from the plant). However, increased yields due to higher CO_2 concentrations may be offset by heat stress due to higher temperatures (Long and Ort 2010), and N limitation in non-N-fixing crop species may result in decreased protein and mineral densities—for example, in wheat (Fernando et al. 2012) and rice (Seneweera 2011), reducing nutritional value.

3.2. Adapting to Climate Change

The mainstream response to addressing the agrifood system's contributions to climate change is to emphasize increasing investments in developing new technologies. Examples include genetic engineering (Peterhansel and Offermann 2012), including patenting plant genes with potential adaptation to changing climate for creating TGVs (ETC Group 2008), and ramping up efforts to access increasingly overdrawn and overallocated production inputs, like water, as in California's plan for the already overstressed Sacramento–San Joaquin River Delta, and massive dam building in China, India, and elsewhere. However, this approach is likely to increase negative externalities, including in other locations and into the future.

The good news is that there are also opportunities within the agrifood system for alternative approaches to adapt to the effects of climate change that also reduce the agrifood system contribution to climate change (Smith and Olesen 2010). Many of these approaches are part of the overall strategy of adapting to the limits of HCC in general, limits that include increasing scarcity of water, fossil fuels, good soils, and plant nutrients like phosphorus (for example, through conventional plant breeding) (Fess et al. 2011).

Alternative approaches often emphasize existing low-cost methods, such as producing shade-grown coffee, which researchers believe can diminish the negative effects of climate change, such as extreme temperature and precipitation, and so reduce farmers' economic and ecological risk (Lin et al. 2008). Seeking crop species and varieties that are better adapted to new climate conditions is another alternative approach. Crop diversity can reduce variation in yield in terms of time and space, increasing resilience in the face of higher variability in temperature and precipitation, and can also provide economic benefits (Lin 2011). Increasing the water-use efficiency of food production to adapt to increasing water scarcity and evapotranspiration rates will also decrease GHGE by decreasing the energy used to deliver and recycle irrigation water.

4. THE AGRIFOOD SYSTEM'S EFFECTS ON CLIMATE AND THE POTENTIAL FOR MITIGATION

One of the three fundamental changes that occurred with the Neolithic revolution was an increase in human management of ecosystems to provide for domesticated crop plants via the diversion of resources (e.g., land, water, nitrogen, and phosphorus) from natural cycles to human-dominated cycles (chapter 2). Some scientists believe that since the early years of the Neolithic, both foragers and early farmers had a major effect on climate via land-use change (LUC) in the form of conversion of an area from a natural ecosystem to a human-dominated ecosystem, mainly via clearing and burning of forests—and that land use per capita decreased about ten times during the last eight thousand years as farmers switched to more intensive cultivation (Kaplan et al. 2011, Ruddiman and Ellis 2009).[5] In more recent times, industrial agriculture has amplified this trend with the application of excessive amounts of water and inorganic nutrients that saturate the soil and lead to high rates of leakage, especially of P and N, out of the production system (Drinkwater and Snapp 2007); these leaked nutrients, in turn, become major contributors to environmental problems, such as pollution and global warming, as well as to social problems, such as malnutrition and inequitable access to resources.

The increasing economic and spatial concentration of the mainstream agrifood system has led to increasing distances between inputs and food production, and between production and consumption (chapter 9). This increases GHGE, not only because of greater transportation distances but much more from increases in resources needed for production (in part to compensate for increased waste), processing, packaging, and storage. In addition, the trend toward diets with a higher proportion of animal and processed foods increases GHGE. As a result, the agrifood system is a major contributor to global warming.

Even though the agrifood system is one of the largest contributors to GHGE (Scialabba and Muller-Lindenlauf 2010)—or even the largest (Goodland and Anhang 2009)—it has received relatively little attention as a target for reducing GHGE compared with sectors such as energy, transportation, and built infrastructure. Below I discuss in more detail the effects of the agrifood system on climate and the potential for mitigating the negative effects through reducing food waste and changing diet.

4.1. Estimating the Agrifood System Contribution to Climate Change

Research on the climate impact of agrifood systems is a relatively new field, with many methodological challenges for gathering and analyzing data. Life-cycle assessment (LCA) was developed for industrial processes and has become an important tool for quantifying the effects of the agrifood system (Roy, Nei et al. 2009). LCA is beginning to make valuable contributions to our understanding; however, it requires large amounts of data that can be difficult to gather, and these have not been collected routinely if at all in the past. LCA also requires defining boundaries in space, time, and food systems' structure that determine what is included and what is excluded in the assessment.

For example, the U.S. Environmental Protection Agency (EPA) estimates that GHGE from agriculture make up 6.1 percent of U.S. anthropogenic GHGE (EPA 2010). This figure is relatively low because it is based on spatial and system boundaries limited to direct agricultural production and therefore do not include postharvest processing, transportation, and preparation, or preproduction components, such as energy for delivering irrigation water or for manufacturing and delivering fertilizers and other chemicals. It is also limited to current GHGE—for example, not including previous LUC. At the global level, similarly limited to agricultural production, the World Bank estimates 14 percent of anthropogenic GHGE come from agricultural production alone, (World Bank 2010:195), while the IPCC estimated 10 to 12 percent including 47 percent of CH_4 and 58 percent of N_2O (Smith et al. 2007:499).

Other estimates have included other parts of the agrifood system. Weber and Matthews used an input-output LCA methodology that extended the boundaries spatially and systemically beyond production to include GHGE from production and transport of inputs such as fertilizer, but did not include postretail components (such as transport to the home, preparation, and postconsumer waste) and did not include past contributions such as LUC. Their estimate was 8.1 MT CO_2e (metric tons CO_2 equivalents) U.S. per household in 1997 (Weber and Matthews 2008), 12.7 percent of the total GHGE per household in that year (calculations based on EPA 2010).

If LUC is included and temporal boundaries are expanded to the beginning of agriculture, the proportional impact of the agrifood system increases dramatically (see below). Goodland and Anhang took this approach in estimating GHGE for just the animal portion of the agrifood system. In terms of system boundaries, they included GHGE from the loss of vegetation resulting from the original conversion of natural vegetation to pasture, and they did not credit existing pasture with net carbon sequestration, as some estimates have done (Goodland and Anhang 2012). This dramatically increased the estimate—they found that the animal portion of the agrifood system *alone* accounted for 51 percent of all anthropogenic GHGE globally (Goodland and Anhang 2009:12–3). Of course, it can be difficult to identify the "natural" ecosystem in many areas, as some ecosystems have been affected by agriculture since the Neolithic and all areas have been affected indirectly by the agrifood system. However, coming to agreement about a definition of ecosystems that are not primarily shaped by recurrent, significant human management should be possible. For example, Goodland and Anhang contrast pasture with forest land cover only in areas where the latter existed before contemporary land clearing.

The defined boundaries of an agrifood system also frame the scope for mitigating its effects on climate change and are necessarily dependent on assumptions, which are often controversial. For example, if past LUC, such as the removal of native forests and other vegetation, is included in a definition, then restoration of this vegetation by reducing animal production is more likely to be considered a possible mitigation strategy, and if the GHGE from composting and landfills is not included, then reducing food waste is less likely to be considered a mitigation strategy.

4.2. The Mitigation Potential

As we saw when calculating HCC in chapter 1, the agrifood system has many points at which efficiencies could be improved, and these could, in turn, decrease GHGE (Smith et al. 2007). This approach to mitigating climate change is preferred by the mainstream because it accommodates current or even increasing rates of growth—for example, by increasing the efficiency of GHGE per unit of GDP (Niu et al. 2011). Resource-delivery efficiency (RDE) could be increased by decreasing the amount of packaging or the distance that production inputs and food travel (Williams and Wikstrom 2011). Resource-use efficiency (RUE) could be improved by: developing TGVs to increase yields (Ronald 2011); adjusting the amounts and timing of water or nitrogen fertilizer application to crops so that a larger proportion would be taken up by plants; plant breeding to increase efficiency of use by the plant (Fess et al. 2011); or managing livestock and breeding new animal varieties to increase efficiency of GHGE per unit of animal food production (Gill et al. 2010).

Unfortunately, this was also the approach adopted at the Durban climate conference in 2011 and the Rio+20 conference held in June 2012, both of which were dominated by industrial nations and multinational corporations and their proposals for creating a "green economy," which was vociferously condemned by advocates of alternative approaches emphasizing environmental and social sustainability (chapter 4).

While these efficiencies are undeniably important, they focus on the supply side and are often seen as mitigating the effects of increasing production by increasing efficiency through reductions in GHGE per unit of production. From the mainstream perspective, dominated by the assumption that economic growth is the only way to solve the world food crisis, increasing these efficiencies is coupled with continuing to increase food production and consumption, which would mean that atmospheric GHG concentration would continue to increase, although possibly at a slower rate, depending on the balance of increasing efficiency and continued growth (see section 1.0).

However, if we assume that we are at or have already exceeded HCC, then increasing efficiencies of GHGE or other measures of environmental damage are not an adequate solution;

we need to decrease production and consumption in order to decrease HI by *decreasing the net demand for resources*. And there is ample opportunity to do this, by increasing the human use efficiency (HUE) of food by reducing food waste (EPA 2012c), improving diets (Carlsson-Kanyama and González 2009, Lagasse and Neff 2010), and reducing disease (Egger and Dixon 2009). For example, it was estimated that in 2002 the United States produced 3,774 kcal per person per day, but the needed average is only about 2,100 kcal per person per day; this difference is attributable to overeating and food being discarded after processing and distribution (Eshel and Martin 2006). Thus, the energy efficiency (energy required in food / energy in food produced) = 2,100 / 3,774 = 0.556, implying a large potential for reducing GHGE in the agrifood system by decreasing overconsumption, and thus demand.

This approach is favored by those advocating an alternative approach to agrifood system sustainability, one based on evidence that HI is already exceeding HCC, because this approach has much greater potential for decreasing total GHGE and will be required to decrease atmospheric GHG concentration. I will discuss in more detail the demand-side approaches of reducing consumption of animal products and reducing food waste later in this chapter.

In the next two sections I briefly describe the carbon and nitrogen cycles and how they have affected climate change, as well as how better management can mitigate climate change.

4.3. The Agrifood Carbon Cycle

Agrifood systems are a critical interface between carbon (C) in the soil and carbon in the atmosphere. Through photosynthesis, plants pump C from the atmosphere into the soil by fixing CO_2 into carbohydrates that store solar energy for later use, and by growing roots into the soil that exude carbon compounds there in the process of taking up water and nutrients. When plants die, some of the carbon in their cells becomes incorporated into the soil as they decompose. Most of the C fixed by plants is pumped back into the atmosphere through plants themselves respiring, microorganisms decomposing plant remains, and animals eating the plants. However, only a small portion (around 4 percent) of the C fixed by plants becomes part of a slow pool of soil C that can remain in the soil for more than about twenty years, and an even smaller portion will become part of the passive pool of soil organic matter that can remain in the soil for hundreds or thousands of years, although this pool accounts for the majority of soil organic matter and plays a vital role in making nutrients available to plants (Brady and Weil 2010:378ff., Brown et al. 2012, EPA 2012b).

Plant organic matter is essential for creating the physical, chemical, and biological properties of soil needed for plant growth and food production. This organic matter provides food for soil microorganisms, and those microorganisms in turn hold nutrients in their cells, making them available to plants when they die. Organic matter also buffers soil pH, helping to keep it at levels that increase the availability of nutrients to plants. Microorganisms and macroorganisms, such as earthworms, break down organic matter and produce polysaccharides that stick soil particles together so they can form aggregates that create soil structure, providing spaces essential for moving water and O_2 into soil and CO_2 out.

Changes in Greenhouse Gases from ice-Core and Modern Data

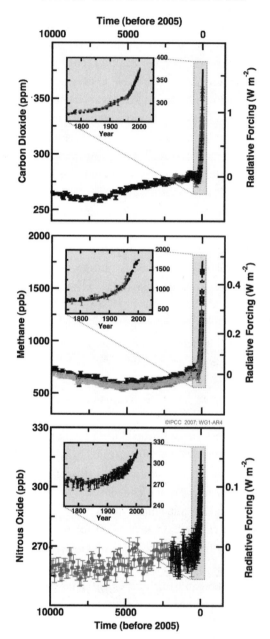

FIGURE 8.2. Atmospheric concentrations of carbon dioxide, methane, and nitrous oxide over the last ten thousand years (large panels) and since 1750 (inset panels). Measurements are shown from ice cores (symbols with different colors for different studies) and atmospheric samples (red lines). The corresponding radiative forcings are shown on the right-hand axes of the large panels. Figure SPM-1 from IPCC 2007b.

There are four major ways in which the agrifood system increases the flux—or rate of flow—of GHG into the atmosphere, and there are known ways in which each of these can be significantly reduced. Figure 8.2 shows the atmospheric concentrations of CO_2 and CH_4 over the last ten thousand years and the dramatic increase with the advent of the industrial revolution (the "hockey stick" curve), and their contributions to global warming measured as radiative forcing on the Earth's surface.

First, LUC in the form of clearing trees and other vegetation for pasture or crop production is a large contributor to CO_2 flux into the atmosphere, but it is often not included in calculations of the agrifood system's contribution to GHGE. Foley et al. calculated that globally, 26 percent of the Earth's 3.38 billion ha of ice-free land is in pasture, and 35 percent (350 million ha) of cropland is devoted to producing animal feed (rather than human food). This equals a total of about 76 percent of the world's 4.91 billion ha of agricultural land that is devoted to animal production (Foley et al. 2011:337–338). Animal production continues to increase and is one of the major causes of LUC. In the Amazon, rain forest has been cleared to make way for cattle pasture and soy, mostly grown for animal feed, and in Indonesia rain forest has been cut down to plant oil palm and soy. Reducing animal production and increasing perennial plant volume on former pasture and feed crop areas could remove much C from the atmosphere.

Second, soil organic matter in land that is converted to cultivation is usually oxidized at a faster rate than it is returned to the soil, leading to a net C flux to the atmosphere. Industrial agriculture, with its heavy use of chemical fertilizers produced, delivered, and applied using large amounts of fossil carbon, reduce the apparent need to add organic materials to soil, exacerbating decreases in soil organic carbon and reducing soil quality for crop production. For example, one hundred years of data from experimental plots at the University of Illinois on N additions and organic carbon in crop residues and soil found that synthetic N fertilization promotes loss of soil organic matter when the addition of N greatly exceeds removal, even with large amounts of organic crop residue incorporated in the soil (Khan, Mulvaney et al. 2007). Loss of aggregates due to tillage and depletion of organic matter leads to lower microbial activity and further decline in aggregates and aeration (Fliessbach et al. 2007, Mäder et al. 2002a) (chapter 6, section 3.5). The same problem occurs in nonindustrial agriculture in much of the Third World, but for a different reason: crop residues are in demand for fuel, feed, and building materials. So there are two major problems in the agricultural C cycle—too much C in the atmosphere and too little in the soil.

Sequestering additional C in the soil can contribute significantly to improving yields (Lal 2010). Many factors affect C sequestration, including initial C content, soil characteristics, and crop rotations (Govaerts et al. 2009); clearly, we still need a better understanding of these dynamics. Reducing tillage, sometimes called conservation or no-till agriculture, which is the main method for increasing soil C, has mixed results. For example, of seventy-eight cases of reduced compared with conventional tillage examined in one review, seven had lower C soil content, forty higher, and in thirty-one there was no difference (Govaerts et al. 2009). Baker et al. reviewed studies and found that while conservation compared with conventional tillage was correlated with increased soil organic carbon in the upper 30 cm, it was also correlated with decreased soil organic carbon in the next 30 cm (Baker et al. 2007). One solution is to return to using local, short-term carbon cycles with organic fertilizers (Birkhofer et al. 2008, Drinkwater and Snapp 2007, Scialabba and Muller-Lindenlauf 2010), and small-scale agriculture using local organic inputs has been advocated as the best and most effective means for doing this (Lin et al. 2008). Kell has proposed breeding crops for deeper roots, which could move significant amounts of C into the soil (Kell 2011, 2012).

Third, CO_2 emissions result from the burning of large amounts of fossil fuels. The industrial revolution was based on the discovery that energy could be extracted from the concentrated long-term carbon stores in coal, oil, and natural gas, which resulted in relatively large increases in the flow of carbon back into the atmosphere, in excess of that being removed by plants. Today, industrial agrifood systems use large amounts of energy in production and transportation of inputs, cultivation of crops, and food transportation, processing, packaging, and storage. One study of the U.S. agrifood system found that it consumed more than 14 percent of all energy used in the United States, projected to rise to nearly 16 percent, although this was only a partial LCA and did not account for all energy used (Canning et al. 2010:iii, 20). It is not surprising that there is a correlation between the prices of oil and food.[6] Reducing the energy intensity of the agrifood system is one obvious way to reduce CO_2 emissions.

A *fourth* major source of atmospheric carbon flux from agrifood systems is in the form of CH_4 from animals and animal waste, paddy rice fields, anaerobic decomposition in landfills, and inadequately managed composting. The largest portion is from animals, and most of this is from cattle. I address mitigation of this source of C flux in the section on diet change below.

4.4. The Agrifood Nitrogen Cycle

Nitrogen (N) is essential for life as a key component of protein, DNA, and vitamins, and humans and other animals ultimately depend on plants for their source of N. While the atmosphere is 78 percent N, in the form of dinitrogen (N_2), this form is nonreactive or unavailable to plants, and life depends on conversion of N_2 into reactive N (Nr) in a process termed N fixation, which occurs in nature primarily by N-fixing microorganisms, and to a lesser extent by lightning.

Plant tissue is ingested and broken down in animal bodies to constituent amino acids, which are absorbed directly, or into Nr compounds, which are resynthesized or excreted. Yet Nr in the diet is unevenly distributed globally—areas with Nr-intensive agriculture have high rates of malnutrition (obesity, diabetes), whereas a lack of Nr is also correlated with malnutrition (undernutrition) and increased disease mortality (Galloway, Townsend et al. 2008). Humans have increased the supply of Nr since the Neolithic by growing certain crops, including leguminous plants such as beans and peas, which have symbiotic relationships with bacteria that fix N in nodules on their roots, rice in flooded paddies where the floating fern *Azolla* contains N-fixing cyanobacteria, and sugarcane with an N-fixing bacterium that grows in its roots without nodules (Brady and Weil 2010:411). In the first fifteen years of the twenty-first century, the discovery of how to fix atmospheric N_2 into Nr (in the form of ammonium) and its development into the industrial Haber-Bosch process producing large quantities economically, although energy intensively, greatly increased human interference in the N cycle (Leigh 2004, Smil 2004).

However, according to Galloway et al., approximately 75 percent of the Nr created globally by humans is used in food production, and approximately 70 percent of that Nr is created via the energy-intensive Haber-Bosch process (Galloway, Aber et al. 2003:345). Today, this process is responsible for approximately 50 percent of all Nr on Earth, which has made the current size of the human population possible, contributing about 40 percent of protein in the human diet globally. From 1995 to 2005 annual cereal production increased 20 percent (1897 to 2270 Tg per year) and meat production increased 26 percent (207 to 260 Tg per year); the Nr necessary for these increases was supplied by the increase in production by the Haber-Bosch process from 100 to 121 Tg N per year (20 percent) over the same period (Galloway, Townsend et al. 2008).[7]

The major storage component of N in soil is organic matter that releases Nr to the soil solution through microbial decomposition as NH_{4+} (ammonium) and NO_{3-} (nitrate), which can be taken up by plants and metabolized. As mentioned in the previous section, excess Nr in the soil speeds up microbial breakdown of organic matter (Khan, Mulvaney et al. 2007), leading to the release of CO_2, as well as loss of soil structure and water-holding capacity, eventually reducing yield. More N fertilizer is commonly applied to crops than can be stored in organic matter (including living plants), and therefore it is quickly made unavailable to plants via NH_{4+} fixation on clays, volatilization into the atmosphere as ammonia (NH_3), leaching from the soil as NO_{3-} in water solution, or denitrification of NO_3^- by soil bacteria, which releases N_2, nitric oxide (NO), and nitrous oxide (N_2O) to the atmosphere.

The most important result of human interference in the N cycle for climate change is the production of N_2O, a GHG roughly three hundred times as effective in warming the Earth's surface as CO_2, and the large increases in Nr in the agrifood systems have led to large increases in N_2O.[8] Figure 8.2 shows the increasing atmospheric concentration of N_2O during the last ten thousand years and the effect of this increase on global temperature, estimated in W per m^2 (IPCC 2007b:3). Note that the curve has the same timing and shape as the "hockey stick" curves for CO_2 and CH_4. From 1860 to 2005, Nr creation increased from about 15 Tg to 187 Tg N per year, with human intervention in the nitrogen cycle increasing dramatically with the invention of the Haber-Bosch process.

There is substantial uncertainty in estimates of amounts and locations of Nr (Galloway, Townsend et al. 2008) and its quantitative effects on climate. However, there is no doubt that human disturbance of the N cycle has made a major contribution to climate warming and will continue to do so—both directly, by increasing N_2O flux to the atmosphere, and indirectly, by speeding the breakdown of O_3 in the troposphere, although human disturbance also has some relatively minor cooling effects via aerosols and reduction of O_3 in the stratosphere.

Human management has transformed the N cycle from a relatively closed system (where inputs and outputs are a small fraction of the nutrients that cycle internally) to an open system (where inputs and outputs are a much larger proportion). N is recognized as the most important case in agricultural production of high levels of input combined with low use efficiencies, resulting in high proportion of leakage into natural cycles in the soil, water, and atmosphere (Spiertz 2010), causing large and increasing environmental damage (Good and Beatty 2011). While natural deposition rates of Nr from the atmosphere globally are around 0.5 kg N per ha per year or less, there are now large areas where human disturbance has resulted in average rates exceeding 10 kg per ha per year, and this number may double by 2050, or in some areas approach 50 kg Nr per ha per year, with rates at the landscape level possibly much higher. These Nr deposition rates far exceed the levels known to negatively impact ecosystems (Galloway, Townsend et al. 2008:889–890).

Currently, the United States, China, northern India, and western Europe are regions of high N fertilizer use and inefficiency (Foley et al. 2011). However, Third World countries of the tropics will experience the most Nr growth in the near future, and, because many tropical regions are *relatively* Nr-rich (i.e., other nutrients are more limiting [Foley et al. 2011], more N fertilizer application could cause rapid N losses to the environment (Galloway, Townsend et al. 2008). Increasing biofuel production will also increase N effects on climate change because biofuels (primarily corn in the United States and sugarcane in Brazil) are N intensive. Sugarcane production is especially inefficient, incorporating only about 30 percent of applied Nr into the crop (Galloway, Townsend et al. 2008).

One of the main ways to reduce N_2O emissions is to increase the efficiency of N use—that is, to increase the value of the ratio (N in crop biomass)/(N added to field)—and field studies have shown that this approach has great potential. For example, in the Yaqui Valley of Mexico, reduced amounts of N applied to wheat, synchronized with crop demand, reduced N_2O as well as NO_3^- by an order of magnitude (Ahrens et al. 2008). This did not decrease the quantity or quality of yield, however, and farmers' after-tax profits increased 12 to 17 percent, though their risk could also increase (Matson et al. 1998).

In China, increasing rates and decreasing efficiency of Nr application in production have resulted in acidification of soils and decreased yields (Guo et al. 2010). It is estimated that that only 21 to 25 percent of applied N is taken up by plants in Chinese agriculture, with much of that returning to the environment in human excrement (Cui et al. 2013). Studies have shown that the potential to increase efficiency without decreasing yield can reduce Nr fertilizer use by 30 to 60 percent (Cui et al. 2013), and the government has embarked on a major effort to do so (Ju et al. 2009, Qiu et al. 2012).

5. AGRIFOOD SYSTEMS, CLIMATE, AND BEHAVIOR CHANGE

As discussed above, most proposed solutions to the problem of anthropogenic climate change include the assumption that *increased* production will be required to keep up with projected growth in demand (see the introduction). The mainstream perspective on the agrifood system assumes that the current socioeconomic structure is a given and that current consumption trends will continue—therefore, supply-side solutions, such as input substitution or increasing efficiency, are the focus. One example of such a solution is a proposal for reducing GHGE from electricity generation in California by reducing fossil fuel use, which assumes as a benefit that it would not "require change in life-style" (Williams et al. 2012:58).

While improved technology can be an important part of the solution, the disproportionate amount of attention and resources it is given are unjustified, in part because increasing the efficiency of production while absolute levels of production remain the same or continue to grow, even slowly, does not address the core challenge of anthropogenic climate change: *rapidly limiting GHGE and reducing GHG concentrations in order to limit global average temperature increase* while increasing equity of access to food and other basic needs (see section 1). Rather, the key to responding to this challenge is likely a better understanding of how humans perceive and interpret their social and biophysical environments in order to support reductions in consumption and to fundamentally change the dominant economic and institutional structures that encourage increasing material consumption and production (Fischer et al. 2012). This is the fundamental challenge of "the zone," as discussed in chapter 1, and of the role of values discussed in chapters 3 and 4.

As we have seen, LCA is an important approach to quantifying the input and output effects of the agrifood system. However, its focus on energy and material resources means that other related and important social and cultural variables are not typically included in LCAs, although they are significant contributors to the effects of the agrifood system on climate change (and the effects of climate change on the agrifood system) (Eshel 2010, Weber and Matthews 2008).

The major driver of climate change is increasing per capita consumption, or, in industrial countries, what could be termed overconsumption (magnified by population growth), which is supported by wasteful and inefficient production, processing, and distribution. This points to the need for changes to the agrifood system that would be physically easy to implement and technologically simple, yet would have very big effects—true "low-hanging fruit." Two of these are *reducing waste and changing diets*, the latter by reducing consumption of processed foods, animal products, fats, and refined carbohydrates, and increasing consumption of fresh fruits and vegetables. These changes would not only reduce GHGE significantly, by reducing consumption of water, energy, and all other resources used in the agrifood system, but would also improve nutrition and health, further reducing GHGE by reducing the resources needed for dealing with excess morbidity and mortality.

5.1. Changing Value-Based Assumptions

The potential for high economic, environmental, and social benefit-cost ratios that could result from changes in diets and reduction of food waste point to the need to directly address ways of encouraging cognitive change to reduce net food consumption via social, cultural, and institutional changes (Gadema and Oglethorpe 2011, Markowitz 2012, Rand et al. 2009). Yet most studies of willingness to change behavior in response to climate change do not focus on diet or food waste reduction, as pointed out by Brody et al. (2012) for the United States, although these components are receiving more attention than in the past (Reay et al. 2012).

Why don't these solutions to global warming get more attention? One reason seems to be that they are unattractive from a mainstream economic perspective because they would reduce production, slow growth, and decrease private profit. This perspective supports the common assumption that cognitive and behavioral change is too difficult to be a realistic alternative. I described the example in chapter 4 of the National Research Council report (NRC 2010:39) on limiting GHGE that recognizes the importance of demand-side drivers (population and economic growth) but then eliminates them from further consideration for "practical" reasons.

A second major and related reason is that these solutions require changes in behaviors perceived by many as lowering the quality of life over the short term, and most people will not be motivated by the prospect of longer-term benefits that are more equitably distributed to society. This, of course, points to the major challenge of reversing dependence on economic growth. It has been suggested that we need to cease relating reductions in GHGE to reductions in GDP, since GDP is seen as a measure of well-being, and instead substitute a more proximate indicator, such as happiness (FitzRoy et al. 2012).

Addressing these barriers to change will involve reexamining mainstream economic assumptions about high resource-supply elasticities and low potential for changing human motivation and behavior (Gowdy 2005, Gowdy and Krall 2009, Gowdy 2008, Jackson 2011). Research has been conducted on the role of knowledge in managing nutrient-cycle externalities (de Kok et al. 2009, Llorens et al. 2009, Malley et al. 2009), but little work has been done on understanding how to change knowledge (including values) among individuals, communities, and organizations (both public and private) that will result in behaviors that reduce those externalities. However, there are some public health examples from the United States that indicate that behavior change is possible, including the success of campaigns to reduce smoking and increase use of seat belts and child safety seats, and behavior change is a major area of research in public health (Ory et al. 2002) that may have valuable insights for supporting behavior change for climate change mitigation. An important component of behavioral change to address climate change and support agrifood systems is awareness of inaccuracies in our empirically based assumptions.

5.2. Changing Empirically Based Assumptions

Understanding the empirical situation is necessary for understanding the value situation (Hansen et al. 2012). Therefore, as discussed in chapter 3, changing empirically based assumptions about the causes of the climate change crisis and its potential solutions can lead to changes in value-based assumptions, specifically values that would support a move toward waste reduction and diet change.

A fundamental concept for understanding how climate change works, and the need for effective action to mitigate it, is the relation between stocks and flows (see chapter 4). Research with university graduate students has shown the large extent to which this concept

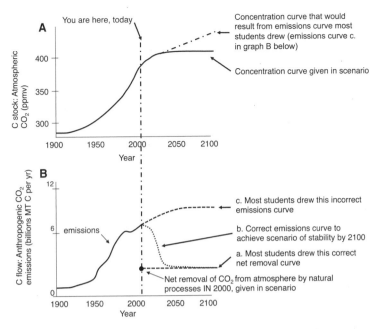

FIGURE 8.3. Relationship between stabilizing concentration and emissions of greenhouse gases. Based on (Sterman 2008). A. Atmospheric CO$_2$ stocks and scenario of stabilized levels at 400 ppm by the year 2100 as presented to students. B. Atmospheric CO$_2$ emissions: (a) net removal by natural processes curve, (b) correct emissions curve to achieve stabilization described in scenario, (c) incorrect curve as drawn by most students.

is misunderstood even by those with extensive formal education in relevant fields. This is an example of how people tend to not be consistent or analytical thinkers and how our common-sense heuristics can be wrong (Kahneman 2011).

Social scientist John Sterman and colleagues presented a scenario problem about GHG and climate change to 212 graduate students at the Massachusetts Institute of Technology (MIT), three out of five of whom had degrees in science, technology, engineering, or mathematics (STEM) and most of the rest of whom were trained in economics. Over 30 percent held a prior graduate degree, 70 percent of these in STEM (Sterman and Booth Sweeney 2007). Sterman et al. first presented the current situation: the net amount of atmospheric CO$_2$ removed by natural processes equals about half of the total anthropogenic CO$_2$ emissions. As a result of this imbalance, concentrations of CO$_2$ in the atmosphere are increasing, from preindustrial levels of about 280 parts per million (ppm) to about 370 ppm in 2000. Next, the students were asked to consider a scenario in which the concentration of CO$_2$ in the atmosphere gradually continues to rise to 400 ppm, about 8 percent higher than the level in 2000, and then stabilizes at that level by the year 2100 (fig. 8.3, A) (IPCC 2007a, Sterman 2008).

Sterman and colleagues then asked the students to sketch lines in the second graph (fig. 8.3, B) showing their estimate of (a) likely, future net GHG removal by natural processes, and future changes in emissions necessary to achieve graph A's stable CO_2 levels. The result was that 84 percent of the MIT students drew lines for emissions (c) that violated the principles of accumulation. They failed to understand that emissions (flows) would have to be reduced to stabilize CO_2 concentration (stocks). The lines that most drew would result in a continuing increase in CO_2 concentrations. (See chapter 4 for a discussion of stocks and flows.)

These results suggested that the students interpreted information about the system dynamics of stocks and flows using a "pattern-matching heuristic, assuming that the output of a system should 'look like'—be positively correlated with—its inputs" (Sterman 2008:532; see also Cronin et al. 2009). The use of this inappropriate heuristic could explain why the majority of respondents to surveys about climate change advocate a "wait and see" or "go slow" approach, which is incapable of dealing with dynamic systems that have long lag times, such as climate change and other systems that affect agrifood system sustainability. Sterman concludes that such errors in understanding are not likely to be correctable by supplying more information, but that "we need new methods for people to develop their intuitive systems thinking capabilities" (Sterman 2008:532).

Perhaps even more conceptually difficult for the average person to understand is how we could achieve a GHG concentration *lower* than the current level, such as Hansen et al.'s well-known 350 ppm target (discussed in section 1) (Hansen, Sato, Kharech et al. 2008). Therefore, we might also ask respondents to consider a scenario in which the current (2013) 400 ppm concentration of CO_2 in the atmosphere decreases to 350 ppm and stabilizes at that level by 2100. We could then ask respondents to sketch lines in the second graph of their estimate of the future change in emissions necessary to achieve a stable CO_2 concentration of 350 ppm. In this case, the correct answer would show emissions (flows) going *below* net natural removal at the point that concentration began to decrease, and then flows would have to equal net natural removal for a stable reduced concentration.

An understanding of these simple, basic empirical relationships between stocks and flows in climate dynamics could help people to see that rapid, significant reduction in GHGE is needed to ensure a sustainable future. This could also help to change values about what we as a global society should do, including making low-cost behavioral changes to reduce food waste and modify our diet to achieve rapid reductions in GHGE.

6. REDUCING FOOD WASTE

Recent research has documented the high levels of resource consumption and waste in the agrifood system (Cuellar and Webber 2010, FAO 2012c, Gustavsson et al. 2011, Stuart 2009). For example, in the United States, approximately 25 percent of embodied energy in produce was wasted in 2004, or 615 trillion BTUs, about 30 percent of the total agrifood system waste (Cuellar and Webber 2010:Table 7). And in 2009, 33.4 MT of food was wasted, which was the largest component of all "materials discarded" (i.e. not composted) in U.S. landfills (EPA 2011).

Every portion of food available to be eaten that is instead wasted is also a waste of all the energy and resources used to produce, process, transport, store, and prepare that food. Food wasted at later stages of the food chain wastes proportionately more because more has been invested in it. As with the agrifood system in general, and as we saw specifically with LCAs earlier, estimates of the amount of waste depend on making assumptions about where the boundaries of the system are drawn in space and time. However, there seems to be general agreement that the proportion of food wasted globally is over 30 percent and much higher for some foods (Gustavsson et al. 2011) and in some countries (Gunders 2012). This represents an equally huge proportion of resources used (e.g., oil, coal, natural gas, water, phosphorus) and environmental damage, like GHGE, that could be avoided.

6.1. Food Waste in the Industrial and Third Worlds

Why do such high proportions of waste persist when there is a world food crisis? The explanation differs for Third World and industrial communities and countries. In the former, a majority of the average 6 to 11 kg per person per year of food waste occurs at the field level and especially the postharvest storage level as a result of pests and pathogens (Gustavsson et al. 2011, Lundqvist et al. 2008). In the industrial world, food waste is much higher, 95 to 115 kg per person per year (Gustavsson et al. 2011), with waste at the production level proportionately much lower, and largely a result of crops that are not sold (whether harvested or not), because of strict aesthetic quality standards and economic inefficiencies—for example, because farmers cannot find suitable markets.

In industrial countries, and in more well-off households in general, the largest proportion of waste is at the retail, food preparation, and consumption levels because food is relatively cheap, in terms of time and money spent on acquiring and preparing it (Gustavsson et al. 2011). A value-based assumption persists in this environment that there should never be a shortage of food, which encourages waste occurring at this level because of the marketing strategy of always having a full stock of all foods (Gustavsson et al. 2011). Because the proportion of disposable income that poor households and Third World countries spend on food is much higher, food is more valuable, and waste is relatively lower at retail and household levels.

Reducing food waste in both the Third World and industrial countries requires changes in knowledge and behavior along with supportive public policies, yet the mainstream emphasizes the need for new technology and infrastructure instead. For example, in the United States and the United Kingdom, governments are promoting better packaging and improved transportation infrastructure to reduce waste in consumption, and in Africa development projects encourage farmers to become more commercial, using more purchased inputs and selling more food to reduce waste in storage (Gustavsson et al. 2011).

6.2. Prospects for Motivating Reduced Food Wastage

There are some campaigns in industrial countries to encourage behavioral change at different points in the agrifood system in order to reduce waste (EPA 2011, Love Food Hate Waste 2012), but without concrete policy and business support these campaigns are unlikely to make much progress. To the extent that they emphasize alternative uses of wasted food—such as composting, energy production, and donation to the hungry—they may actually reinforce wasting by rationalizing it.

Overconsumption, especially of certain foods, can lead to many of the major public health problems currently afflicting the industrial world, including malnutrition, obesity, and associated noncommunicable diseases. As such, this overconsumption is also a form of food waste and so is an opportunity for synergy between efforts to improve public health and efforts to reduce waste in the agrifood system. Researchers are documenting both psychological (Wansink et al. 2005) and biochemical (Lustig 2006) factors that encourage overconsumption of food. Understanding and addressing these factors through actions and policies as diverse as reducing portion size in home meals to limiting sucrose content in processed foods could help improve the poor energy efficiency in current U.S. diets mentioned in section 4.2 (energy required in food / energy in food produced = 0.556). I discuss the idea of changing diets, and especially food choice, in the next section.

7. CHANGING DIETS

There has been increasing realization that current and forecasted diets are high in resource-intensive foods. These foods generate relatively more GHGE before consumption (Popp et al. 2010), and, because they are unhealthy, they also generate more GHGE after consumption, in the form of resources consumed by excess morbidity and mortality (Egger and Dixon 2009, Walker et al. 2005). Therefore, food choice has the potential to significantly reduce GHGE, especially by reducing the proportion of animal products (Eshel 2010, Gonzalez et al. 2011, Roy et al. 2012) and other foods produced, processed, or transported in energy- and resource-intensive ways. This also includes reducing the more energy-intense methods used for fruits and vegetables, such as growing in heated greenhouses and transportation by air (Stoessel et al. 2012).

How could the tremendous potential for diet change to reduce GHGE be harnessed in an equitable way within and between nations? From our discussion of externalities in chapter 4, it will be obvious that from environmental and social perspectives on sustainability, any accounting of changes in the agrifood system should take into account all effects of an activity, and there should be public agreement about the boundaries in space and time that will be used. In this section I show how changing the diet to reduce GHGE and improve nutrition could increase the well-being of the whole population by reducing negative effects. These negative effects are currently externalized by the mainstream economic system to maximize profits, and they vigorously oppose efforts to eliminate them—for example, in the United States, by taxing soda, limiting junk food advertising to children, and promoting lower meat consumption (described below).

7.1. Evolution of the Human Diet

Histories of the human diet tend to begin with the Paleolithic, a critical time period because the advent of hunting by humans may be an important causal factor in the emergence of large brain size, and the Paleolithic environment probably also selected for physiological and cognitive adaptation to periodic scarcity of food (see chapter 2, section 3.2). There is evidence that we are evolutionarily adapted to desire nutrient-dense foods, because this trait was adaptive during the period in human history when we were foragers and high-quality foods were not always abundant in time and space (Armelagos 2010). However, for about the last seven million years, the diet of our human ancestors has been predominantly vegetarian (Ungar and Sponheimer 2011), and gut microbiota in humans today reflect this, leading some scientists to argue that humans are biologically adapted to a primarily vegetarian diet (Dunn 2012).

There is a well-documented correlation between increasing wealth, often measured as per capita GDP, and increasing consumption of animal and high-caloric-density foods (Popkin 2007). This correlation is referred to as the nutritional transition (Popkin 2006), and it has resulted in a global crisis of human and environmental health—both overnutrition and its associated noncommunicable diseases, such as obesity and diabetes, and undernutrition—as well as massive environmental degradation, including climate change (Armelagos 2010, Egger and Dixon 2009, Walker et al. 2005). Often, the mainstream describes this trend as the result of people "diversifying" their diets, or in the case of animal foods, increasing the protein in their diets, as they obtain the resources to buy these more expensive foods.

In 2010, consumption of fruits and vegetables (both fresh and in other forms) was 50 percent and 36 percent, respectively, below recommended levels (for a 2000 kcal diet), according to the USDA's *Dietary Guidelines* (calculations based on USDA 2010:51), while consumption of fats, sugars, and refined grains was 42, 147, and 93 percent above recommended levels, respectively. Increasing the proportion of fruits and vegetables in the diet by replacing some of the calories and much of the volume currently contributed by fats, sugar, and refined grain products with fresh produce would significantly improve nutrition and could reduce GHGE. Doing so would improve nutrition because fresh fruits and vegetables have a higher concentration of the nutrients (vitamins and minerals) most often lacking in the U.S. diet (Buzby et al. 2006), as well as higher fiber content and lower caloric density, which addresses the obesity epidemic (Rolls et al. 2010). Doing so would decrease GHGE because the GHGE for produce at the production level (the source of the most agrifood GHGE) is much lower on average than for other food groups (Eshel and Martin 2006, Weber and Matthews 2008).

Today, even mainstream nutritional associations like the American Dietetic Association recognize that vegetarian and vegan diets can easily provide adequate protein (Craig and Mangels 2009). In addition, the bulk of accumulating evidence suggests that vegetarian and especially vegan diets can be much *more* healthy than diets containing animal products, and vegan diets are now being widely recommended by health professionals in industrialized nations like the United States (HSPH 2011, PCRM 2011). The main bodies of evidence are from epidemiological (Rankin et al. 2012) and clinical (Esselstyn 1999, 2001, 2010, Sinha et al. 2009) studies. Perhaps the largest epidemiological study looking expressly at the relationship between diet and health was the China study (Campbell and Campbell 2005), from which the most statistically significant positive correlation was between the increasing consumption of animal foods and the rates of chronic and noncommunicable diseases, including cardiovascular disease, cancers, and diabetes.

7.2. Nutrition and Malnutrition Today

One of the biggest obstacles to diet change is the existing agrifood system structure, dominated by a mainstream economic perspective that assumes that the most important way to evaluate the agrifood system is monetarily. This often means promoting diets that are unhealthy, a trend that was first brought to widespread attention by the late pediatrician and nutritionist Derrick Jelliffe, who dubbed the result "comerciogenic malnutrition" (Jelliffe 1972). Thus, the "health" of agrifood industries is totally disconnected from the health of people who obtain their food from those industries. For example, the U.S. government blatantly equates increased consumption of unhealthy foods with a healthy food economy, as evidenced by this statement from the USDA: "Trends in the soft drinks and beverage sector are often an indicator of consumer ability to purchase higher value foods, and foreign investment in the beverage sector often functions as a bellwether for the health of local food industries" (USDA ERS 2011b).

Not only does the dominance of economic criteria mean that good nutrition is ignored, but malnutrition is actively promoted. Marion Nestle has documented the powerful influence of the agrifood industry on food choices in the United States (Nestle 2001, 2002, 2006), which uses advertising tactics similar to those of the tobacco industry (Brownell and Warner 2009), and the foods they promote are directly linked to poor health and may be addicting in ways similar to that of drugs like tobacco (Gearhardt, Grilo et al. 2011, Gearhardt, Yokum et al. 2011). In response to an epidemic of childhood obesity and associated disease among children in the United States, a soda tax was proposed that was vigorously opposed by the beverage and entertainment industry (which profits from advertising), which spent USD $40 million successfully lobbying against it in 2009 (Wilson and Roberts 2012).

Mainstream economic players, which by definition have a vested interest in business as usual, will oppose any change. For example, when the USDA suggested that its employees could contribute to climate change mitigation by participating in the Meatless Monday movement, the National Cattlemen's Beef Association immediately attacked, calling the suggestion an "animal rights extremist campaign" to end meat consumption.[9] The USDA responded immediately, removed this advice, and apologized.[10]

The U.S. federal food stamp program, or SNAP (Supplemental Nutrition Assistance Program), whose purpose is provide nutritious food, is in fact subsidizing the mainstream agrifood system, and attempts to modify SNAP—for example, to make soda or candy ineligible for purchase with food stamps—have been vigorously opposed by the industry in the name of "consumer choice" (Simon 2012). At the same time, the industry attempts to take advantage of increased consumer interest in nutrition by promoting new or "improved" food products, which critics often see as deceptive "nutriwash" because these products have minimal if any improvement in nutritional value (Simon 2007:68).

As we saw in chapter 4, food insecurity, malnutrition, and hunger have reached record levels not only in the Third World, but in the industrial world as well—including in rich industrial nations like the United States. The mainstream agrifood system contributes to this problem by promoting diets that contribute to malnutrition and related chronic diseases, such as obesity and diabetes. For example, average food energy consumption in the United States has increased by 533 calories per person per day since 1970 (USDA ERS 2010a), comprising mostly the empty calories from fats and oils, refined grains, and especially sweeteners (Putnam et al. 2002), which provided more calories per day than any other food group in 2007 (USDA ERS 2010a). Increasing consumption of these empty-calorie foods may be accompanied by lower consumption of more nutrient-rich foods, such as fruits and vegetables (Wells and Buzby 2008), leading to the combination of malnutrition and overweight so common in the industrialized world.

Low-income and food-insecure people are especially vulnerable to poor nutrition and health problems. They have unique risk factors that make healthy eating especially challenging, such as lack of access to healthy food (FRAC 2010). For example, in the United States in 2005, 40.1 percent of low-income adults in California with fair or poor health status lived in food-insecure households, compared with 24.8 percent of adults with good, very good, or excellent health (Harrison et al. 2007). Low-income residents also spend a greater proportion of their income on food—up to 25 percent for the lowest income bracket, compared with the U.S. average of just below 10 percent (USDA ERS 2010b). Healthy, nutritious foods are becoming more difficult for the U.S. consumer to obtain, with the price of fruits and vegetables increasing by 40 percent over the last twenty-five years. At the same time, the price of sweets, fats and oils, and soft drinks has declined, supported by government subsidies, leaving many, especially low-income and food-insecure citizens, more likely to purchase calorie-dense, nutrient-poor foods (Schoonover and Muller 2006).

7.3. The Potential Contribution of Dietary Change to Climate Change Mitigation

As we have seen, animal production, including pasture and growing and processing corn and soy for feed, is a major source of GHGE: CO_2 from respiration and fossil fuel use, CH_4 from enteric fermentation in the digestive system of ruminant food animals (primarily cattle, sheep, and goats), and N_2O from N fertilizers and manure (Zervas and Tsiplakou 2012). Foley et al. estimated that 35 percent of all crops globally are grown for animal feed (and 3 percent are grown for biofuels and other nonfood uses); therefore, "the potential to increase food supplies by . . . shifting 16 major crops to 100 percent human food could add over a billion tonnes to global food production (a 28 percent increase), or the equivalent of 3×10^{15} food kilocalories (a 49 percent increase)" (Foley et al. 2011:338, 340). As discussed above, animal pasture accounts for 26 percent of the Earth's ice-free terrestrial area. Production of animal products for human consumption is also a major consumer of the Earth's freshwater and is another reason that shifting global diets are having a huge impact on food production resources. For example, it takes 3,500 liters of water to produce one kg of rice, while producing one kg of beef requires 15,000 liters of water (Hoekstra and Chapagain 2008).

Animal production is a prime example of how the carbon and nitrogen cycles have been disrupted by human activity, increasing the space and time between the production of carbohydrates, protein, and other food ingredients by plants and their return to the soil. At the center of this disruption in the industrial world are CAFOs (concentrated animal feeding operations), which concentrate large numbers of animals—especially cattle, pigs and chickens—in one place. CAFOs require food and water to be imported and produce huge amounts of feces and urine that are impossible to use locally, and so need to be shipped elsewhere or are managed on site, often leading to the escape of GHGs and other pollutants (Gurian-Sherman 2008). This is the situation Wendell Berry famously commented on: "The genius of America farm experts is very well demonstrated here: they can take a solution and divide it neatly into two problems" (Berry 1977:62). In contrast, on smaller, decentralized, diversified farms, animal manure is a resource to be incorporated into the soil to enhance its physical and biological qualities and supply nutrients to growing crops.

CAFOs are a priority for the EPA, primarily because mismanagement has resulted in contamination of local and regional water systems, where runoff from CAFOs carries excess nitrogen, phosphorus, bacteria, pesticides, antibiotics, hormones, and trace elements including metals (EPA 2008b). CAFOs are also a significant source of CH_4 from manure, but the EPA does not regulate CH_4, though it has the authority to do so (Verheul 2011). In the United States, CH_4 emissions from animal manure, while comprising only 7 percent of total CH_4 emissions in 2009, has increased 56 percent since 1990, while overall CH_4 emissions have increased by only 1.7 percent (calculated from data in EPA 2012a).

With growing awareness of the contribution of the animal agrifood system to GHGE, reducing the role of animals in the agrifood system is quickly moving from "an interesting and little explored area for mitigating climate change" (Carlsson-Kanyama and González 2009) to attracting attention and advocacy from a number of scientists and organizations (Lagasse and Neff 2010), including Rajendra K. Pachauri, the chair of the IPCC since 2002 (Pachauri 2008). Proponents see this as a desirable change, since it would also provide major benefits in terms of improved human health and animal welfare. It could also have other environmental benefits, such as reducing water use and pollution and converting grazed lands to denser, perhaps more natural vegetation, which could sequester more C and even lead to greater biodiversity. The diet with the most positive impact would be one that comprised 100 percent plant-based foods, produced using methods minimizing N and energy inputs, and with minimal processing and transportation. Diets often described as vegetarian or ovo-lacto vegetarian (which include dairy and eggs) can still have significant impact because dairy products are from ruminant animals whose production generates a lot of GHGE (Carlsson-Kanyama and González 2009, Eshel and Martin 2006). Even vegan diets are no guarantee of minimal GHGE, since they can include highly processed and packaged foods, as well as foods produced conventionally with high GHGE from nitrogen fertilizer, land clearing, and so on.

A number of studies have shown a direct relationship between improved diet and reduced GHGE (Carlsson-Kanyama and González 2009, Eshel and Martin 2006, Lagasse

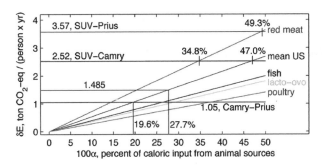

FIGURE 8.4. The greenhouse burden (considering CO_2, N_2O, and CH_4) exerted by various plant- and animal-based diets. Note that δE = difference in C footprint between vegan and mixed diet. Figure 3 from Eshel and Martin 2006.

and Neff 2010, Rolls et al. 2010, Weber and Matthews 2008), including for a vegetarian diet in California (Marlow et al. 2009). While increasing the fresh fruits and vegetables in the U.S. diet to meet USDA dietary guidelines would require a significant increase in the amount of produce grown (Buzby et al. 2006), total land in cultivation could decrease if production of the high-fat, high-carbohydrate, and animal-derived foods in the diet were also reduced (Buzby et al. 2006). It is important to note that under current market and production subsidy regimes in the United States, less energy-dense, healthier diets would also be more expensive, and therefore might increase food insecurity (Townsend et al. 2009) unless other changes were also made.

Eshel and Martin calculated the difference in GHGE between a vegan diet and diets containing different proportions of different types of animal products (fig. 8.4). For example, the "standard" U.S. diet containing 27.7 percent of calories from animal products produces 1.485 MT CO_2e per person per year more than a vegan diet (Eshel and Martin 2006). Eshel and Martin compared the GHGE of diets with different proportions of animal products with choice of automobile. For the standard U.S. diet (keeping the proportion of different animal products constant), they found that at 19.6 percent animal products the difference in GHGE from a vegan diet was equivalent to that between driving a Toyota Camry and a Toyota Prius (1.05 MT CO_2e per person per year), and at 47.0 percent animal products the difference was equivalent to that between driving a Toyota Camry and a GMC Suburban (2.52 MT CO_2e per person per year). This is an important comparison, because the kind of car people drive is so much more visible than the kind of food they eat, and automobile choice has become an icon of how climate friendly people's lifestyles are.

7.4. Prospects for Motivating Diet Change to Reduce Greenhouse Gas Emissions

Eating animals and animal products is an important part of many groups' food and agriculture history and culture, and processed and fast foods are synonymous with modernity, and are aggressively promoted by the mainstream agrifood industries. The result is that the knowledge and behaviors of consumers and policy makers are strongly adapted to the current meat system (de Boer and Aiking 2011). The emotional tie to this part of the diet can make it quite difficult to objectively evaluate its relation to climate change, even among the most vehement promoters of behavior change to mitigate global warming. For example,

Bill McKibben, founder of 350.org and one of the most prominent campaigners for reducing GHGE—and even more radically, for reducing GHG concentrations—sees local agrifood systems, including small businesses, as part of the solution but seems unaware of the importance of diet. In writing about the difficulties that the Farmers Diner in Barre, Vermont, experienced in converting to more local food, he stated, "You can't have a diner without bacon," and waxed poetic about the high-animal-content offerings of the restaurant's menu (McKibben 2010:137). James McWilliams sees this reluctance of the environmental community to promote a more climate-friendly diet as due to veganism's relative lack of sensationalism, aesthetic appeal, and personal choice component, compared with issues such as eliminating fossil fuel development (McWilliams 2012).

Understanding the determinants of food choice values and behavior is a field of scientific study that is just beginning to engage with fundamental research in psychology and physiology (Koster 2009). Yet we know that dietary preferences can change. One study found significant reductions in meat consumption or increase in fruit and vegetable consumption as a result of subjects reading positive or negative information (Allen and Baines 2002). As advocacy for diet change increases, some organizations, such as the World Preservation Foundation, see this as the most effective way of mitigating climate change (WPF 2011), and they are exploring ways to change cultural preferences to encourage plant-based diet choices (Taylor 2012).

NINE

The Big Solutions

Localizing Agrifood Systems

1. INTRODUCTION

You are in the produce section of your grocery store, you want to buy carrots, and you have a choice. There are two bins of carrots, the carrots in each bin look the same and are the same price, but there are different signs in front of each bin. One sign reads "Just harvested, delicious carrots, locally grown within 40 km of this store"; the other says simply "Fresh, delicious carrots." Which do you choose? Do you choose the "locally grown" carrots and feel good about it? Why? What are your conscious and unconscious assumptions about what "locally grown" means?

If you buy the local carrots, do you assume that doing so is also a vote in favor of small-scale farmers in your community, in favor of good working conditions and a living wage for farm workers, and in favor of production methods that are environmentally friendly? If you do, you are part of growing number of people for whom *local* has become a proxy for a more *sustainable* agrifood system. Or maybe you don't buy the local carrots, assuming that the craze for local is meaningless, or even economically irrational, and perhaps even an irritating distraction.

Like many words we use to describe the agrifood system, "local" is used as an indicator that implies goals that are often assumed, like socially just working conditions, environmentally friendly production, and community-supporting economics. The problem is that the implied goals are difficult to verify, because there is usually little or no documentation to support them, and those of us who want them to be true can be misled (chapter 3). Another major problem is that those who either don't care or who oppose those goals can use labels like "locally grown" to manipulate the marketplace for their own gain and undermine those very goals. It is always important to make sure that indicators are really doing what we assume they are doing—serving as valid proxies for goals that are more difficult to measure.

So, how do we know if local is a good indicator of our goals for a more sustainable agrifood system, and therefore if localization is a good strategy for reaching them? If we do decide to work for localization, how do we make sure that it really is moving toward our goals? This chapter is about how we can begin to answer these questions. Here we come full circle to the theme of economic globalization and its critics that was briefly discussed in chapter 2 in terms of the ongoing global revolution in agriculture, and which was key for understanding the debate about the causes of and solutions to the world food crisis discussed in chapter 4. We have seen the effects of globalization on plant breeding and biotechnology, agricultural and natural ecosystems, and social systems in chapters 5 through 7, and its effect on greenhouse gas emissions (GHGE) and diet in chapter 8. In this chapter I consider the potential of the grassroots localization movement as an initial phase of developing a comprehensive alternative to the mainstream agrifood system.

Localization also provides a critical case study of how different values and goals for the future can lead to very different interpretations and actions, despite superficial similarities. The battle over localization is a microcosm of the battle over who gets to set the goals of our agrifood system and select the paths to reach them, a battle embedded in the larger economic, military, environmental, and cultural struggle for the future of the planet. This is why we must all keep asking the key questions and examining our empirical and values-based assumptions before accepting indicators such as local as substitutes for our goals.

In the first half of this chapter, focusing on the industrial world, I discuss four major challenges to grassroots localization: (1) the dominance of the mainstream global agrifood system, (2) the local trap that grassroots advocates can get caught in, (3) greenwashing by mainstream agrifood system players, and (4) straightforward opposition by localization bashers. In the second half of this chapter, I address three questions that need to be asked about agrifood systems to successfully reach the goals of localization, as illustrated by the example of Santa Barbara County, California: (1) How local is the agrifood system in terms of indicators that we use to measure goals? (2) What would be the benefits of agrifood system localization as measured by indicators in terms of goals? and (3) How can using indicators that are more closely linked to goals help us to achieve those goals? And how can we exploit the potential for synergies between different goals of localization?

2. LOCALIZATION AS AN ALTERNATIVE TO GLOBALIZATION

The agrifood system has become increasingly dominated by long-distance import and export (Saltmarsh and Wakeman 2004:7). It has been driven by economies of scale, relatively cheap energy (Martinez et al. 2010), and supportive economic and political systems. Critics claim that this globalized, centralized, industrial agrifood system, while highly productive in terms of quantity, contributes to many major social and environmental problems,

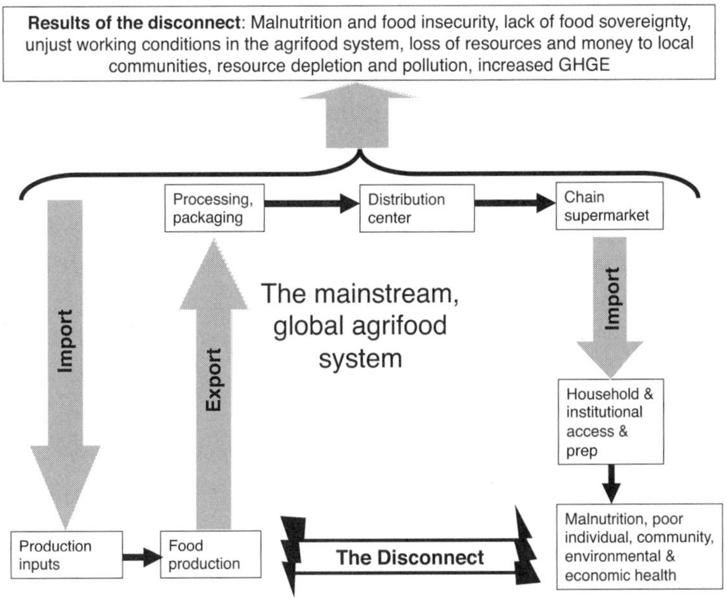

Results of the disconnect: Malnutrition and food insecurity, lack of food sovereignty, unjust working conditions in the agrifood system, loss of resources and money to local communities, resource depletion and pollution, increased GHGE

FIGURE 9.1. The disconnect in the mainstream agrifood system. © 2013 David A. Cleveland.

including the current world food crisis, and soaring levels of environmental damage (IAASTD 2009). More and more, localization is being promoted as a comprehensive way of solving agrifood system problems by reconnecting the globalized system spatially, temporally, and structurally.

2.1. The Agrifood System Disconnect

In the industrial world, and increasingly in the Third World, mainstream agrifood systems are characterized by three spatial and structural disconnects: (1) between the places where food is grown and the places where it is eaten, (2) between the places where food is grown, processed, transported, and consumed and the places where the resources (e.g., water, energy, nutrients, labor) used are from, and (3) between the eating of food and its fundamental roles of biological, psychological, and cultural nourishment (fig. 9.1).

The economic emphasis sees globalization in the agrifood system as the result of industry responding to supply and demand via the invisible hand of the market—for example, "by importing food from around the world and modifying their products and retail formats to better meet consumer needs" (USDA ERS 2011b). This description is based on the assumptions of the mainstream economic perspective (chapter 4) and ignores the role of the industry itself, via advertising and monopolization, in generating demand. For example, between 1999 and 2009 there was a pattern of faster growth in demand for packaged convenience foods and imported foods in those countries with the fastest "penetration of modern food retailers" (Tandon et al. 2011:19).

In industrial world countries like the United States, most food is distributed through centralized networks, and most people obtain food from large grocery stores, fast food vendors, restaurants, or institutions such as college dining halls, workplace cafeterias, and health care facilities, which typically source their food from regional or global distributors who in turn buy from large-scale producers (Martinez et al. 2010). Similarly, the retail grocery options for those who purchase food for home preparation in the United States are rapidly being consolidated: in 1992 the top five retailers captured 27 percent of retail sales, and by 2009 they claimed 60 percent of sales, with one retailer—Walmart—holding a 30 percent market share of all grocery sales in that country (Wood 2013).

Unlike in the industrial world, where localization focuses on reestablishing local agrifood systems, in much of the Third World many small-scale farmers and local agrifood systems still exist, although they have also been affected by globalization, many beginning with European colonialism in the late fifteenth century. Today, corporate globalization of the Third World is officially promoted by the policies of international organizations, such as the World Bank and the World Trade Organization, and by many industrial nations—for example, through the *Alliance for a Green Revolution in Africa* (AGRA) program discussed in chapter 5. The New Alliance for Food Security and Nutrition, announced by the G8 leaders in May 2012, assumes that agrifood systems in Africa can improve only if global multinational corporations are in charge (White House 2012).

Today, globalization, aided by free trade arrangements, is advancing rapidly in the Third World in terms of food marketing (Tandon et al. 2011), farm production (Reardon et al. 2009), and, recently, land acquisition (Cotula, Vermeulen, Mathieu et al. 2011, see chapter 7). The rapid rise of supermarkets in the Third World has generally taken sales away from traditional markets and producers, and while some supermarkets are adopting a greater emphasis on healthier foods, the overall strategy is profit maximization at the top through encouraging greater consumption of the most profitable food items (Hawkes 2008). On the other hand, in some cases supermarkets may increase the variety of foods available to consumers at a cost similar to that of traditional markets (Hawkes 2008), as in New Delhi (Minten et al. 2010). Supermarkets have had mixed effects on small-scale Third World farmers (Reardon et al. 2009), but they generally exclude poorer farmers in marginal areas, such as rainfed farmers in Kenya (Neven et al. 2009) and small-scale farmers in Ghana (Field et al. 2010).

Some evidence also suggests that international trade agreements have negatively affected consumers and small-scale farmers. For example, the North American Free Trade Agreement (NAFTA) has led to increased imports of subsidized maize from the United States to Mexico, with import dependency (proportion of domestic consumption met by imports) just for maize in Mexico rising from 7 to 34 percent from 1990–1992 to 2006–2008 (Wise 2010:18). During the same period, the real producer price for Mexican maize farmers fell 66 percent, amounting to an estimated loss of USD $6.5 billion to those farmers (Wise 2010), which undermined many positive aspects of traditional agriculture, including social relations and knowledge, and forced many small-scale farmers to migrate (Fitting 2011). There

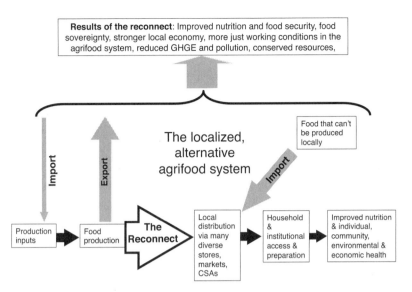

FIGURE 9.2. Reconnecting the agrifood system. © 2013 David A. Cleveland.

is now evidence that consumption in Mexico of U.S. ingredients and foods whose imports are facilitated by NAFTA are contributing to rising obesity there (Clark et al. 2012).

2.2. Reconnecting the Agrifood System

Localization faces many financial, legal, logistic, and infrastructural obstacles (fig. 9.2). In the industrial world, changes in consumer shopping and eating habits, farmer production and marketing strategies, and community and government action would be required to change regulations for land and water use, farm labor, marketing, and food safety. Recognizing the regulatory barriers to effective localization, advocates are actively working to make the system friendlier to local agrifood systems, especially in urban areas, by creating resources to deal with legal, organizational, and infrastructural issues (Hodgson et al. 2011, Sanders and Shattuck 2011, SELC 2013, Wooten and Ackerman 2011).

The localization movement in the Third World is more about rejuvenating and strengthening remaining local systems, resisting the pressures that are either eliminating local systems or pushing them into the global marketplace, and ensuring that not only food security but also food sovereignty are regained. Local and regional efforts are often supported by international NGOs, including La Vía Campesina and GRAIN (see chapter 4).

2.3. Defining Grassroots Localization

As a grassroots alternative to globalization, localization questions the assumptions made as part of globalization, which are predominantly those of mainstream economics, as discussed in chapter 4. Proponents of localization, on the other hand, often assume that it will improve many of what they feel are the negative consequences of the industrialization and commoditization of the agrifood system as part of the growing dominance of markets and multinational corporations in a globalized world. I take the grassroots vision of localization as my reference point, because it embodies much of what I have discussed as environmental and social emphases of an alternative perspective and contrasts most starkly with the economic emphasis of the existing mainstream approach.

Localization advocates emphasize that economic goals should be defined within social and environmental goals—financial profits should be secondary to both the long-term health of people and the environment and social justice for farm workers and others in the agrifood system, often including humane treatment of farm animals (e.g., Food Circle 2012). Specific goals of localization that are frequently mentioned include reducing GHGE from burning fossil fuels in transportation, improving nutrition by increasing the availability and therefore the consumption of fresh fruits and vegetables, and socially and economically reinvigorating local communities by increasing their control over the system (Martinez et al. 2010). Indeed, localization has emerged as the dominant theme of "sustainable" agrifood systems from environmental and social perspectives. While some proponents of the mainstream economic perspective also support some aspects of localization, most do not support its basic goals.

The discourse on local food systems is often accompanied by an emphasis on "community" and the idea that localization reinforces community cohesion (DeLind 2010, DeLind and Bingen 2008). Community in this context is generally understood as a tighter connection between growers and consumers—a connection that has been almost entirely lost in the mainstream agrifood system. Food sovereignty is also a major component of statements in support of localization, as in this one from the Food Circles Networking Project: "Control over our food should rest with us . . . communities and the people in those communities" (Food Circle 2012). And strengthening communities may be related to reducing the economic, physical, and social scale of agrifood systems. The late rural sociologist Thomas Lyson and colleagues have contrasted community-oriented or "civic agriculture" with commodity agriculture in the United States, and found that the former—measured by six indicators from the Census of Agriculture—is associated with "sustainable" agriculture, defined as "increased profitability, decreased resource use, and stable or increasing farm numbers" (Lyson and Barham 1998). They also found that smaller-scale farming is positively related to community well-being, due to "the level of civic engagement and the strength of the economically independent middle class" (Lyson et al. 2001). Further analysis of data from 433 agriculture-dependent counties in the United States found that those with laws limiting nonfamily corporate farming scored higher on indicators of social welfare, including lower levels of relative poverty and unemployment and greater cash gains from farming (Lyson and Welsh 2005).

"Localization," like "sustainability," has become highly contested terrain as its popularity has grown, attracting criticism from a range of different perspectives. In the next section I discuss how localization has been defined and measured, and how those who share the goals of grassroots localization can get caught in the local trap—assuming that indicators of localization, especially food miles, necessarily promote their goals. In the following section, I discuss two critiques from the mainstream. One critique is implicit; it adopts some of the same indicators, like food miles, but uses these to distract attention from its goals, which are often opposed to those of grassroots localization advocates. The other mainstream critique is explicit; it attacks the empirical and value assumptions of grassroots localism head-on.

3. IMPLEMENTING LOCALIZATION

In this and the following section, I focus on some key challenges to implementing localization in terms of selecting and measuring indicators and assessing progress toward reaching goals for which it is an indicator (see chapter 3, and fig. 3.5); following that, I give some specific examples.

3.1. Assessing Localness

Assessing the localness of an agrifood system and opportunities for further localization is a challenging task. The positive, monetary value directly generated by the agrifood system is tracked by businesses, governments, and organizations such as the FAO. In contrast, negative externalities such as GHGE and loss of local economic, social, and environmental resources are tracked less extensively or not all; this is also the case for many of the goals of grassroots localization—such as improved individual, community, and environmental health. In addition, there has been little interest until lately in tracing the frequently complex paths connecting the components of the global agrifood system.

This means that a comprehensive assessment of the agrifood system is very challenging. It also means that progress toward the goals of grassroots localization are difficult to quantify, measure, and monitor. Progress can most easily be quantified in terms of indicators like the miles that production inputs or food travels; direct marketing arrangements such as farmers markets, CSAs (community supported agriculture), and local distribution networks; or the amount of local food served in community schools and hospitals (in the United States, Martinez et al. 2010; in the Third World, Pretty et al. 2006; in Europe, Renting et al. 2003). However, these variables do not directly measure many of the underlying social, environmental, and ethical goals of localization. One of the most often used, intuitively appealing, and relatively easy to quantify indicators is food miles. Whether and to what extent food miles are a measure of sustainability has become the center of an often heated debate about localization, especially in industrial countries.

As we will see in this and the following section, reducing food miles *is necessary and may be sufficient* for some of the common goals of localization, such as increasing personal interactions between farmers and eaters, and *may be necessary but not sufficient* for many goals such as strengthening the local economy or conserving small-scale periurban farms and historical rural landscapes. However, it is *neither necessary nor sufficient* for many of the most important goals of localization, such as improving human nutrition, supporting farm worker rights, or reducing negative environmental impacts, although if done in appropriate ways it can make important contributions to achieving these goals as well.

3.2. Food Miles Pros and Cons

Spatial scale is the most concrete, intuitively appealing, and easiest aspect of agrifood systems scale to quantify, compared with structural or economic scale, and it is a commonly used indicator for gauging the negative impact of the mainstream agrifood system and the progress toward alternatives. "Food miles" (distance from farm to retail, or "direct transport") has become a popular way of measuring spatial scale; a 2003 estimate that conventional fresh produce sold in Iowa traveled an average of almost 1,500 miles (2,414 km) (Pirog and Benjamin 2003) is often cited as representing the entire U.S. food system in general. Consequently, reducing food miles is also a popular indicator for the goals of localization, including reducing GHGE, improving produce quality and freshness, and improving community food access, nutrition, and economic health.

One problem is defining what food miles and spatial scale per se mean. A case study in the United Kingdom found that there was both innocent and intentional confusion about whether "local" referred to food produced locally, within a given geographically defined distance, or to "locality" foods—foods that could be produced at a distant location but had an "identifiable geographical provenance" (Ilbery and Maye 2006:355, 365–366). A survey of grocery stores in the Willamette Valley of Oregon found a wide range of definitions for local, including spatial criteria that often included several states (Dunne et al. 2011).

A frequent focus of food miles research is the potential for spatially defined populations to be fed from within arbitrarily defined foodsheds, or regions where the food consumed by a particular population is produced, defined primarily in terms of food miles (e.g., Zajfen 2008). A number of recent studies in the United States have focused on the potential for minimizing the size of foodsheds—for example, estimating the potential for feeding the city of San Francisco from farms within a 160 km radius of the Golden Gate (Thompson et al. 2008); another seeks to calculate "the minimum distance possible" between farms and population centers within New York State (Peters et al. 2009); and a third is aimed at "minimizing the distance between population centers and available cropland" in eight midwestern states (Hu et al. 2011).

3.3. Food Miles as an Indicator

Despite the prominence of food miles as an indicator of localization, the assumption that decreasing food miles will always, or even usually, result in progress toward the goals of localization and thus sustainability has been increasingly challenged in favor of a more rigorous assessment (Edwards-Jones et al. 2008).

In terms of environmental goals, there is more and more recognition that the spatial extent of production and consumption captures only a portion of the environmental consequences of agrifood systems. Comprehensive life-cycle assessment (LCA) is increasingly seen as a more valid method (Weber and Matthews 2008), though also a more challenging one, with most work to date done in Europe (Brodt et al. 2008) (see chapter 8, section 4.1). For example, Weber and Matthews' input-output LCA found that food miles accounted for only about 4 percent of the U.S. agrifood system GHGE in 1997, while 83 percent was from food production and processing (2008:3511).

A study in the United Kingdom evaluating food miles and carbon emissions for a wide range of locations for products sold by a major UK importer and retailer found those to be poorly correlated (Coley et al. 2011). If consumers drive more than 7.4 km to purchase organic produce directly from a farm, the GHGE are greater than "the emissions from the system of cold storage, packing, transport to a regional hub and final transport to customer's doorstep used by large-scale vegetable box suppliers" (Coley et al. 2009:154). A comparison between locally produced food in the United Kingdom and imports from New Zealand found that apples and onions produced in and shipped from New Zealand were more energy efficient than those grown in the United Kingdom (Saunders and Barber 2008).

Similarly, localization is often promoted because it is assumed to improve nutrition, although there are few data to support this. While localization can support improved nutrition, there is no necessary causal relationship. For example, 100 percent spatial localization of an agrifood system for fruits and vegetables may not have any effect on nutritional status unless local residents have access to and eat those fruits and vegetables; at the same time, nutrition could be comparably improved if residents obtained and ate imported fruits and vegetables (Cleveland, Radka, and Müller 2011b).

While advances in LCA are improving the potential for environmental assessment of the sustainability of localization, social assessment is not included in LCA. Therefore, a focus on evaluating sustainability using food miles as an indicator with LCA risks omitting important variables, including the supralocal, often global connections between all "local" agrifood systems, although some advocates do recognize this (Feagan 2007). For example, one study estimated that changes in food purchasing in western Europe based on a linear relationship with food miles would decrease welfare in three exporting countries—New Zealand, Madagascar, Malawi—by 0.30, 0.12, and 0.28 percent, respectively as a result of reduced exports (Ballingall and Winchester 2010). This illustrates the need to *think globally when advocating localization.*

Another key factor that is often omitted is the extent to which "local" agrifood systems depend on imported migrant labor. In California, agriculture has always relied on noncompetitive labor from Chinese, Filipino, and Mexican migrants, as well as from migrants from within the United States—such as people escaping the dust bowl in the 1930s and African Americans fleeing racist exclusion in the southeastern United States in the first half of the twentieth century (Arax and Wartzman 2003). While these migrants have been essential for California's highly productive but often labor-intensive agriculture, they may have a negative or positive effect on the sending communities. Recently, most farm workers in California are immigrants from Mexico and Central America. Therefore, localization of the agrifood systems in California may be at the price of delocalization of communities in Mexico and Central America. A complete LCA would require inclusion of the effects of migrant labor on both the communities people migrate from as well as those they migrate to.

On the other hand, there are data supporting a causal link between food miles and some of the goals of localization. For example, there appear to be synergies between direct marketing, organic production, and small size of farms. Smaller farms are most likely to do direct marketing; for example, in 2007, small farms (<$50,000 in direct sales annually) accounted for 84 percent of all direct sales in the United States (Martinez et al. 2010:20). A larger proportion of small-scale farmers are organic—of 250 direct-marketing farmers surveyed in California, 19 percent sell some organic products (18.8 percent sell only organic), compared with 2 percent of all California farmers (Kambara and Shelley 2002:14). A 2003 survey of 1,014 organic farmers in the United States found an average farm size of 277 acres and a median size of 40 acres (Walz 2004:26), compared with 441 acres and 120 acres, respectively, for all farms in the United States (USDA NASS 2003).

To the extent that local food systems include more environmentally sustainable production, like organic practices, they could decrease GHGE through changes in inputs and resource management, although research suggests a complex situation highly influenced by context-specific variables (Mondelaers et al. 2009:1111, Scialabba and Muller-Lindenlauf 2010). For example, reducing energy-intensive inputs, such as manufactured pesticides and fertilizers, can reduce CO_2 emissions; substitution of concentrated synthetic N fertilizers with carefully managed organic fertilizers can reduce N_2O emissions; management of animals and animal waste can reduce CH_4 emissions; and soil management can potentially increase C sequestration. Moreover, organic agriculture could produce an adequate food supply even if yields are marginally lower than those of conventional agriculture; organic agriculture also generally has higher efficiency in terms of energy and nutrients and is likely to be more environmentally sustainable over the long term (Mäder et al. 2002b) (chapter 6, section 3.5). A review of the literature supports the hypothesis that organic can "feed the world" (Badgley et al. 2007; also see chapter 6, section 3.5).

3.4. The Local Trap

The danger of using food miles as an indicator of the deeper goals for sustainable agrifood systems is that food miles can *become* the goal, causing grassroots localization advocates to fall into a "local trap" (Born and Purcell 2006). The local trap is an instance of drinking green Kool-Aid (chapter 3).

For example, the Iowa localization movement has focused on the spatial definition of local without questioning the extent to which this is related to other potential aspects of localization, such as economic control or production practices (Hinrichs 2003:43). Localization advocates caught in this trap are criticized by those who support grassroots localization but question the connection between definitions of "local" based on indicators (e.g., food miles) that are not inclusive enough (e.g., do not include production methods or diet) and so are not articulated with many of the larger goals of localization (e.g., reducing GHGE; see McWilliams 2009).

Born and Purcell's solution is to realize that scale, per se, is socially constructed and has no inherent properties (Born and Purcell 2006). Therefore, they argue that to understand agrifood systems, we need to look not at scale, but at the motives and goals of those controlling the agrifood system. I think they go too far in dismissing scale as not useful—scale *does* have some inherent characteristics. For example, as mentioned above, the physical proximity between agrifood system participants creates social interactions often valued by those participants that are impossible at larger spatial scales. Scale is also functionally related to other important qualities—for example, the time required to deliver fresh produce from a farm to a store or school, other factors being equal, will be greater for larger-scale systems. Finally, goals and motives can be difficult to characterize and measure with consistency, and some indicators that are easier to measure will be useful. Thus, as with using indicators to assess the success of any move toward increasing agrifood system sustainability (chapter 3, and fig. 3.5), it is necessary to constantly check to make sure that the indicators used are adequate and meaningful.

4. THE MAINSTREAM RESPONSE

In addition to those who advocate for the goals of localization but criticize its methods and assumptions, including the uncritical use of food miles as an indicator, as leading into a local trap, there are two other major critiques, both of which come from a mainstream economic perspective. The first one critiques localization implicitly by using localwashing and co-opting the goals of the grassroots localization movement, based on the assumption that those goals are irrelevant or undesirable.[1] The second explicitly attacks localization, arguing that it is economically irrational and elitist.

4.1. Localwashing and Co-optation

The first critique comes from those who see localization as a market trend and opportunity for profit, without being concerned with, or perhaps not accepting, the goals that would bring fundamental change to the mainstream agrifood system—perpetrating "local-wash" (Roberts 2011). Because localization is an idea popular with consumers and with critics of the mainstream system, who are also generating significant market demand, localization can become a new label for an old bottle of wine. The most visible localwashers are the mainstream retailers, distributors, and producers who emphasize food miles, intentionally seeking to conflate it with the goals sought by many grassroots advocates because they assume doing so will increase sales. For example, corporations like Walmart have moved aggressively toward sourcing local food primarily because "buying locally grown produce is a hot marketplace trend, with customers increasingly reaching for staples such as tomatoes and corn that grew in local soil" (Walmart 2012).

Governments also often emphasize the mainstream economic aspects of regional or local food hubs, as in this USDA definition: "A regional food hub is a business or organization that actively manages the aggregation, distribution, and marketing of source-identified food products primarily from local and regional producers to strengthen their ability to satisfy wholesale, retail, and institutional demand" (Barham et al. 2012:4). The USDA has partnered with the Wallace Center at Winrock International in the National Good Food Network (NGFN). The NGFN emphasizes the concept of "value chains," which they assume can be created within the current agrifood system structure. These value chains can include main-stream corporate players like the Sysco Corporation, one of the world's largest food distribution companies, which had sales of $39 billion in 2011. The Sysco/Wallace project was motivated by Sysco's realization that it was losing market share by not carrying more diverse and "sustainable" products (Cantrell 2010), yet in this project, most of the goals of grassroots localization—like improved nutrition, social justice for farm workers, and support for small farms—are absent or treated superficially.

Also in the United States, the American Farmland Trust, in collaboration with the California Department of Food and Agriculture, published a vision for the future that promotes "regional markets" as part of supporting California's conventional agrifood industry (AFT 2012). Its goal for water, for example, is not to decrease consumption, but to increase efficiency, and its goal for farm workers is not to improve their working conditions and benefits, or to address the negative impacts on migrants' home communities, but to ensure a steady supply of cheap labor for California agriculture.

Localwashing is also on the rise in the Third World. "Local" has also become the target of venture capitalists and of immigrants from the finance and investment industry (LFS 2013), an objective promoted by the U.N. Food and Agriculture Organization (Chakrabarti and Graziano da Silva 2012). Organizations like the Sustainable Food Lab promote projects that are advertised as supporting local Third World farming communities by helping them to supply multinational food corporations' desire to tap the market in local and "ethical" foods, based on the unsubstantiated assumption that integrating these communities into the mainstream global food system will provide long-term benefits (SFL 2011).

A case study of Bayer CropScience's vertical integration of private agrifood networks in India suggests that the corporation's Food Chain Partnership model selected only farmer participants who had the resources to generate profit for the companies involved, and marginal farmers who most need help were left out (Trebbin and Franz 2010). The selected farmers were restricted to growing crops designated by Bayer, were given information only about Bayer products, and had their traditional knowledge ignored. Monsanto's corporate social responsibility projects have targeted the same type of farmers, and even these have been dropped when they conflict with the company's primary goal of profit making (Glover 2007). These examples don't necessarily mean that investors or corporations intend to undermine public health, communities, small-scale farmers, farm workers, or the environment, but neither is support of these their goal, and therefore what happens to people and environments is secondary to profits. This is another reason for making value- and empirically based assumptions about the agrifood system clear and available for social discussion.

4.2. The Localization Bashers

An explicit critique of the localization movement from a mainstream economic perspective is being promoted by the localization bashers, who find the arguments for localization irrational, romantic, ignorant, and elitist. They see the market as the ultimate arbiter of values and ignore environmental and social data and values (for example, Desrochers and Shimizu

2012, Lusk and Norwood 2011). We saw this critique in terms of the Third World in previous chapters, such as in chapter 5, where I described Paul Collier's critique of advocates for small-scale agriculture in Africa as deluded romantics (Collier 2008).

Those who control and profit from the currently dominant centralized agrifood system may see localization as a threat and resist it—for example, by using food safety regulations in ways that discriminate against small-scale farmers, direct marketing, and local distribution hubs (DeLind and Howard 2008).

5. THE SANTA BARBARA COUNTY CASE STUDY

So far I have presented an overview of the idea of localization as a grassroots alternative to globalization, a way to reconnect the disconnected parts of our agrifood system. Now I want to illustrate the implementation of localization by asking three questions that need to be answered to successfully reach the goals of sustainability via localization, illustrated by the example of Santa Barbara County, California: (1) How local is the agrifood system in terms of indicators that measure goals? and (2) What would be the benefits of agrifood system localization as measured by indicators in terms of goals? In the following section I address the third question: How can using indicators more closely linked to our goals help us to achieve these goals, and to exploit the potential for synergies between different goals of localization?

I worked with students to investigate localization as a solution to the agrifood system disconnect, using Santa Barbara County (SBC) as a case study of California and the United States.[2] Our objective was to provide information needed for discussion of the agrifood system and the potential for localization to increase sustainability, following the outline described in chapter 3 (see figs. 3.1 and 3.5). Our definition of sustainability was based on the goals of grassroots localization described above, and we used food miles as the indicator.

SBC makes a great case study for localization and its potential, because it is a major agricultural county in terms of resource use, employment, and production. Agricultural production in the county in 2008 was valued at $1.14 billion, which placed it in the top 1 percent of all counties in the United States and fourteenth of the fifty-eight counties in California (the most agriculturally productive state in the United States). SBC also has many people and organizations working successfully to improve the system, including a thriving localization movement. Like other agricultural counties, it also has high levels of environmental and social problems, including water shortages, water and air pollution, farmland conversion, poor conditions for farm workers, food insecurity, diabetes, and obesity.

My students and I focused on produce (fruit and vegetables) because produce (1) dominates SBC agriculture economically, with 82 percent of total value in 2009 (calculated from SBC ACO 2010); (2) dominates direct sales of agricultural products in SBC—accounting for 79 percent of the total agricultural sales at the Santa Barbara Certified Farmers Markets from November 2008 to September 2009; and (3) contains nutrients widely believed to be lacking in U.S. diets due to underconsumption of fresh fruit and vegetables (Buzby et al.

2006), and many SBC residents have low nutritional status and are food insecure (CCPHA 2007, CFPA 2010).[3] In addition, fruits and vegetables are among the most traded crops; in 2010, for example, 17 percent of California fruits and vegetables were exported out of the country, accounting for 65 percent of all California foreign agricultural exports (calculations based on AIC 2012:Table 1).[4] Vegetables and fruits produced in SBC in 2008 totaled 1.07 million MT (metric tons; 2.36 billion pounds). We predicted that SBC would be well on its way to localizing its agrifood system, at least in terms of consumption of produce.

5.1. How Local is the Agrifood System in Terms of Indicators that Measure Goals?

Our first question was: How local is the SBC agrifood system using the indicator of food miles? More specifically: What is the proportion of "local" produce in SBC, defined as produce that was both grown and consumed in SBC, not including SBC produce shipped outside the county to a regional warehouse and then imported back to SBC?[5] Since no data on this have been collected, we gathered data on the dollar value of sales (by wholesalers or retailers) or purchases (by retailers or consumers) of SBC-grown produce by farmers' markets, community supported agriculture programs (CSAs), U-pick operations, grocery stores, farm-to-school programs, institutional purchases, and food assistance programs.

We converted these dollar amounts to weight, and used various methods to estimate the proportion that was grown in SBC. This told us the amount of SBC-grown produce that was consumed directly in SBC—our local produce.[6] Together with data for the amount grown in SBC (readily available from the county or the USDA), and the amount consumed in SBC (estimated using national-level USDA data), we were able to estimate how local the SBC produce agrifood system was.

We and many others were shocked to find that less than 1 percent of produce grown in SBC was consumed in SBC and less than 4 percent of produce consumed in SBC was grown in SBC (table 9.1). The amount of SBC-grown produce consumed directly in SBC is so small that even if the actual amount were three times our estimate, it would still be less than 11 percent of total estimated produce consumption in SBC, and less than 1.5 percent of all SBC-grown produce.

5.2. What Would Be the Benefits of Agrifood System Localization?

Given the answer to our first question, we then asked: What would be the effect of dramatically increasing localization? More specifically: How would complete localization (measured in food miles) of the SBC agrifood system for produce help to achieve two of the common goals of grassroots localization—decreasing GHGE and improving nutrition? Complete localization means that all produce consumed in SBC would be grown in SBC and delivered to consumers without leaving the county. SBC currently produces approximately nine times more fruits and vegetables than are consumed in the county (table 9.1), and production takes place outdoors year round, so completely localizing the produce agrifood system could be done without increasing the quantity of production. In addition, production for SBC consumption would not require large export-oriented farms to change their operation. Farms

TABLE 9.1. Santa Barbara County Produce: Annual Production, Consumption, Export, and Import

Category of produce	Weight MT, unless otherwise indicated.	
	Current conditions	With 100% localization (all consumed in SBC grown in SBC)
1 Grown in SBC[a] (MT)	1,068,957	1,068,957
2 Consumed in SBC[b] (MT)	118,348	118,348
3 Consumed in SBC as % grown in SBC	11.07%	11.07%
4 Grown and consumed in SBC[c] (MT)	3,871	118,348
5 Grown and consumed in SBC as % grown	0.36%	11.07%
6 Exported from SBC[d] (MT)	1,065,086	950,609
7 Exported as % grown	99.64%	88.93%
8 Imported to SBC[e] (MT)	114,477	0
9 Imported as % consumed	96.73%	0.00%
10 Grown and consumed in SBC as % consumed	3.27%	100.00%

Source: (Cleveland et al. 2011b), with minor corrections and edits.
Note: Annual estimates are based on data for 2008 and 2009. All weights were converted to primary (farmgate) weights.
[a] Data from (SBC ACO 2009), with amount for backyard harvest added.
[b] Calculated as: [population of SBC 2008 (US Census Bureau 2012)] × [average annual consumption fruits & vegetables converted to primary weight (USDA ERS 2010)]
[c] From table S-3 in Cleveland et al. 2011b.
[d] = (row 1) − (row 4)
[e] = (row 2) − (row 4)

under 50 acres in size accounted for 13,744 acres, or 14.7 percent, of harvested cropland in SBC in 2007. If these farms produced fruits and vegetables with the average yield of fruits and vegetables in SBC in 2008, they could produce 114 percent of the estimated consumption of produce in SBC in 2008 (118,348 MT) (calculations based on data in SBC ACO 2009, USDA NASS 2009).

We found that under the current export-import agrifood system, produce imported into SBC accounts for approximately 100,000 MT CO_2e, with about 11,000 MT of this for direct transport (table 9.2).[7] To estimate the effect of complete localization by substituting imported produce with SBC-grown produce, we calculated the CO_2 emissions for direct transport within SBC of the quantity of produce currently imported, assuming all deliveries would be done in light trucks, and calculated a net emissions savings of 8,925 MT CO_2e per year. This was a savings per household of only 0.058 MT, or approximately 9 percent of the average U.S. household's GHGE for produce, but only about 0.7 percent of a household's total agrifood system GHGE (8.1 MT, per calculations of Weber and Matthews 2008:3511)[8] and less than 0.1 percent of *total* U.S. GHGE CO_2e per person (table 9.2, fig. 9.3).[9]

It is not surprising that total localization would result in such a small reduction in GHGE for SBC. The most recent LCA estimate is that direct transport of produce (food miles)

TABLE 9.2. Greenhouse Gas Emissions (GHGE) from Import and Export of Produce in the Santa Barbara County Agrifood System

GHGE CO_2e total for produce in the United States, 1997[a,c] (MT CO_2e)	68,555,632
GHGE CO_2e total for direct transportation of produce in the United States, 1997[a, b, c] (MT CO_2e)	7,521,442
Households in the United States, 1997 (Weber and Matthews 2008)	101,000,000
GHGE MT CO_2e household[-1] total for produce in the United States, 1997[a] (MT CO_2e hh[-1])	0.679
GHGE MT CO_2e household[-1] for direct transportation of fruit and vegetables in the United States, 1997[a] (MT CO_2e hh[-1])	0.074
Households ("housing units") in SBC, 2008 (US Census Bureau 2012)	151,763
Proportion of produce consumed that was imported to SBC, 2008	96.5%
Total GHGE MT CO_2e total for produce consumed in SBC that was imported into SBC (for proportion imported in 2008) (MT CO_2e)	99,393
Total GHGE MT CO_2e for direct transportation of produce consumed in SBC that is imported into SBC (for proportion imported 2008) (MT CO_2e)	10,905
Total produce imported into SBC for consumption, 2008 (MT) (see table 10.1)	114,477
Average km traveled (round trip) from farm to retail for produce grown and consumed directly in SBC[d]	65
Direct transport of produce (MT yr[-1]) within SBC (km) (assume 1 MT average load)	7,440,991
MT CO_2 km[-1] for direct transportation of produce in SBC, assuming light-duty trucks (below 6,000 pounds or 2.72 MT), averaged for model years 1995–2008; n.b., CO_2, not CO_2e (EPA 2008)	0.00028118
CO_2 for direct transportation when all produce grown and consumed in SBC (MT yr[-1]) (assume most GHGE from light truck transport is CO_2)	2,092
Potential net savings CO_2e for maximum localization of SBC (MT CO_2e yr[-1]) for reduced imports	8,812
Potential savings household[-1] (MT CO_2e yr[-1]) for maximum localization of SBC (imports only)	0.058
Potential savings household[-1] as proportion of U.S. average total household agrifood system GHGE in 1997, imports only (Weber and Matthews 2008:3511)	0.0072
Potential savings household[-1] as proportion of total household GHGE for all produce	0.0855
Potential savings person[-1] yr[-1] as proportion of U.S. average total GHGE CO_2e person[-1] in 1997, 24.2 MT (EPA 2010:2–7, US Census Bureau 2012)	0.0009
Potential savings person[-1] yr[-1] as proportion of U.S. average total GHGE CO_2e person[-1] in 2008, 22.9 MT (EPA 2010:2–7, US Census Bureau 2012)	0.0010

Source: (Cleveland et al. 2011b), with minor corrections and edits.

[a] Based on the life-cycle assessment in (Weber and Matthews 2008) of GHGE for food consumed by U.S. households in 1997, in CO_2 equivalents (CO_2e). Weber and Mathews estimated that GHGE from direct transport increased by 5 percent between 1997 and 2004 (Weber and Matthews 2008:3512).

[b] Direct transportation is final delivery transportation or delivery from farm to retail store, and is one component of the supply chain that also includes indirect transportation, wholesaling and retailing, passenger transportation, and production.

[c] Weber and Matthews (2008) used U.S. Department of Commerce commodity flow data for three commodity groups: vegetable and melon farming, fruit farming, and canned or dried fruits and vegetables.

[d] Round-trip estimate based on data for farmers market vendors in the western United States (Ragland and Tropp 2009:44–45).

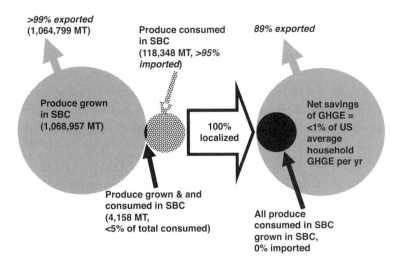

FIGURE 9.3. The effect of completely localizing the Santa Barbara County agrifood systems for produce, as of about 2008. Data from (Cleveland, Radka et al. 2011) with minor corrections. © 2011 David A. Cleveland.

accounts for only 2.5 percent of total agrifood system GHGE in the United States (Weber and Matthews 2008:3511).

On the other hand, there is great potential for localization to improve nutrition in SBC. While SBC is generally considered to be an affluent county and has a median household income of $60,645, slightly below the California average and seventeenth highest of all fifty-eight California counties (USDA ERS 2010a), it also has high levels of food insecurity and malnutrition. In 2007, 53 percent of SBC adults were overweight or obese and 7 percent were diagnosed with diabetes (CDC 2007), and in 2010, 37 percent of fifth, seventh, and ninth graders were overweight or obese, placing SBC twenty-eighth of the fifty-eight California counties (1 = lowest prevalence) (calculated from data in CCPHA 2011). However, simply substituting SBC-grown produce for imported produce will not automatically have a positive effect on nutrition and health because there are many intervening cultural, social, economic and geographic obstacles, including national agrifood policy (Eshel 2010). SBC ranks forty-seventh of the fifty-eight California counties in terms of food security (1 = lowest proportion insecure), with 39.5 percent of the population food insecure (CFPA 2010).[10] Difficulty obtaining healthy food contributes to poor nutrition—SBC has three times as many fast-food restaurants and convenience stores as supermarkets and produce vendors (CCPHA 2007). So while localization as defined here may have some social benefits, it does not necessarily address the problems of poor nutrition and lack of access to healthier foods.

6. HOW CAN WE ACHIEVE THE BENEFITS OF LOCALIZATION?

Our Santa Barbara research showed that even in an agricultural county that produces an abundance of fresh fruit and vegetables year round and has an active local food and nutrition movement, the agrifood system is still very much a part of the global disconnect. Our research also demonstrated that food miles are not always a good indicator of the goals of grassroots localization, as they are frequently assumed to be.

This leads us to the third question that needs to be asked about localization: How can using indicators more closely linked to our goals help us to achieve these goals? That is: How can we avoid the localization trap (drinking green Kool-Aid), greenwashing and co-optation? And how can we exploit the potential for synergies between different goals of localization?

6.1. Case Study of Successful Localization

The results of our SBC study pointed to the need for research and action on how localization can be accomplished in a way that directly supports the underlying goals of grassroots localization advocates. One of the projects we have carried out is analysis of Farmer Direct Produce (FDP), a successful local food hub first established in 2006, selling in southern SBC, with most of the produce sourced within the county (Cleveland, Müller et al. n.d.). Its success was a result of support by the University of California's Santa Barbara campus Residential Dining Service (RDS), initiated by student activism beginning in 2004, and supported by the creation of sustainability goals for UC and UCSB food systems. However, the most important ingredient appears to have been the commitment of the FDP owners and the RDS staff to creating alternatives to the mainstream agrifood system, especially in rejecting the assumptions of the mainstream economic emphasis in favor of definitions of sustainability based on social and environmental emphases.

RDS cultivated an unconventional business relationship with FDP: flexible deadlines, personal relations, and an understanding of the constraints and needs of farmers. This relationship has been essential for the success of RDS in increasing the sustainability of their produce purchases. Procuring local produce for its dining commons was also a new experience for RDS, and they were therefore willing to work through the logistical difficulties involved in order to achieve their goals for sustainability. This meant that as a young and growing business FDP was able to make mistakes without the fear of losing RDS as a client. By June 2010, UCSB Dining Services had gone from procuring all of its produce from large distributors who sourced conventionally grown fruits and vegetables from around the world, to procuring 73 percent of *all* produce within a 150-mile radius, already surpassing the UC sustainable food policy goal of 20 percent by 2020. In addition, 26 percent of all produce purchased was grown within 150 miles of UCSB by small family farmers, and was either pesticide free or certified organic.

FDP owners Wesley Sleight and Anna Breaux have a commitment to local food and farmers that has been key to their success in achieving dependability and quality with the produce they deliver (fig. 9.4). Business transactions have been based on personal relationships, oral contracts, and a loyalty to business partners, and farmers value the personal relationship they have with the FDP owners. Despite barriers and disadvantages, farmers choose to participate in local hubs to some extent, even when it means not maximizing their profits, in order to achieve their social goals of selling their food locally. While profit is an important

FIGURE 9.4. Wesley Sleight and Anna Breaux, Farmer Direct Produce, outside their warehouse in Santa Barbara County, California. Photo © David A. Cleveland. Used with permission of subjects.

criterion, personal and philosophical motives are also important when choosing sales outlets, and some farmers feel that participating in local food hubs is the "right thing to do" in order to support their local food system.

However, as FDP, RDS, and the farmers have become more efficient in working with one another, there have also been increased savings and profits all around. For example, one of the possibilities that is being explored is FDP coordinating with farmers to grow the specific crops RDS uses in the amounts required, which would reduce waste and uncertainty for all participants. Local food hubs such as the one described in this chapter can provide examples of the profit motive being put back into balance with social values and norms and the biophysical limits of our agrifood system, of moving beyond the reification of economic growth as the primary source of social stability and wealth as the primary determinant of human happiness (Jackson 2009b). Thus, local hubs can be an important force within the food system for realigning goals toward social and environmental criteria for success (Friedmann 2009).

6.2. Lessons Learned

Based on the concepts I have presented in this book, what are the most important approaches for moving toward greater agrifood systems sustainability in the face of the global food crisis? I think there are at least five key strategies.

First, we can think critically about how to define the goals for our food systems from the global to the very local, and what assumptions they are based on. If grassroots localization and smaller-scale agrifood systems are the overarching goals, this implies a fundamental change in the way agrifood systems function, including replacing the ruling assumptions of the mainstream economic emphasis and the dominant economic criteria of individual profit maximization and continued economic growth.

Second, we can constantly assess actions in terms of the progress they are making toward our goals, including evaluating what our indicators can and cannot do for that process (see fig. 3.5). That is, we need to continually monitor the outcomes of our actions and the relationship between those outcomes and our goals. This will help us uncover self-deception that can give rise to the localization trap, as well as expose greenwashing and co-optation. Of course, analyzing and testing assumptions underlying indicators is part of this process.

Third, we can look for the options that would provide the most efficient and effective way to meet our goals as quickly as possible, such as changing diet and reducing food waste, instead of focusing on high-tech solutions requiring large investments and infrastructure and long development and implementation periods, such as creating national or large regional food distribution hubs versus local food distribution hubs.

Fourth, we can look for opportunities for synergies by evaluating existing data, and by conducting strategic research to test hypotheses about potential synergies—for example, the synergy among direct marketing, organic production, and small farm size mentioned above, and between reduction in GHGE and improved nutrition with changes toward a more plant-based diet (chapter 8). There will also be numerous instances when conflict between goals can be resolved only by making major changes, and in the short term these goals will therefore have to be negotiated and modified. For example, given the current economic structure, improving working conditions for farm workers could result in higher prices for fruits and vegetables, which would make it more difficult for low-income households to improve their diets and for small-scale organic farmers to compete. Resolving these basic conflicts may require a fundamental rethinking and restructuring of the economic system (see chapter 8, section 1.0)

Fifth, we can recognize the global linkages that are necessarily present in any local system. For example, improving working conditions for migrant farm workers in affluent communities could exacerbate the loss of local agrifood systems in the workers' home communities. In this case, the mainstream economic policies, including trade agreements such as NAFTA, which made life so difficult in those home communities, would need to be changed as part of localization.

7. BALANCING ON A PLANET

The case study of the current state and potential for localization in the SBC agrifood system is an example of the kind of critical thinking I have been describing throughout this book. It reinforces the idea that finding sustainable solutions to our food crises will be a balancing act. It will be a balancing act among the different emphases, with their contrasting and often conflicting understandings of the problems and solutions, at the community, regional, and global levels. It will also be a balancing act in a more personal way, in that it will challenge each of us to maintain a balance between seeking an objective, critical assessment of the situation and seeking changes in line with our values about what defines a sustainable agrifood system—to be engaged but not attached. To me, it is this combination of critical engagement and nonattached advocacy for our values that defines a viable path for our problem solving.

As I stated in the introduction, I believe that this way of thinking about and improving our agrifood system will be essential if we are to address the human biological, socioeconomic, and cultural need for food while supporting the ecological processes on which our agrifood system depends. Such a path is characterized by intellectual honesty and integrity as well as by compassion. This will require a willingness to be explicit about our own values and empirically based assumptions and to assess them just as critically as the assumptions fundamental to perspectives different than our own—in other words, to engage in what I have referred to as critical thinking. This process and the opportunities that it opens up for positive synergies in our agrifood system have been the subject of this book.

We can see the need for honest critical thinking about the challenges faced by our agrifood system when we place these challenges within the deep history of and contrasting perspectives on human impact and the Earth's human carrying capacity (chapter 1); between demand for and supply of food (chapter 2); between traditional and modern knowledge (chapter 3); among economic, environmental, and sociocultural emphases in the definition of sustainability (chapter 4); and among yield, yield stability, and diversity in plant breeding (chapter 5).

Critical thinking provides greater clarity for identifying or developing positive synergies among different tactics that can provide relatively rapid results, which are now so essential, especially between traditional and modern knowledge and technologies. Such synergies include those described in this chapter and throughout this book: between diversity at all levels of our agroecosystems (chapter 6); between individual short-term and community long-term welfare in managing resources (chapter 7); between the need for food and its effect on climate change (chapter 8); and among different goals of localization, as discussed in this chapter.

As I described in the preface, I spent many sleepless nights in northern Ghana decades ago trying to understand the inequities and inconsistencies I was observing when looking at the local agrifood system in Zorse in the context of the larger world. The deep knowledge and hard work of the people in Zorse were part of a strategy that was increasingly influenced by individuals, societies, and biophysical processes beyond their control, if not their awareness. Based on my further understanding of the deep history of life on Earth, humans, and agrifood systems since then, I have concluded that the mainstream approach to solving the world food crisis is fundamentally flawed, and that we need to focus on exploring alternatives. Given future trajectories based on a business-as-usual scenario, we need to do this very quickly. Human thinking and behavior are the sources of the current global crises, including in our agrifood systems, and it is still possible that we can address those crises by changing how we think and behave. Indeed, I believe that this is critical if these crises are to be averted for the long term—relying only on the Neolithic formula of new technologies and better management of crop evolution, agroecosystems, and society is not the answer. By understanding ourselves and the impact our thinking and actions have, we will be better able to define and work together toward a sustainable food and agriculture system.

Acronyms, Abbreviations, and Symbols

Acronym, Abbreviation, Symbol	Definition
AE	application efficiency, regarding irrigation water
AGRA	Alliance for a Green Revolution in Africa
B.C.E.	[years] before common era
BP	before present
BTU	British thermal unit(s), a unit of energy, = 0.25 kcal
C	carbon, or consumption (as in H = NCT)
CAFO	concentrated animal feeding operation
CBD	Convention on Biological Diversity
C.E.	[years in] common era
CE	conveyance efficiency, regarding irrigation water
CGIAR	Consultative Group on International Agricultural Research; grew out of and subsequently became the main promoter of the Green Revolution
CGRs	crop genetic resources
CH_4	methane
CIMMYT	Centro Internacional de Mejoramiento de Maíz y Trigo (International Maize and Wheat Improvement Center), headquartered outside of Mexico City, one of the CGIAR centers
CO_2	carbon dioxide
CO_2e	carbon dioxide equivalent, a measure of the global warming potential (GWP) of all the greenhouse gases included in specific measurements. The GWP is the warming effect of greenhouse gases relative to that of CO_2, taking into account their lifetime in the atmosphere and their relative effectiveness in absorbing thermal infrared radiation.
ÇPM	common property management
CPR	common-pool resource
CV	coefficient of variation
DNA	deoxyribonucleic acid
DR	discount rate
Ev	macroevolution, or change in allele frequencies over a number of generations resulting in new species, in contrast to mEv
FAO	United Nations Food and Agriculture Organization

Acronym, Abbreviation, Symbol	Definition	Acronym, Abbreviation, Symbol	Definition
FK	farmer knowledge	N	nitrogen, or population size (as in HI = NCT)
FV	farmer crop variety, or future value (when calculating DR)	N_2	dinitrogen (an inert gas)
GDP	gross domestic product, or the actual market value of all goods and services produced in a country during a designated time period	NH_4^+	ammonium
		NO_3^-	nitrate
		N_2O	nitrous oxide
		NPP	net primary product
GHG	greenhouse gas(es)	Nr	reactive nitrogen
GHGE	greenhouse gas emission(s)	O_2	oxygen
GWP	global warming potential, e.g. of GHG, see CO_2e	O_3	ozone
		P	phosphorus, or (in statistics) probability
GxE	genotype by environment		
h^2	narrow-sense heritability, $= V_A/V_P$	P_0	population at time zero
		P_t	population at time t
H_2O	water	PES	payment for ecosystem services
HCC	human carrying capacity	ppb	parts per billiion
HI	human impact, or harvest index	ppm	parts per million
HUE	human use efficiency, used in HCC calculation	PV	present value (when calculating DR)
IAASTD	International Assessment of Agricultural Knowledge, Science and Technology for Development	R	genetic gain, i.e., change in allele frequencies between generations
		r_G	genetic correlation
IPCC	Intergovernmental Panel on Climate Change	RDA	recommended dietary allowance
IPR	intellectual property right	RDE	resource-delivery efficiency, HCC calculation
K	potassium	RMP	risk management process
LCA	life-cycle assessment	RNA	ribonucleic acid, tRNA: transfer RNA; mRNA: messenger RNA
LER	land equivalent ratio		
ln	natural logarithm	RUE	resource-use efficiency
LUC	land-use change	S	selection pressure
mEv	microevolution, or result of change in allele frequencies between generations (R), in contrast to macroevolution (Ev)	SK	scientist knowledge
		T	technology (as in HI = NCT)
		TDM	total aboveground dry matter
		TFC	total factor cost
MFC	marginal factor cost	TFR	total fertility rate, or the average number of children born per woman in a population, assuming she lives through the end of her reproductive period
mRNA	messenger RNA		
MV	modern crop variety		
MVP	marginal value product		

Acronym, Abbreviation, Symbol	Definition
TGV	genetically engineered, transgenic crop variety
tRNA	transfer RNA
TVP	total value product
UN	United Nations Organization
US EPA	United States Environmental Protection Agency
USD	United States dollars
USDA	United States Department of Agriculture
V_E	environmental variation

Acronym, Abbreviation, Symbol	Definition
V_G	genetic variation
V_{GxE}	genotype-by-environment variation
V_P	phenotypic variation
W m^{-2}	watts per square meter, a measure of radiative forcing
WUA	water users' association
WUE	water use efficiency, regarding irrigation water
yr	year

Metric Units and Metric-English Unit Conversions

"English" measurements are U.S. standard.

AREA

1 kilometer (km)^2 = 100 hectare (ha)

1 ha = 10,000 meter (m)^2 = 0.01 km^2 = 2.47 acre (a) = 107,639 feet (ft)^2

1 m^2 = 10,000 cm^2 = 10^6 millimeter (mm)^2 = 10.76 ft^2

DISTANCE

1 km = 1,000 m = 0.621 mi

1 m = 100 cm = 1,000 mm = 3.28 ft

1 centimeter (cm) = 10 mm = 0.394 inch (in)

1 micrometer (µm) = 0.0001 mm

VOLUME

1 m^3 = 1,000 liters (l) = 264. 2 gallons (gal) = 35.31 ft^3

1 liter = 1,000 milliliters (ml) = 1,000 cm^3 = 0.264 gal = 33.8 fluid ounces (fl oz) = 1 mm
depth of water from rain or irrigation on 1 m^2 = 61.02 in^3 = 0.035 ft^3

WEIGHT

1 metric ton (MT) = 1,000 kilograms (kg) = 1.102 (US short) tons (T)

1 kg = 1,000 gram (gm) = 2.205 pounds (lb)

1 gm = 1,000 milligrams (mg) = 10^6 micrograms (mcg) = 0.035 ounces

1 megagram (Mg) = 1 MT

1 teragram (Tg) = 10^{12} gm = 1 million metric tons

ENERGY

1 Cal (with uppercase *C*) = 1,000 cal (with lowercase *c*) = 1 kcal

Temperature in °C (centigrade) = (Temperature in °F - 32) x 5/9

Freezing point of water = 0°C = 32°F

Boiling point of water = 100°C = 212°F

NOTES

PREFACE

1. I give all measurements in metric units; see appendix 2 for conversions.

2. American readers use the term "corn" to refer to *Zea mays,* but in most of the rest of the world it is called "maize," or versions of this in different languages, derived from the language of a Native American group in the Caribbean that was one of the first that the Europeans invaded. "Corn" in British English refers to the grain of any cereal crop.

INTRODUCTION

1. For simplicity, I use the term "traditional" to mean "traditionally based." "Traditionally based" implies ongoing adaptation and is a more accurate description because all traditionally based communities are constantly changing in response to external and internal changes, including adopting and modifying aspects of industrial societies.

2. For simplicity, I use the term "empirical assumption" to mean "empirically based assumption"—that is, an assumption about what the biophysical and sociocultural world *is,* based on interpretation of perceptions of that world. I use the term "value assumption" to mean "value-based assumption"—that is, an assumption about how the biophysical and sociocultural world *should be,* based on values. Most scientists agree that testing empirical assumptions as hypotheses results in improved assumptions, not in the "truth." See chapter 3 for more discussion of assumptions and knowledge in general.

3. This phenomenon has been documented by psychologists. Daniel Kahneman describes it as a contest between two different ways our minds work: "System 1," which is largely unconscious and quickly creates a coherent story using only the information currently available, without questioning the quality or completeness of this information, and "System 2," our conscious analytical mind. System 2 is lazy and often accepts the suggestions of System 1, though if forced to it can actively seek additional and higher-quality information needed for better analyses (Kahneman 2011). In other words, we have to exert conscious effort to engage our analytic System 2, or System 1 may lead us to make bad decisions.

4. Throughout this book I use the term "Third World" to refer what some call the "Global South," and the United Nations and other mainstream development agencies call "developing countries," and the term "industrial world" to refer to the "Global North," or "developed countries."

CHAPTER 1

1. The 1977 film *Powers of Ten*, made by the office of designers Charles and Ray Eames, is a famous example of this imaginary travel across scales (www.youtube.com/watch?v = ofKBhvDjuyo).

2. A light year is the distance light travels in one year, or about 9.461×10^{12} km.

3. The theory of selection for density-dependent growth rates has been repeatedly challenged since its initial conceptualization in the 1960s, but it continues to be elaborated (Bassar et al. 2013), and experimental studies have provided evidence of both intra- and interspecies effects of density on growth rates—for example, r and K selection under high and low predation (Lande et al. 2009).

4. See http://en.wikipedia.org/wiki/Ambalappuzha_Sri_Krishna_Temple, or www. indiadivine.org/audarya/hare-krishna-forum/491548-ambalapuzha-palpayasam.html. This legend is one among many about the origin of *paal payasam*.

5. Malthus was what today we would call a libertarian, in that he opposed the French Revolution and advocated repeal of public support for the poor based on his belief that human progress depends not so much on reforming political institutions, but on "the slow process of education . . . and the exercise of individual prudence in personal affairs" (Winch 1992:ix).

6. Assuming exponential growth at the current annual rate (0.012), it would take only three hundred years to reach this size, forty times larger than the current population (7.1 billion).

7. The Tallensi, also of northern Ghana, also have long periods of postpartum abstinence, which they explicitly stated was to control fertility to protect the health of the next youngest (Fortes 1949:20). Researchers have noted postpartum abstinence practices among many other groups in Africa, and that periods of abstinence may be growing shorter due to social changes, which could increase fertility (Benefo 1995).

8. The Upper Region was the northernmost region of Ghana at the time of my research. In 1983 it was divided into the Upper East Region, where Zorse is located, and the Upper West Region.

9. This section was inspired by Joel Cohen's (1995a) calculation of HCC.

10. This is an instance of the first law of thermodynamics, introduced in section 2.0, and discussed further in chapter 4.

11. Negative exponents are a shorthand way of indicating "per." For example, 348 m³ water person⁻¹ yr⁻¹) = 348 m³ water / person / year = 348 m³ water per person per year.

CHAPTER 2

1. This description is based on (Shostak 1981).

2. Still, Darwinian evolution was also misinterpreted in a social context as a unilineal process leading to European males as the most advanced form, even by prominent evolutionists including T. H. Huxley (Desmond 1997).

3. Some researchers have suggested that the variation in fire in Australia during the last seventy thousand years was due primarily to climatic variation, not foragers (Mooney et al. 2011).

4. However, Harlan says that "crop" seems to define a range from wild to fully domesticated (Harlan 1992:63). Some crops retain the ability to survive without human care and can escape cultivation to establish feral populations, which can be facilitated by crossing with wild relatives—for example, radish, sunflower, rye, wheat, and sorghum (Bagavathiannan and Van Acker 2008). Other exceptions exist, especially among some perennial fruit crops, more accurately described as "semidomesticates," for example, olive (Baldoni et al. 2006, Breton et al. 2006).

5. The description of evolution given here is quite simplified. For example, some phenotypic traits resulting from the environment are heritable, and there are units of selection at higher levels than the individual organism. I will discuss these in more detail in chapter 5.

6. The term "evolution" was widely used in Darwin's time to refer to the evolution (origin) of species. Population geneticists came to use the term to mean change in allele frequencies

between generations, also referred to as microevolution or subspecific evolution, to differentiate it from macroevolution or transspecific evolution, which results in new species (Endler 1986:5–7). I use the term "evolution" to refer to change in allele frequencies over generations in a general sense, and the more specific terms when needed, especially in chapter 5.

7. See chapter 5 for more details on FVs and MVs and the differences between them.

8. There is some confusion over terms in the literature, with indirect artificial selection sometimes defined as "natural" selection (Simmonds 1979:14–15), as the same as conscious selection (Allard 1999), or as entirely "unconscious" selection (Poehlman and Sleper 1995:9).

9. I discuss polyculture in more detail in chapter 6, including how to evaluate its yields in comparison with monocultures using the land equivalent ratio.

CHAPTER 3

1. This is probably the most commonly cited definition of sustainability. It comes from the World Commission on Environment and Development (WCED 1987:Chapter 2), known as the Brundtland Commission after Chair Gro Harlem Brundtland, which was convened in 1983 and published a report four years later (WCED 1987). Brundtland is the former prime minister of Norway and has served as the director-general of the World Health Organization. She now serves on the Blue Ribbon Advisory Board of PepsiCo, widely acknowledged to be one of the world's foremost manufacturers and promoters of unhealthy food (Norum 2008)—which brings up the critical issue of whether we can move toward a more sustainable agrifood system by working with corporations and others when they promote systems that may be unsustainable according to our definition. We will discuss this later in this and subsequent chapters.

2. The neuroscientist Sam Harris has promoted the controversial idea that human values can be objectively evaluated in terms of human biology (Harris 2010). This is an intriguing idea, but even if there are universally "good" ideas, behaviors, and events, we are still far from agreement on their biological indicators or how they can be objectively measured.

3. Scientists whose research questions the assumptions of TGV supporters are often fiercely attacked as doing "bad science" or as guilty of scientific "misconduct" (Waltz 2009). Prominent proponents of TGVs like Miller et al. (2008) take this position, as does the U.S. government, as described by Essex (2008). The debate around TGVs is discussed in more depth in chapter 5.

4. I use the term "farmer knowledge" to refer to the knowledge of farmers that is based on local tradition, yet this knowledge is always influenced to a greater or lesser degree by knowledge from other places, including Western scientific knowledge, and it is constantly changing over time due to both local and outside influence. FK as I use it is often included under the label "indigenous" knowledge, but this is term is fraught with controversy; Dove (2006) and Ellen and Harris (2000) provide excellent analyses of political debates over the status of "indigenous" knowledge and the inherent ambiguity in defining it.

5. IFPRI is a CGIAR center (see appendix 2); the report was a product of the Agricultural Science and Technology Policy project, whose funders include the Gates Foundation (CGIAR 2011).

6. For example, in the United States, Weber and Matthews estimate that food miles account for only about 4 percent of total agrifood system GHGE (2008). I discuss food miles and GHGE in more detail in chapter 9.

7. The phrase "drinking green Kool-Aid" in relation to climate change mitigation is discussed in McIntyre 2011.

CHAPTER 4

1. The G8 comprises eight leading industrial countries that have been meeting annually since 1975; it currently includes Britain, Canada, France, Germany, Italy, Japan, Russia, and the United

States (www.g8.utoronto.ca/what_is_g8.html). The G8 Summit is held yearly, and since 2008 it has issued statements specifically addressing the world food crisis.

2. The G8 appears to have become aware of the bad press that resulted from its members' indulgence in opulent meals while discussing solutions to the world food crisis, and it now apparently enforces strict privacy on its meals; I have found it impossible to locate information on meals at subsequent G8 summits. Even at the time, some G8 officials had doubts about such ostentatious consumption.

3. See chapter 8 for an example of strong sustainability versus weak sustainability in the case of climate change.

4. It is important not to confuse ecological economics with environmental economics, which is usually a mainstream economic approach to dealing with the environment based on monetizing all environmental and social factors so that they can be dealt with in the marketplace (Gowdy and Erickson 2005:209). A third economic approach to sustainability has been dubbed "sustainability economics"; this approach purports to find common ground between ecological and environmental economics but seems to be heavily biased toward mainstream economic assumptions, since it is defined as the "environmental sustainability of economic performance and growth, and assumes that whether natural capital can be replaced by human-made capital depends on technological progress" (Bartelmus 2013:xiv, 122).

5. These findings have been supported by experiments with nonhuman animals (e.g., rats; see Bartal et al. 2011), but interpretation of these and human experiments supporting altruism have been challenged (e.g., Rankin 2011, West and Domingos 2012).

6. This is the same exponential formula used to calculate population growth in chapter 1, here using different notation; $FV \approx P_t$, $PV \approx P_o$, and e, r, and t have the same meaning. The use of an exponential DR is based on the assumption that discount rate of change is constant through time, and this has been challenged. Experimental and field evidence suggests that a hyperbolic function more realistically models people's actual change of preferences through time; DRs are higher in the near future and lower in the more distant future, making it possible that preferences could be reversed with time. Hyperbolic discounting has been used to model food choice (Shapiro 2005) and attitudes about the effects of global warming (Karp 2005).

7. This is from the chair's summary of a meeting of G8 development ministers before the summit in 2008 (G8 2008b). Other information is from the presummit statement on food security (G8 Summit 2008) and from a statement issued at the end of the summit (G8 2008a). The Millennium Development Goals, established by the United Nations in 2000, include "eradicate extreme poverty and hunger" and "ensure environmental sustainability," and they are embraced by the United Nations and other mainstream organizations (www.un.org/millenniumgoals/index.shtml).

8. La Vía Campesina states that it is an "international movement which brings together millions of peasants, small and medium-size farmers, landless people, women farmers, indigenous people, migrants and agricultural workers from around the world. It defends small-scale sustainable agriculture as a way to promote social justice and dignity. It strongly opposes corporate driven agriculture and transnational companies that are destroying people and nature" (http://viacampesina.org/en/index.php/organisation-mainmenu-44).

9. Because the population was larger in 2010–2012, the proportion of undernourished actually decreased from 15 percent to 12.9 percent from 1990–1992 to 2010–2012 (FAO 2012d). The FAO defines "undernourishment" as "chronic hunger" or "the status of persons, whose food intake regularly provides less than their minimum energy requirements. The average minimum energy requirement per person is about 1800 kcal per day. The exact requirement is determined by a person's age, body size, activity level and physiological conditions such as illness, infection, pregnancy and lactation" (FAO 2013b).

10. The USDA defines food insecurity as a situation in which "food intake of one or more household members was reduced and their eating patterns were disrupted at times during the year because the household lacked money and other resources for food." By 2010, the numbers had not changed substantially; 14.5 percent were food insecure at least some time during the year, and 5.3 percent had very low food security (Coleman-Jensen et al. 2011).

11. GRAIN (Genetic Resources Action International) is an international NGO based in Spain whose stated purpose is to "support small farmers and social movements in their struggles for community-controlled and biodiversity-based food systems" (www.grain.org). It received the Right Livelihood award in 2011.

12. For the FAO definition, see (FAO 2013b).

13. For example, see (Firestone 2012).

14. The CGIAR is an international network of agricultural research centers focused on plant breeding that grew out of and was subsequently the main promoter of the Green Revolution. It has a "strong relationship with the World Bank," which also serves as the trustee of the CGIAR Fund (http://cgiar.bio-mirror.cn/who/structure/index.html), and as an institution it takes a mainstream approach to agrifood system sustainability (see Cleveland 2006).

CHAPTER 5

1. FVs include older and newer traditional varieties selected by farmers, and sometimes MVs that have been adapted to farmers' environments by farmer and natural selection (sometimes referred to as "creolized" or "degenerated" MVs), as well as progeny from crosses between landraces and MVs. FVs are sometimes referred to as "landraces," but this inaccurately implies that they are selected by the "land"—that is, by the environment and not by humans (Zeven 1998).

2. In ribonucleic acid (RNA), uricil replaces thymine, and in mitchodrial RNA there are other differences. Three of the sixty-four codons are stop codons that signal the end of synthesis.

3. Two important epigenetic processes are histone modification and DNA methylation. Epigenetics is a rapidly increasing area of investigation in crop science—for example, in understanding variation in environmental adaptation in rice (Ding et al. 2012, Zhang et al. 2012).

4. Many domesticated crops have more than one maternal and paternal chromosome in each set of homologous chromosomes, a condition known as *polyploidy*. Polyploidy is another source of diversity and has been important in the evolution of crop plants, including tetraploid and hexaploid wheats.

5. The crossing-over occurs at a stage in meiosis when each of the homologous chromosomes have duplicated, forming four chromatids; crossing-over and recombination can occur between any two of these chromatids.

6. Cross-pollination can occur between separate plants in species with three different types of flowers: *hermaphroditic* species, such as tomato, bean, and wheat, have *perfect* flowers consisting of both male and female parts; *monoecious* species, such as squash and maize, have separate female and male flowers on the same plant; and *dioecious* species, such as carob and pistachio, have plants that have either female or male flowers only.

7. Broad sense heritability (H) is the proportion of V_p due to genetic variance (V_G / V_p), while narrow sense heritability (h^2) is the proportion of V_p due to additive genetic variance (V_A / V_p)—that is, the proportion of V_G directly transmissible from parents to progeny, and therefore of primary interest to breeders.

8. Expression of S in standard deviation units (the standardized selection differential) permits comparison of selections among populations with different amounts or types of variation (Falconer and Mackay 1996). The standardized selection differential = iσ, where i = selection intensity that depends on proportion of the parent population selected, and σ = phenotypic standard devia-

tion of the parental population (Allard 1999:101–102, Falconer and Mackay 1996:189, Simmonds and Smartt 1999:193).

9. See (Soleri, Smith, and Cleveland 2000), discussed further below.

10. Type 1 stability can also defined in terms of variance (s^2) across environments defined by average yield ($\bar{X}$), and type 2 stability as *coefficient of variation* ($CV = s/\bar{X}$); the value of these statistics is inversely related to stability—that is, a higher value of s^2 or CV indicates greater instability than a lower value. These definitions are commonly used in aggregate stability analyses by economists, who disagree about whether s^2 or CV is a better measure of stability, a debate that is analogous to the one in plant breeding over the use of type 1 or type 2 stability. Even when absolute stability (s^2) increases with increases in yield, relative stability (CV) will remain the same or decrease if increases in average yield are large enough. Thus, a choice between these two measures can be seen as a choice of whether to emphasize yield stability or yield. Anderson and Hazell suggest that even when yield increases are great enough so that the CV does not increase as s^2 increases, the risks for poorer households can still increase (Anderson and Hazell 1989b:347).

11. Simmonds, however, believed that it has not had much influence on breeding decisions because the data needed are not usually available until after the variety is close to being or already finished (Simmonds and Smartt 1999:185–186).

12. However, some scientists working with farmers have emphasized that they are still in control, and prefer the term "client oriented breeding" (Witcombe et al. 2006)

13. Recall that the basic model for variation in phenotype is $V_P = V_G + V_E + V_{GxE}$. A null hypothesis is a statement that there is no difference between the things being compared.

14. Agrifood biotechnology has sometimes been defined to include any human effect on other species—for example, making bread with yeast—and agbiotech proponents often equate genetic modification with genetic engineering, so that any plant whose genetic structure has been influenced by humans is a GMO and crops domesticated in the Neolithic are equated with TGVs (Fedoroff 2003). The intent in making this equivalence appears to be to obfuscate scientific inquiry into differences required for a robust risk assessment, and Gepts shows that describing TGVs as "genetically modified" is inaccurate and unscientific (2002).

CHAPTER 6

1. Aggregates are clumps of soil that give it its structure, with spaces for the movement of water and air into the soil, and CO_2 from respiration of microorganisms and plant roots out of the soil. Soils with more abundant and stable aggregates are more supportive of plant growth.

2. There are, however, alternate explanations of the metabolic quotient in young soils that may be important for better understanding the Mäder et al. study. For example, Banning et al. state that "if it is assumed that the efficiency of amino acid utilization is representative of the efficiency of organic C utilization *in situ*, then high metabolic quotients in young rehabilitation soils do not represent lower cellular efficiency but higher respiration potential as a result of higher C bioavailability" (2008).

3. "Race" is a taxonomic category within a species, dividing up the species into groups more similar to each other—for example, maize varieties are grouped into races.

CHAPTER 7

1. There is a tendency to conflate properties of natural resources in relation to potential human use, management strategies for them, and interaction between resources and management strategies. Hardin did this in his famous essay (1968), as discussed below. The services to humans provided by resources have also been classified using Ostrom's scheme (Fisher et al. 2009).

2. Privately managed is not the same as privately owned. *Usufructory* rights are use rights that may appear to outsiders as private ownership, as was my experience in Zorse—the day-to-day management of fields there appeared to me the same as privately owned property as I knew it in the United States. The crucial difference is that the user has rights of use as long as she follows the rules, but she does not have the right to give or sell the land.

3. Ostrom defines a fourth resource category—toll or club goods (Ostrom 2010)—but I will not discuss this class of goods here.

4. "The Prize in Economics 2009—Press Release." Nobelprize.org, 12 October 2009 (www .nobelprize.org/nobel_prizes/economics/laureates/2009/press.html).

5. See (FAO 2012a).

CHAPTER 8

1. The IPCC (Intergovernmental Panel on Climate Change) is the leading international scientific body assessing climate change. It was established in 1988 and operates under the auspices of the UN (www.ipcc.ch/publications_and_data/publications_and_data_reports.shtml#.Ufj5ZFNiHjE). It has issued four major assessments, the most recent being *Climate Change 2007*, the Fourth Assessment Report (see www.ipcc.ch/organization/organization.shtml#.UemM3lNiHjE). Its Fifth Assessment is due for release from September 2013 to October 2014.

2. Note that these targets are for CO_2 concentrations and do not include other GHGs; thus, in terms of effects on the climate, CO_2 concentrations are being used as an indicator for the combined effect of all GHGs, expressed in terms of CO_2e, or carbon dioxide equivalent, which I will refer to later in this chapter. CO_2e is a measure of the total warming potential of the greenhouse gases included in a specific measure in terms of their global warming potential (GWP) compared with that of CO_2, which is set at 1.0. The GWP is the warming effect of greenhouse gases relative to that of CO_2, taking into account their lifetime in the atmosphere and their relative effectiveness in absorbing thermal infrared radiation. However, CO_2e does not include the effect of other factors, such as aerosols, which have a cooling effect. The two most abundant GHGs after CO_2 are methane (CH_4), which has a GWP more than twenty times higher than CO_2, but lasts only about a decade in the atmosphere, and nitrous oxide (N_2O), which has a GWP three hundred times that of CO_2 over a hundred-year period, and remains in the atmosphere even longer (see www.epa.gov /climatechange/ghgemissions/gases.html). Thus, in 2011, while the global average CO_2 concentration was 392 ppm (ftp://ftp.cmdl.noaa.gov/ccg/co2/trends/co2_annmean_mlo.txt), the concentration of GHGs in CO_2e was 473 ppm (www.esrl.noaa.gov/gmd/aggi/)

3. The "greenhouse effect", although a standard term in scientific literature, is actually a misnomer because global climate warming is not the result of the same process that heats a greenhouse.

4. While water vapor is the most abundant and important GHG in the atmosphere, humans have only a relatively small influence on its concentration (IPCC 2007a:135), although agriculture may affect water holding capacity as a result of increased albedo on cropland (Eshel and Martin 2009).

5. There is an upward tick about eight thousand years ago in the IPCC graph of CO_2 concentrations (fig. 8.2). There is debate about the extent to which increases in CH_4 and CO_2 concentrations are the result of land clearing and development for agriculture, especially during the following millennia, but a contribution by anthropogenic activities is generally acknowledged (Wanner et al. 2008).

6. However, one statistical analysis found that while a USD $1 per barrel increase in crude oil prices leads to food commodity prices increasing by between USD $0.09 to $1.65 per MT, most of the effect is due to the use of crops for biofuel, rather than the direct use of oil in food production (Ciaian and Kancs 2011).

7. Global natural biological nitrogen fixation in agricultural systems was estimated by Galloway et al. to be approximately 40 Tg N per year in 2005 (Galloway et al. 2008:889).

8. Smaller amounts are also created by combustion of fossil fuels.

9. See "USDA supports Meatless Monday campaign: NCBA questions USDA's commitment to U.S. cattlemen," National Cattleman's Beef Association, 2012 July 25. http://www.beefusa.org /newsreleases1.aspx?NewsID = 2560, accessed 2012 August 20.

10. See Amy Harmon, "Retracting a plug for Meatless Mondays," New York Times, 2012 July 25. http://www.nytimes.com/2012/07/26/us/usda-newsletter-retracts-a-meatless-mondays-plug. html, accessed 2013 August 7.

CHAPTER 9

1. Localwashing is analogous to greenwashing. Jaffee and Howard have noted these two responses in the case of mainstream marketing of organic and fair-trade foods, with the goal of profit maximization (Jaffee and Howard 2010:388); see chapter 3.

2. This section is based on our research in (Cleveland, Radka, Müller et al. 2011)

3. While produce dominates SBC agricultural production economically, it occupies a small proportion of agricultural land. In 2007, 94,000 acres were harvested for fruit and vegetables, or 13 percent of all agricultural land in the county, the remainder being almost all rangeland (calculated from data in USDA NASS 2009).

4. Data are not available on any exports out of SBC, or out of California to other states; there are only estimates of exports from California to other countries. The increase in both exports and imports illustrates the disconnect of the global system. In 2010, 24 percent of California agricultural production was exported, including 61 percent of raspberries and blackberries, 42 percent of oranges and orange products, 36 percent of cauliflower, 17 percent of broccoli, and 11 percent of strawberries. In 2005, the United States imported an estimated 38.5 percent of all fresh fruits and 13.6 percent of vegetables consumed (Johnson 2012:13), and during the twenty years from 1990 to 2009, the value of U.S. imports of both types of produce (in fresh, dried, frozen, and preserved forms) each increased 261 percent, while exports increased 200 percent for fruits and 37 percent for vegetables (based on Johnson 2012:2,13). For example, in 2009, 2,288 million pounds of fresh strawberries were produced in the United States, of which 271 million pounds were exported, while 187 million pounds of strawberries were imported; for fresh broccoli, the numbers are 1,909 and 229,261 million pounds, respectively (USDA ERS 2011a).

5. While difficult to quantify, some produce that is exported to distribution centers outside of SBC is subsequently imported to grocery stores and other transaction points in SBC, and several farmers told us of such occurrences. SBC-grown produce that is exported and then imported is similar to produce grown outside the county and imported in important ways (e.g., freshness, packaging, GHGE), and we did not include this in our estimate of "local" produce.

6. Details of methods are in (Cleveland, Radka, Müller et al. 2011b). The methods could be easily used by any local group to assess its own area's localness; the most difficult part is the field-work required to document locally grown and sold food.

7. To estimate GHGE in terms of CO_2e for direct transport (farm to retail) of produce imported into SBC, we summed CO_2e for all three categories of produce from Weber and Matthews's LCA, the most recent of the U.S. agrifood system (2008); calculated the CO_2e per household per year; and multiplied that quantity by the number of households in SBC in 2008 (US Census Bureau 2010).

8. Weber and Matthews used 1997 as the reference year for their LCA of the U.S. agrifood system, the most complete analysis to date.

9. Savings per person was less than 0.1 percent for 2008 as well (calculated using data from EPA 2010:2–7, US Census Bureau 2010) (table 9.2).

10. These figures are based on surveys of low-income adults (those residing in households with incomes less than 200 percent of the federal poverty level) (CFPA 2010). In addition, 17.6 percent, or fifteen thousand, of these residents are affected by "very low food security," meaning a condition of severe food insecurity characterized by disruption of eating patterns and reduction in food intake (Harrison et al. 2007).

REFERENCES

AASHE (Association for the Advancement of Sustainability in Higher Education). 2010. STARS History and System Development. http://stars.aashe.org/pages/faqs/4102, accessed 2010 October 10.

Abbo, S, Lev-Yadun, S, and Gopher, A. 2010. Yield stability: An agronomic perspective on the origin of Near Eastern agriculture. Vegetation History and Archaeobotany 19(2):143–150.

Abdullahi, AA, and CGIAR. 2003. Adoption and impact of dry-season dual-purpose cowpea in the Nigerian semiarid region. http://impact.cgiar.org/pdf/223.pdf, accessed 2013 August 1.

ABSPII (Agricultural Biotechnology Support Project II). 2010. Consortium Partners. Cornell University. www.absp2.cornell.edu/consortiumpartners, accessed 2011 September 5.

Achebe, C. 1959. Things Fall Apart. New York: Fawcett.

Adams, RM. 1978. Strategies of maximization, stability, and resilience in Mesopotamian society, settlement, and agriculture. Proceedings of the National Academy of Science 122:329–355.

Adesina, AA 1992. Village-level studies and sorghum technology development in West Africa: Case study in Mali. Pages 147–168 in Moock, JL, and Rhoades, RE, eds. Diversity, Farmer Knowledge, and Sustainability. Ithaca, NY: Cornell University Press.

Adolphs, R. 2009. The social brain: Neural basis of social knowledge. Annual Review of Psychology 60:693–716.

AFT (American Farmland Trust). 2012. California Agricultural Vision: From Strategies to Results. www.cdfa.ca.gov/agvision/docs/Ag_Vision_Progress_Report.pdf, accessed 2013 January 5.

AGRA (Alliance for a Green Revolution for Africa). 2012. About the Alliance for a Green Revolution in Africa. www.agra-alliance.org/section/about, accessed 2012 May 4.

Agrawal, A. 1995. Dismantling the divide between indigenous and scientific knowledge. Development and Change 26:413–439.

Ahrens, TD, Beman, JM, Harrison, JA, Jewett, PK, and Matson, PA. 2008. A synthesis of nitrogen transformations and transfers from land to the sea in the Yaqui Valley agricultural region of Northwest Mexico. Water Resources Research 44: article no. W00A05.

AIC (Agricultural Issues Center, University of California). 2012. Estimating California's Agricultural Exports. Davis, California: University of California. http://aic.ucdavis.edu/pub/exports.html, accessed 2012 May 1.

Alexandratos, N., and Bruinsma, J. 2012. World agriculture towards 2030/2050: The 2012 revision. Working Paper 12–03. FAO. http://www.fao.org/docrep/016/ap106e/ap106e.pdf, accessed 2013 January 5.

Allard, RW. 1999. Principles of Plant Breeding. 2nd ed. New York: John Wiley & Sons.

Allard, RW, and Bradshaw, AD. 1964. Implications of genotype-environmental interactions in applied plant breeding. Crop Science 4:503–508.

Allen, MW, and Baines, S. 2002. Manipulating the symbolic meaning of meat to encourage greater acceptance of fruits and vegetables and less proclivity for red and white meat. Appetite 38(2):118–130.

Altieri, MA, Funes-Monzote, FR, and Petersen, P. 2012. Agroecologically efficient agricultural systems for smallholder farmers: Contributions to food sovereignty. Agronomy for Sustainable Development 32(1):1–13.

Anand, S., and Sen, A. 2000. Human development and economic sustainability. World Development 28(12):2029–2049.

Andersen, R. 2008. Governing Agrobiodiversity: Plant Genetics and Developing Countries. Farnham, Surrey, UK: Ashgate.

Anderson, JR, and Hazell, PBR. 1989a. Introduction. Pages 1–10 in Anderson, JR, and Hazell, PBR, eds. Variability in Grain Yields: Implications for Agricultural Research and Policy in Developing Countries. Baltimore, MD: Johns Hopkins University Press.

———. 1989b. Synthesis and needs in agricultural research and policy. Pages 339–356 in Anderson, JR, and Hazell, PBR, eds. Variability in Grain Yields: Implications for Agricultural Research and Policy in Developing Countries. Baltimore, MD: Johns Hopkins University Press.

Anderson, MK. 2005. Tending the Wild: Native American Knowledge and the Management of California's Natural Resources. Berkeley: University of California Press.

Angermeier, PL, and Karr, JR. 1994. Biological integrity versus biological diversity as policy directives. BioScience 44:690–697.

Anseeuw, W, Alden Wily, L, Cotula, L, and Taylor, M. 2012. Land Rights and the Rush for Land: Findings of the Global Commercial Pressures on Land Research Project. Rome: ILC. www.landcoalition.org/sites/default/files/publication/1205/ILC GSR report_ENG.pdf, accessed 2013 April 18.

Aragón-Cuevas, F, Taba, S, Hernández Casillas, JM, Figueroa C, JdD, Serrano Altamirano, V, and Castro García, FH. 2006. Catálogo de Maíces Criollos de Oaxaca. Libro Técnico Núm. 6. Oaxaca, Mexico: INIFAP-SAGARPA.

Arax, M, and Wartzman, R. 2003. The King of California: JG Boswell and the Making of a Secret American Empire. New York: PublicAffairs.

Armelagos, GJ 2010. The Omnivore's Dilemma: The evolution of the brain and the determinants of food choice. Journal of Anthropological Research 66(2):161–186.

Arrow, K, et al. 1995. Economic growth, carrying capacity, and the environment. Science 268:520–521.

Asrat, S, Yesuf, M, Carlsson, F, and Wale, E. 2010. Farmers' preferences for crop variety traits: Lessons for on-farm conservation and technology adoption. Ecological Economics 69(12):2394–2401.

Atahan, P, Itzstein-Davey, F, Taylor, D, Dodson, J, Qin, J, Zheng, H, and Brooks, A. 2008. Holocene-aged sedimentary records of environmental changes and early agriculture in the lower Yangtze, China. Quaternary Science Reviews 27(5–6):556–570.

Atlin, GN, and Frey, KJ. 1990. Selecting oat lines for yield in low-productivity environments. Crop Science 30(3):556–561.

Atlin, GN, McRae, KB, and Lu, X. 2000. Genotype × region interaction for two-row barley yield in Canada. Crop Science 40:1–6.

Badgley, C, Moghtader, J, Quintero, E, Zakem, E, Chappell, MJ, Aviles-Vazquez, K, Samulon, A, and Perfecto, I. 2007. Organic agriculture and the global food supply. Renewable Agriculture and Food Systems 22(2):86–108.

Bagavathiannan, MV, and Van Acker, RC. 2008. Crop ferality: Implications for novel trait confinement. Agriculture Ecosystems and Environment 127(1–2):1–6.

Baker, JM, Ochsner, TE, Venterea, RT, and Griffis, TJ. 2007. Tillage and soil carbon sequestration—What do we really know? Agriculture, Ecosystems and Environment 118):1–5.

Baldoni, L, Tosti, N, Ricciolini, C, Belaj, A, Arcioni, S, Pannelli, G, Germana, MA, Mulas, M, and Porceddu, A. 2006. Genetic structure of wild and cultivated olives in the central Mediterranean basin. Annals of Botany 98(5):935–942.

Ballingall, J, and Winchester, N. 2010. Food miles: Starving the poor? World Economy 33(10):1201–1217.

Banning, NC, Grant, CD, Jones, DL, and Murphy, DV. 2008. Recovery of soil organic matter, organic matter turnover and nitrogen cycling in a post-mining forest rehabilitation chronosequence. Soil Biology and Biochemistry 40(8):2021–2031.

Bänziger, M, Betr n, FJ, and Lafitte, HR. 1997. Efficiency of high-nitrogen selection environments for improving maize for low-nitrogen target environments. Crop Science 37:1103–1109.

Barbier, EB. 2012. The green economy post Rio+20. Science 338(6109):887–888.

Bardhan, P, and Dayton-Jones, J. 2002. The drama of the commons. Pages 87–112 in Ostrom, E, Dietz, T, Dolšak, N, Stern, PC, Stonich, S, and Weber, EU, eds. The Drama of the Commons. Washington, DC: National Academy Press.

Barham, J, Tropp, D, Enterline, K, Farbman, J, Fisk, J, and Kiraly, S. 2012. Regional Food Hub Resource Guide. Washington, DC: USDA, Agricultural Marketing Service. http://ngfn.org/resources/ngfn-database/knowledge/FoodHubResourceGuide.pdf, accessed 2013 July 2.

Barlow, M. 2008. Blue covenant: The alternative water future. Monthly Review: An Independent Socialist Magazine 60(3):125–141.

Barnett, TP, Adam, JC, and Lettenmaier, DP. 2005. Potential impacts of a warming climate on water availability in snow-dominated regions. Nature 438(7066):303–309.

Barnosky, AD, et al. 2011. Has the Earth's sixth mass extinction already arrived? Nature 471(7336):51–57.

———. 2012. Approaching a state shift in Earth's biosphere. Nature 486(7401):52–58.

Barrett, CB, Bellemare, MF, and Hou, JY. 2010. Reconsidering conventional explanations of the inverse productivity-size relationship. World Development 38(1):88–97.

Barrett, SCH. 2002. The evolution of plant sexual diversity. Nature Review Genetics 3(4):274–284.

Bartal, IBA, Decety, J, and Mason, P. 2011. Empathy and pro-social behavior in rats. Science 334(6061):1427–1430.

Bartelmus, P. 2013. Sustainability Economics: An Introduction. London: Routledge.

Baskin, Y. 1994. Ecologists dare to ask: How much does diversity matter? Science 264:202–203.

Bassar, RD, Lopez-Sepulcre, A, Reznick, DN, and Travis, J. 2013. Experimental evidence for density-dependent regulation and selection on Trinidadian guppy life histories. American Naturalist 181(1):25–38.

Bayer. 2009. Interview with Dr. Wolfgang Plischke, Sustainable Development Report 2009. www.sustainability2009.bayer.com/en/interview-with-prof.-dr-wolfgang-plischke.aspx, accessed 2010 October 15.

Becker, CD., and Ostrom, E. 1995. Human ecology and resource sustainability: The importance of institutional diversity. Annual Review of Ecology and Systematics 26:113–133.

Benefo, KD. 1995. The determinants of the duration of postpartum sexual abstinence in West-Africa: A multilevel analysis. Demography 32(2):139–157.

Benz, BF, Sanchez-Velasquez, LR, and Santana Michel, FJ. 1990. Ecology and ethnobotany of *Zea diploperennis:* Preliminary investigations. Maydica 35:85–98.

Bergelson, J, and Purrington, CB. 2002. Factors affecting the spread of resistant *Arabidopsis thaliana* populations. Pages 17–31 in Letourneau, DK and Elpern Burrows, B, eds. Genetically Engineered Organisms: Assessing Environmental and Human Health Effects. Boca Raton, FL: CRC Press.

Berger, Christopher C, and Ehrsson, HH. 2013. Mental imagery changes multisensory perception. Current Biology 23(14):1367–1372.

Berlin, B. 1992. Ethnobiological Classification: Principles of Categorization of Plants and Animals in Traditional Societies. Princeton, NJ: Princeton University Press.

Berry, W. 1977. The Unsettling of America: Culture and Agriculture. San Francisco, CA: Sierra Club Books.

Berthaud, J, Clément, JC, Emperaire, L, Louette, D, Pinton, F, Sanou, J, and Second, S. 2001. The role of local-level gene flow in enhancing and maintaining genetic diversity. Pages 1–23 in Cooper, HD, Spillane, C, and Hodgkin, T, eds. Broadening the Genetic Base of Crop Production. Wallingford, Oxon, UK: CABI.

Birkhofer, K, et al. 2008. Long-term organic farming fosters below and aboveground biota: Implications for soil quality, biological control and productivity. Soil Biology and Biochemistry 40(9):2297–2308.

Blaikie, P. 2006. Is small really beautiful? Community-based natural resource management in Malawi and Botswana. World Development 34(11):1942–1957.

Blakeney, M. 2011. Recent developments in intellectual property and power in the private sector related to food and agriculture. Food Policy 36:S109-S113.

Block, J. 2012. A reality check for organic food dreamers. Wall Street Journal. 2012 December 23. http://online.wsj.com/article/SB10001424127887323297104578174963239598312.html, accessed 2013 January 5.

Bloom, P, and Weisberg, DS. 2007. Childhood origins of adult resistance to science. Science 316(5827):996–997.

Bocquet-Appel, JP. 2011. When the world's population took pff: The springboard of the Neolithic demographic transition. Science 333(6042):560–561.

Boehm, C. 2012. Ancestral hierarchy and conflict. Science 336(6083):844–847.

Borges, JL. 1960. Del rigor en la ciencia. In El hacedor, p. 137. Buenos Aires: Emecé Editores.

Borlaug, NE. 1999. How to feed the 21st century? Pages 509–519 in Coors, JG and Pandey, S, eds. The Genetics and Exploitation of Heterosis in Crops. Madison, WI: ASA-CSSA-SSSA.

———. n.d. The Green Revolution: Peace and Humanity. 1970 Nobel Peace Prize lecture. Mexico City: CIMMYT.

Born, B, and Purcell, M. 2006. Avoiding the local trap: Scale and food systems in planning research. Journal of Planning Education and Research 26(2):195–207.

Borojevic, S. 1990. Genetic composition and adaptability of the cultivar. Pages 322–349 in Principles and Methods of Plant Breeding. Amsterdam: Elsevier.

Boserup, E. 1965. The Conditions of Agricultural Growth. Chicago: Aldine.

———. 1990. Economic and Demographic Relationships in Development. Baltimore, MD: Johns Hopkins University Press.

Boster, JS. 1985. Selection for perceptual distinctiveness: Evidence from Aguaruna cultivars of *Manihot esculenta.* Economic Botany 39:310–325.

Boulding, K. 1968. The economics of the coming spaceship Earth. Pages 275–287 in Boulding, K, ed. Beyond Economics: Essays on Society, Religion and Ethics. Ann Arbor: University of Michigan Press.

Bourdieu, P. 2000. Pascalian Meditations. Translated by Richard Nice. Cambridge, MA: Polity Press. First published 1997 as Méditations pascaliennes by Éditions du seuil.

Bowles, S. 2011. Cultivation of cereals by the first farmers was not more productive than foraging. Proceedings of the National Academy of Sciences 108(12):4760–4765.

Brady, NC, and Weil, RR. 2010. Elements of the Nature and Properties of Soil. 3rd ed. Upper Saddle River, NJ: Prentice Hall.

Breen, SD. 2010. The mixed political blessing of campus sustainability. Ps-Political Science and Politics 43(4):685–690.

Breton, C, Medail, F, Pinatel, C, and Berville, A. 2006. From olive tree to oleaster: Origin and domestication of *Olea europaea* L. in the Mediterranean basin. Cahiers Agricultures 15(4):329–336.

Brock, DA, Douglas, TE, Queller, DC, and Strassmann, JE. 2011. Primitive agriculture in a social amoeba. Nature 469(7330):393–398.

Brodt, S, Feenstra, G, and Tomich, T. 2008. The Low-Carbon Diet Initiative: Reducing Energy Use and Greenhouse Gas Emissions in the Food System Using Life Cycle Assessment. Summary of a Symposium on Critical Issues and Research Methods. Davis, California: UC Davis Agricultural Sustainability Institute.

Brody, S, Grover, H, and Vedlitz, A. 2012. Examining the willingness of Americans to alter behaviour to mitigate climate change. Climate Policy 12(1):1–22.

Brooks, S. 2011. Is international agricultural research a global public good? The case of rice biofortification. Journal of Peasant Studies 38(1):67–80.

Brown, S, Miltner, E, and Cogger, C. 2012. Carbon sequestration potential in urban soils. Pages 173–196 in Lal, R, and Augustin, B, eds. Carbon Sequestration in Urban Ecosystems. Dordrecht: Springer Netherlands.

Brownell, KD, and Warner, KE. 2009. The perils of ignoring history: Big tobacco played dirty and millions died. How similar is big food ? Milbank Quarterly 87(1):259–294.

Bruns, R, and Peterson, CJ. 1998. Yield and stability factors associated with hybrid wheat. Euphytica 100(1–3):1–5.

Buzby, JC, Wells, HF, and Vocke, G. 2006. Possible Implications for U.S. Agriculture from Adoption of Select Dietary Guidelines. USDA Economic Research Service. www.ers.usda.gov/media/860109/err31_002.pdf, accessed 2012 December 2.

Campbell, D. 2001. Conviction seeking efficacy: Sustainable agriculture and the politics of co-optation. Agriculture and Human Values 18:353–363.

Campbell, TC, and Campbell, TMI. 2005. The China Study: The Most Comprehensive Study of Nutrition Ever Conducted and the Startling Implications for Diet, Weight Loss and Long-term Health. Dallas, TX: BenBella Books.

Canning, P, Charles, A, Huang, S, Polenske, KR, and Waters, A. 2010. Energy Use in the U.S. Food System. Economic Research Report No. (ERR-94). USDA Economic Research Service. www.ers.usda.gov/media/136418/err94_1_.pdf, accessed 2011 January 6.

Cantrell, P. 2010. Sysco's Journey from Supply Chain to Value Chain: 2008–2009 Final Report. Wallace Center at Winrock International. www.wallacecenter.org/our-work/Resource-Library/Innovative-Models/Sysco Case Study 2009.pdf, accessed 2012 February 18.

Cardinale, BJ, et al. 2012. Biodiversity loss and its impact on humanity. Nature 486(7401): 59–67.

Carlsson-Kanyama, A, and González, AD. 2009. Potential contributions of food consumption patterns to climate change. American Journal of Clinical Nutrition 89(5):S1704-S1709.

Carmody, RN, and Wrangham, RW. 2009. The energetic significance of cooking. Journal of Human Evolution 57(4):379–391.

Carpenter, SR, et al. 2009. Science for managing ecosystem services: Beyond the Millennium Ecosystem Assessment. Proceedings of the National Academy of Sciences of the United States of America 106(5):1305–1312.

Carr, DL, Lopez, AC, and Bilsborrow, RE. 2009. The population, agriculture, and environment nexus in Latin America: Country-level evidence from the latter half of the twentieth century. Population and Environment 30(6):222–246.

Catling, DC, and Claire, MW. 2005. How Earth's atmosphere evolved to an oxic state: A status report. Earth and Planetary Science Letters 237(1–2):1–20.

CBD (Convention on Biological Diversity). 2012. Convention on Biological Diversity. www.cbd.int, accessed 2012 June 28.

CCPHA (California Center for Public Health Advocacy). 2007. The Food Landscape in Santa Barbara County: Searching for Healthy Food. Davis: California Center for Public Health Advocacy. www.publichealthadvocacy.org/RFEI/Santa_Barbara_County_Fact_Sheet.pdf, accessed 2009 November 10.

———. 2011. A Patchwork of Progress: Changes in Overweight and Obesity among California 5th, 7th, and 9th Graders, 2005–2010. Davis: California Center for Public Health Advocacy. www.publichealthadvocacy.org/research_patchworkprogress.html, accessed 2012 April 30.

CDC (Centers for Disease Control and Prevention). 2007. Data and Trends. County Level Estimates of Obesity: 2007 Age-Adjusted Estimates of the Percentage of Adults Who Are Obese in California. Washington, DC: Centers for Disease Control and Prevention. http://apps.nccd.cdc.gov/DDT_STRS2/CountyPrevalenceData.aspx, accessed 2010 January 15.

Ceballos, H, Pandey, S, Narro, L, and Perez-Velásquez, JC. 1998. Additive, dominant, and epistatic effects for maize grain yield in acid and non-acid soils. Theoretical and Applied Genetics 96:662–668.

Ceccarelli, S. 1989. Wide adaptation: How wide? Euphytica 40:197–205.

———. 1996. Adaptation to low/high input cultivation. Euphytica 92:203–214.

———. 2012. Plant Breeding with Farmers: A Technical Manual. Aleppo, Syria: International Center for Agricultural Research in the Dry Areas.

Ceccarelli, S, et al. 2003. A methodological study on participatory barley breeding II: Response to selection. Euphytica 133:185–200.

Ceccarelli, S, Erskine, W, Hamblin, J, and Grando, S. 1994. Genotype by environment interaction and international breeding programmes. Experimental Agriculture 30:177–187.

Ceccarelli, S, and Grando, S. 2002. Plant breeding with farmers requires testing the assumptions of conventional plant breeding: Lessons from the ICARDA barley program. Pages 297–332 in Cleveland, DA and Soleri, D, eds. Farmers, Scientists and Plant Breeding: Integrating Knowledge and Practice. Wallingford, Oxon, UK: CABI.

———. 2009. Participatory plant breeding. Pages 395–414 in Carena, MJ, ed. Cereals. New York: Springer US.

Ceccarelli, S, Grando, S, and Hamblin, J. 1992. Relationship between barley grain yield measured in low- and high-yielding environments. Euphytica 64:49–58.

Ceccarelli, S, Grando, S, and Impiglia, A. 1998. Choice of selection strategy in breeding barley for stress environments. Euphytica 103:307–318.

Ceccarelli, S, Grando, S, Tutwiler, R, Bahar, J, Martini, AM, Salahieh, H, Goodchild, A, and Michael, M. 2000. A methodological study on participatory barley breeding I: Selection phase. Euphytica 111:91–104.

CFPA (California Food Policy Advocates). 2010. Santa Barbara County Nutrition and Food Insecurity Profile. www.cfpa.net/2010CountyProfiles/SantaBarbara.pdf, accessed 2010 April 7.

CGIAR (Consultative Group on International Agricultural Research). 2006. Science Council Brief: Summary Report on System Priorities for CGIAR Research, 2005–2015. Washington, DC: CGIAR. www.sciencecouncil.cgiar.org/fileadmin/user_upload/sciencecouncil /Reports/SCBrief_SystPrior.pdf, accessed 2009 March 12.

————. 2011. Subtheme 9.2: Agricultural Science and Technology Policy (GRP 31). Washington, DC: CGIAR. http://cgmap.cgiar.org/docsRepository/documents/MTPProjects/2011–2013 /IFPRI_2011–2013_22.PDF, accessed 2013 August 7.

Chakrabarti, S, and Graziano da Silva, J. 2012. Hungry for investment: The private sector can drive agricultural development in countries that need it most. Wall Street Journal, September 6. http://online.wsj.com/article/SB10000872396390443686004577633308019 0871456.html, accessed 2013 January 16.

Chapman, J. 2008. Summit that's hard to swallow: World leaders enjoy 18-course banquet as they discuss how to solve global food crisis. Daily Mail, July 8. www.dailymail.co.uk/news /article-1032909/Summit-thats-hard-swallow—world-leaders-enjoy-18-course-banquet-discuss-solve-global-food-crisis.html, accessed 2013 August 7.

Chon, SU, and Nelson, CJ. 2011. Allelopathy in Compositae Plants. Pages 727–739 in Lichtfouse, E, Hamelin, M, Navarrete, M, and Debaeke, P, eds. Sustainable Agriculture, vol. 2. Dordrecht, Netherlands: Springer.

Ciaian, P, and Kancs, dA. 2011. Food, energy and environment: Is bioenergy the missing link? Food Policy 36:571–580.

Clark, CW. 1990. Mathematical Bioeconomics: The Optimal Management of Renewable Resources. 2nd ed. New York: Wiley.

Clark, S, and Walsh, H. 2011. Fair trade proving anything but in growing $6 billion market. Bloomberg Businessweek, December 28. www.businessweek.com/news/2011–12–28 /fair-trade-proving-anything-but-in-growing-6-billion-market.html, accessed 2012 January 4.

Clark, SE., Hawkes, C, Murphy, SME, Hansen-Kuhn, KA, and Wallinga, D. 2012. Exporting obesity: US farm and trade policy and the transformation of the Mexican consumer food environment. International Journal of Occupational and Environmental Health 18(1):53–64.

Cleveland, DA. 1980. The population dynamics of subsistence agriculture in the West African Savanna: A village in northeast Ghana. Ph.D thesis. University of Arizona, Tucson.

————. 1986a. Culture and horticulture in Mexico. Culture and Agriculture 7(29):1–5.

————. 1986b. The political economy of fertility regulation: The Kusasi of savanna West Africa (Ghana). Pages 263–293 in Handwerker, WP, ed. Culture and Reproduction: An Anthropological Critique of Demographic Transition Theory. Boulder, CO: Westview Press.

————. 1988. Social Feasibility of the Swabi SCARP. Report submitted to the Harza Engineering Company and the Asian Development Bank, for the Swabi SCARP Feasibility Study, Peshawar, Pakistan. Unpublished report.

————. 1990. Development alternatives and the African food crisis. Pages 181–206 in African Food Systems in Crisis. Part Two: Contending with Change. New York: Gordon and Breach.

————. 1991. Migration in West Africa: A savanna village perspective. Africa [London] 61:222–246.

————. 2001. Is plant breeding science objective truth or social construction? The case of yield stability. Agriculture and Human Values 18(3):251–270.

————. 2006. What kind of social science does the CGIAR, and the world, need? Culture and Agriculture 28(1):4–9.

Cleveland, DA, Bowannie, FJ, Eriacho, D, Laahty, A, and Perramond, EP. 1995. Zuni farming and United Stated government policy: The politics of cultural and biological diversity. Agriculture and Human Values 12:2–18.

Cleveland, DA, Müller, NM, Tranovich, A, Mazaroli, DN, and Hinson, K. n.d. Local food hubs: Case study of a success in Santa Barbara County. Ms. in preparation.

Cleveland, DA, and Murray, SC. 1997. The world's crop genetic resources and the rights of indigenous farmers. Current Anthropology 38:477–515.

Cleveland, DA, Radka, CN, and Müller, NM. 2011a. Localization: Necessary but not sufficient. Huffington Post, May 31. www.huffingtonpost.com/david-a-cleveland/local-food-climate-change-_b_869491.html, accessed 2011 June 2.

Cleveland, DA, Radka, CN, Müller, NM, Watson, TD, Rekstein, NJ, Wright, HvM, and Hollingshead, SE. 2011b. The effect of localizing fruit and vegetable consumption on greenhouse gas emissions and nutrition, Santa Barbara County. Environmental Science and Technology 45:4555–4562.

Cleveland, DA, and Soleri, D, eds. 2002a. Farmers, Scientists and Plant Breeding: Integrating Knowledge and Practice. Oxon, UK: CABI.

———. 2002b. Indigenous and scientific knowledge of plant breeding: Similarities, differences, and implications for collaboration. Pages 206–234 in Sillitoe, P, Bicker, AJ, and Pottier, J, eds. "Participating in Development": Approaches to Indigenous Knowledge. London: Routledge.

———. 2005. Rethinking the risk management process for genetically engineered crop varieties in small-scale, traditionally based agriculture. Ecology and Society 9(1):Article 9.

———. 2007a. Extending Darwin's analogy: Bridging differences in concepts of selection between farmers, biologists, and plant breeders. Economic Botany 61:121–136.

———. 2007b. Farmer knowledge and scientist knowledge in sustainable agricultural development. Pages 211–229 in Sillitoe, P, ed. Local Science versus Global Science: Approaches to Indigenous Knowledge in International Development. Oxford, UK: Berghahn Books.

Cleveland, DA, Soleri, D, Aragón Cuevas, F, Crossa, Jand Gepts, P. 2005. Detecting (trans)gene flow to landraces in centers of crop origin: Lessons from the case of maize in Mexico. Environmental Biosafety Research 4:197–208.

Cleveland, DA, Soleri, D, and Smith, SE. 1994. Do folk crop varieties have a role in sustainable agriculture? BioScience 44:740–751.

———. 1999. Farmer Plant Breeding from a Biological Perspective: Implications for Collaborative Plant Breeding. Economics Working Paper. Mexico, D.F.: CIMMYT.

———. 2000. A biological framework for understanding farmers' plant breeding. Economic Botany 54:377–394.

Cohen, JE. 1995. How Many People Can the Earth Support? New York: W. W. Norton and Company.

———. 2005. Human population grows up. Scientific American 293(3):48–55.

Coleman-Jensen, A, Nord, M, Andrews, M, and Carlson, S. 2011. Household Food Security in the United States in 2010. Economic Research Report Number 125. www.ers.usda.gov/Publications/ERR125/ERR125.pdf, accessed 2013 August 7.

Coley, D, Howard, M, and Winter, M. 2009. Local food, food miles and carbon emissions: A comparison of farm shop and mass distribution approaches. Food Policy 34(2):150–155.

———. 2011. Food miles: Time for a re-think? British Food Journal 113(6–7):919–934.

Collier, P. 2008. The politics of hunger: How illusion and greed fan the food crisis. Foreign Affairs 87(6):67–79.

Comadran, J, et al. 2008. Mapping adaptation of barley to droughted environments. Euphytica 161:35–45.

Commoner, B. 2002. Unraveling the DNA myth: The spurious foundation of genetic engineering. Harper's Magazine, February, 39–47.

Confino, J. 2012. Rio+20: Tim Jackson on how fear led world leaders to betray green economy. Guardian, June 25. www.guardian.co.uk/sustainable-business/rio-20-tim-jackson-leaders-green-economy, accessed 2013 January 8.

Connolly, AJ, Connolly, KP, and Lyons, M. 2012. A seismic change: Land control in Africa. Is this a wake-up call for agribusiness? International Food and Agribusiness Management Review 15(2):171–177.

Connor, DJ, Loomis, RS, and Cassman, KG. 2011 [1992]. Crop Ecology: Productivity and Management in Agricultural Systems. 2nd ed. Cambridge, UK: Cambridge University Press.

Cornucopia Institute. 2012. The Organic Watergate. http://www.cornucopia.org/USDA/Organic-WatergateWhitePaper.pdf, accessed 2013 August 7.

Cooper, HD, Spillane, C, and Hodgkin, T. 2001. Broadening the genetic base of crops: An overview. Pages 1–23 in Cooper, HD, Spillane, C, and Hodgkin, T, eds. Broadening the Genetic Base of Crop Production. Wallingford, Oxon, UK: CABI.

Cooper, M, and Hammer, GL. 1996. Plant Adaptation and Crop Improvement. Wallingford, Oxford, UK: CABI in association with IRRI and ICRISAT.

COP CBD (Conference of the Parties to the Convention on Biological Diversity). 2000. Cartagena Protocol on Biosafety. www.biodiv.org/biosafety, accessed 2013 January 9.

Corenblit, D, Baas, ACW, Bornette, G, Darrozes, J, Delmotte, S, Francis, RA, Gurnell, AM, Julien, F, Naiman, RJ, and Steiger, J. 2011. Feedbacks between geomorphology and biota controlling Earth surface processes and landforms: A review of foundation concepts and current understandings. Earth-Science Reviews 106(3–4):307–331.

Cornes, R, and Sandler, T. 1996. The Theory of Externalities, Public Goods, and Club Goods. 2nd ed. Cambridge: Cambridge University Press.

Costanza, R. 2001. Visions, values, valuation, and the need for ecological economics. BioScience 51:459–468.

Cotula, L, Vermeulen, S, Leonard, R, and Keeley, J. 2009. Land Grab or Development Opportunity? Agricultural Investment and International Land Deals in Africa. London: IIED /FAO/IFAD. ftp://ftp.fao.org/docrep/fao/011/ak241e/ak241e.pdf, accessed 2013 May 9.

Cotula, L, Vermeulen, S, Mathieu, P, and Toulmin, C. 2011. Agricultural investment and international land deals: Evidence from a multi-country study in Africa. Food Security 3:S99-S113.

Craig, WJ, and Mangels, AR. 2009. Position of the American Dietetic Association: Vegetarian diets. Journal of the American Dietetic Association 109(7):1266–1282.

Cronin, MA, Gonzalez, C, and Sterman, JD. 2009. Why don't well-educated adults understand accumulation? A challenge to researchers, educators, and citizens. Organizational Behavior and Human Decision Processes 108(1):116–130.

Cuellar, AD, and Webber, ME. 2010. Wasted food, wasted energy: The embedded energy in food waste in the United States. Environmental Science and Technology 44(16):6464–6469.

Cui, S, Shi, Y, Groffman, PM, Schlesinger, WH, and Zhu, Y-G. 2013. Centennial-scale analysis of the creation and fate of reactive nitrogen in China. Proceedings of the National Academy of Sciences 110(6):2052–2057.

Daily, GC, and Ehrlich, PR. 1992. Population, sustainability, and Earth's carrying capacity. BioScience 42:761–770.

Daly, H, and Cobb, J. 1989. For the Common Good. Boston: Beacon Press.

Daly, HE, and Farley, J. 2004. Ecological Economics: Principles and Applications. Washington, DC: Island Press.

Darwin, C. 1859. On the Origin of Species by Means of Natural Selection. London: John Murray. First facsimile edition 1967 by Atheneum.

Dasgupta, P. 2008. Discounting climate change. Journal of Risk and Uncertainty 37(2–3):141–169.

Davies, MIJ. 2009. Wittfogel's dilemma: Heterarchy and ethnographic approaches to irrigation management in Eastern Africa and Mesopotamia. World Archaeology 41(1):16–35.

Dawes, CT, Fowler, JH, Johnson, T, McElreath, R, and Smirnov, O. 2007. Egalitarian motives in humans. Nature 446:794–796.

de Boer, J, and Aiking, H. 2011. On the merits of plant-based proteins for global food security: Marrying macro and micro perspectives. Ecological Economics 70(7):1259–1265.

de Kok, JL, Kofalk, S, Berlekamp, J, Hahn, B, and Wind, H. 2009. From design to application of a decision-support system for integrated river-basin management. Water Resources Management 23(9):1781–1811.

De Schutter, O. 2009. Seed Policies and the Right to Food: Enhancing Agrobiodiversity and Encouraging Innovation. Report of the Special Rapporteur on the right to food. United Nations General Assembly. www.srfood.org/images/stories/pdf/officialreports/20091021_report-ga64_seed-policies-and-the-right-to-food_en.pdf, accessed 2010 December 29.

De Schutter, O. 2011a. The green rush: The global race for farmland and the rights of land users. Harvard International Law Journal 52(2):503–559.

———. 2011b. How not to think of land-grabbing: Three critiques of large-scale investments in farmland. Journal of Peasant Studies 38(2):249–279.

———. 2011c. Right to food. Université de Louvain, Louvain-la-Neuve, Belgium. www.srfood.org/index.php/en/right-to-food, accessed 2011 November 24.

Dean, RM, Valente, MJo, and Carvalho, AnF. 2012. The Mesolithic/Neolithic transition on the Costa Vicentina, Portugal. Quaternary International 264(0):100–108.

DeLacy, IH, Basford, KE, Cooper, M, Bull, JK, and McLaren, CG. 1996. Analysis of multi-environment trials: An historical perspective. Pages 39–124 in Cooper, M and Hammer, GL, eds. Plant Adaptation and Crop Improvement. Wallingford, Oxford, UK: CABI in association with IRRI and ICRISAT.

DeLind, LB. 2010. Are local food and the local food movement taking us where we want to go? Or are we hitching our wagons to the wrong stars? Agriculture and Human Values 28(2):273–283.

DeLind, LB, and Bingen, J. 2008. Place and civic culture: Re-thinking the context for local agriculture. Journal of Agricultural and Environmental Ethics 21(2):127–151.

DeLind, LB, and Howard, PH. 2008. Safe at any scale? Food scares, food regulation, and scaled alternatives. Agriculture and Human Values 25(3):301–317.

Dembélé, NN, and Staatz, JM. 2000. The response of cereals traders to agricultural market reform. Pages 145–165 in Bingen, RJ, Robinson, D, and Staatz, JM, eds. Democracy and Development in Mali. East Lansing: Michigan State University Press.

Desmond, A. 1997. Huxley: From Devil's Disciple to Evolution's High Priest. New York: Basic Books.

Desmond, A, and Moore, J. 2009. Darwin's Sacred Cause: How a Hatred of Slavery Shaped Darwin's Views on Human Evolution. New York: Houghton Mifflin Harcourt.

Desrochers, P, and Shimizu, H. 2012. The Locavore's Dilemma: In Praise of the 10,000-mile Diet. New York: PublicAffairs.

Diamond, J. 2005. Collapse: How Societies Choose to Fail or Succeed. New York: Viking.

Dietz, T, Dolšak, N, Ostrom, E, and Stern, PC. 2002. The drama of the commons. Pages 3–35 in Ostrom, E, Dietz, T, Dolšak, N, Stern, PC, Stonich, S, and Weber, EU, eds. The Drama of the Commons. Washington, DC: National Academy Press.

Dietz, T, Frey, RS, and Rosa, EA. 2002. Risk, technology, and society. Pages 329–369 in Dunlap, RE and Michelson, W, eds. Handbook of Environmental Sociology. Westport, CT: Greenwood Press.

Ding, JH, Lu, Q, Ouyang, YD, Mao, HL, Zhang, PB, Yao, JL, Xu, CG, Li, XH, Xiao, JH, and Zhang, QF. 2012. A long noncoding RNA regulates photoperiod-sensitive male sterility, an essential component of hybrid rice. Proceedings of the National Academy of Sciences of the United States of America 109(7):2654–2659.

Dove, MR. 1996a. Center, periphery, and biodiversity: A paradox of governance and a developmental challenge. Pages 41--67 in Brush, SB and Stabinsky, D, eds. Valuing Local Knowledge. Washington, DC: Island Press.

———. 1996b. Process versus product in Bornean augury: A traditional knowledge system's solution to the problem of knowing. Pages 557–596 in Ellen, R and Fukui, K, eds. Redefining Nature: Ecology, Culture and Domestication. Oxford, UK: Berg.

———. 2006. Indigenous people and environmental politics. Annual Review of Anthropology 35:191–208.

Dowd, BM, Press, D, and Huertos, ML. 2008. Agricultural nonpoint source water pollution policy: The case of California's Central Coast. Agriculture, Ecosystems and Environment 128(3):151–161.

Drinkwater, LE, and Snapp, SS. 2007. Nutrients in agroecosystems: Rethinking the management paradigm. Advances in Agronomy 92:163–186.

Duarte, CM. 2009. Coastal eutrophication research: A new awareness. Hydrobiologia 629(1):263–269.

Dunn, R. 2012. Human ancestors were nearly all vegetarians. Scientific American, July 23. http://blogs.scientificamerican.com/guest-blog/2012/07/23/human-ancestors-were-nearly-all-vegetarians, accessed 2013 June 30.

Dunne, JB, Chambers, KJ, Giombolini, KJ, and Schlegel, SA. 2011. What does "local" mean in the grocery store? Multiplicity in food retailers' perspectives on sourcing and marketing local foods. Renewable Agriculture and Food Systems 26(1):46–59.

DuPuis, EM, and Gillon, S. 2009. Alternative modes of governance: Organic as civic engagement. Agriculture and Human Values 26:43–56.

Duquette, E, Higgins, N, and Horowitz, J. 2012. Farmer discount rates: Experimental evidence. American Journal of Agricultural Economics 94(2):451–456.

Durack, PJ, Wijffels, SE, and Matear, RJ. 2012. Ocean salinities reveal strong global water cycle intensification during 1950 to 2000. Science 336(6080):455–458.

Duvick, DN. 2002. Theory, empiricism and intuition in professional plant breeding. Pages 239–267 in Cleveland, DA and Soleri, D, eds. Farmers, Scientists and Plant Breeding: Integrating Knowledge and Practice. Oxon, UK: CABI.

Dyson, T. 1996. Population and Food: Global Trends and Future Prospects. London: Routledge.

Eberhart, SA, and Russell, WA. 1966. Stability parameters for comparing varieties. Crop Science 6:36–40.

Edmeades, GO, Bolaños, J, Chapman, SC, Lafitte, HR, and Bänziger, M. 1999. Selection improves drought tolerance in tropical maize populations: I. gains in biomass, grain yield, and harvest index. Crop Science 39:1306–1315.

Edwards-Jones, G, et al. 2008. Testing the assertion that "local food is best": The challenges of an evidence-based approach. Trends in Food Science and Technology 19(5):265–274.

Egger, G, and Dixon, J. 2009. Should obesity be the main game? Or do we need an environmental makeover to combat the inflammatory and chronic disease epidemics? Obesity Reviews 10(2):237–249.

Ehrlich, PR. 1968. The Population Bomb. New York: Ballantine Books.

Ekman, P. 2003. Darwin, deception, and facial expression. Annals of the New York Academy of Sciences 1000:205–221.

Ellen, R. 1999. Models of subsistence and ethnobiological knowledge: Between extraction and cultivation in Southeast Asia. Pages 91–117 in Medin, DL and Atran, S, eds. Folkbiology. Cambridge, MA: MIT Press.

Ellen, R, and Harris, H. 2000. Introduction. Pages 1–33 in Ellen, R, Parkes, P, and Bicker, A, eds. Indigenous Environmental Knowledge and Its Transformations: Critical Anthropological Perspectives. Amsterdam: Harwood Academic Publishers.

Ellis, F. 1993. Peasant Economics: Farm Households and Agrarian Development. 2nd ed. Cambridge, UK: Cambridge University Press.

Ellstrand, NC. 2003. Dangerous Liaisons? When Cultivated Plants Mate with Their Wild Relatives. Baltimore, MD: Johns Hopkins University Press.

Elton, CS. 1958. The Ecology of Invasions by Animals and Plants. London: Chapman and Hall.

Endler, JA. 1986. Natural Selection in the Wild. Princeton, NJ: Princeton University Press.

EPA (Environmental Protection Agency). 2008a. Cars and Light Trucks. Light-duty Automotive Technology, Carbon Dioxide Emissions, and Fuel Economy Trends: 1975 through 2009. www.epa.gov/otaq/fetrends.htm—report, accessed 2010 March 15.

———. 2008b. Compliance and Enforcement Annual Results FY2008: Important Environmental Problems/National Priorities: Concentrated Animal Feeding Operations (CAFOs). www.epa.gov/compliance/resources/reports/endofyear/eoy2008/2008-sp-nat-cafo.html—problem, accessed 2012 May 15.

———. 2010. 2010 U.S. Greenhouse Gas Inventory Report. www.epa.gov/climatechange/emissions/usinventoryreport.html, accessed 2010 April 25.

———. 2011. Food waste. www.epa.gov/waste/conserve/foodwaste, accessed 2013 August 7.

———. 2012a. 2012 U.S. Greenhouse Gas Inventory Report. www.epa.gov/climatechange/emissions/downloads12/US-GHG-Inventory-2012-Main-Text.pdf, accessed 2012 April 28.

———. 2012b. Organics: Yard Trimmings and Food Scraps. In WARM Version 12. www.epa.gov/climatechange/wycd/waste/downloads/Organics.pdf, accessed 2013 February 20.

———. 2012c. Solid Waste Management and Greenhouse Gases: Documentation for Greenhouse Gas Emission and Energy Factors Used in the Waste Reduction Model (WARM). www.epa.gov/climatechange/wycd/waste/SWMGHGreport.html, accessed 2013 February 20.

ESA (Ecological Society of America). 2009. Ecological Impacts of Economic Activities. www.esa.org/pao/economic_activities.php, accessed 2009 September 2.

Escobar, A. 1999. After nature: Steps to an antiessentialist political ecology. Current Anthropology 40:1–30.

Eshel, G. 2010. A geophysical foundation for alternative farm policy. Environmental Science and Technology 44(10):3651–3655.

Eshel, G, and Martin, PA. 2006. Diet, energy, and global warming. Earth Interactions 10:Paper no. 9. journals.ametsoc.org/doi/pdf/10.1175/EI167.1, accessed 2017 January 27.

———. 2009. Geophysics and nutritional science: Toward a novel, unified paradigm. American Journal of Clinical Nutrition 89(5):S1710–S1716.

Esquinas-Alcazar, J. 2005. Protecting crop genetic diversity for food security: Political, ethical and technical challenges. Nature Review Genetics 6(12):946–953.

Esselstyn, CB. 1999. Updating a 12-year experience with arrest and reversal therapy for coronary heart disease (An overdue requiem for palliative cardiology). American Journal of Cardiology 84(3):339–341.

———. 2001. Resolving the coronary artery disease epidemic through plant-based nutrition. Preventive Cardiology 4:171–177.

———. 2010. Is the present therapy for coronary artery disease the radical mastectomy of the twenty-first century? American Journal of Cardiology 5:902–904.

Essex, J. 2008. Biotechnology, sound science, and the Foreign Agricultural Service: A case study in neoliberal rollout. Environment and Planning C-Government and Policy 26(1):191–209.

ETC Group (Action group on Erosion, Technology and Concentration). 2008. Patenting the "Climate Genes" . . . and Capturing the Climate Agenda. www.etcgroup.org/upload /publication/pdf_file/687, accessed 2009 March 30.

Evans, LT. 1993. Crop Evolution, Adaptation and Yield. Cambridge, UK: Cambridge University Press.

———. 1998. Feeding the Ten Billion: Plants and Population Growth. Cambridge, UK: Cambridge University Press.

Eyre-Walker, A, Gaut, RL, Hilton, H, Feldman, DL, and Gaut, BS. 1998. Investigation of the bottleneck leading to the domestication of maize. Proceedings of the National Academy of Science 95:1441–1446.

Faisal, AA, Selen, LPJ, and Wolpert, DM. 2008. Noise in the nervous system. Nature Reviews Neuroscience 9(4):292–303.

Falconer, DS, and Mackay, TF. 1996. Introduction to Quantitative Genetics. 4th ed. Edinburgh: Prentice Hall/Pearson Education.

FAO (UN Food and Agriculture Organization). 1996. Report on the State of the World's Plant Genetic Resources, International Technical Conference on Plant Genetic Resources. Rome: FAO.

———. 2004. Agricultural Biotechnology: Meeting the Needs of the Poor? The State of Food and Agriculture. Rome: FAO.

———. 2010. The Second Report on the State of the World's Plant Genetic Resources for Food and Agriculture. Rome: Commission on Genetic Resources for Food and Agriculture, Food and Agriculture Organization of the United Nations. www.fao.org/docrep/013 /i1500e/i1500e00.htm, accessed 2013 March 6.

———. 2011a. FAO Food Price Index. Rome, Italy: FAO. www.fao.org/worldfoodsituation /wfs-home/foodpricesindex, accessed 2011 June 20.

———. 2011b. The State of Food Insecurity in the World: How Does International Price Volatility Affect Domestic Economies and Food Security? Rome: FAO. www.fao.org/publications /sofi, accessed 2012 March 8.

———. 2012a. AQUASTAT database. Rome: FAO. www.fao.org/nr/water/aquastat/data/query, accessed 2013 August 7.

———. 2012b. FAO Food Price Index. Rome: FAO. www.fao.org/worldfoodsituation/wfs-home /foodpricesindex, accessed 2011 June 20.

———. 2012c. Food Wastage Footprint: An Environmental Accounting of Food Loss and Waste. Concept Note. March 2012. Rome: FAO. http://www.fao.org/fileadmin/templates/nr /sustainability_pathways/docs/Food_Wastage_Concept_Note_web.pdf, accessed 2013 August 7.

———. 2012d. The State of Food Insecurity in the World, 2012. Rome: Food and Agriculture Organization of the United Nations. www.fao.org/publications/sofi, accessed 2013 August 7.

———. 2013a. FAO Food Price Index. Rome: FAO. www.fao.org/worldfoodsituation /foodpricesindex, accessed 2013 June 20.

———. 2013b. FAOSTAT. Rome, Italy: FAO, faostat.fao.org, accessed 2013 January 1.

———. 2013c. Hunger Portal. Rome: FAO. www.fao.org/hunger, accessed 2013 June 15.

FBSBC (Foodbank of Santa Barbara County). 2010. www.foodbanksbc.org, accessed 2010 December 20.

Feagan, R. 2007. The place of food: Mapping out the "local" in local food systems. Progress in Human Geography 31(1):23–42.

Fedonkin, MA. 2009. Eukaryotization of the early biosphere: A biogeochemical aspect. Geochemistry International 47(13):1265–1333.

Fedoroff, NV. 2003. Prehistoric GM corn. Science 302:1158–1159.

Fedoroff, NV, et al. 2010. Radically rethinking agriculture for the 21st century. Science 327(5967):833–834.

Fernando, N, Panozzo, J, Tausz, M, Norton, R, Fitzgerald, G, and Seneweera, S. 2012. Rising atmospheric CO_2 concentration affects mineral nutrient and protein concentration of wheat grain. Food Chemistry 133(4):1307–1311.

Fess, TL, Kotcon, JB, and Benedito, VA. 2011. Crop breeding for low input agriculture: A sustainable response to feed a growing world population. Sustainability 3:1742–1772.

Feynman, J, and Ruzmaikin, A. 2007. Climate stability and the development of agricultural societies. Climatic Change 84(3–4):295–311.

Field, S, Masakure, O, and Henson, S. 2010. Rethinking localization—A low-income country perspective: The case of Asian vegetables in Ghana. Cambridge Journal of Regions Economy and Society 3(2):261–277.

Finlay, KW, and Wilkinson, GN. 1963. The analysis of adaptation in a plant-breeding programme. Australian Journal of Agricultural Research 14:742–754.

Firestone, D. 2012. The "redistribution" of wealth. New York Times, September 19. http://takingnote.blogs.nytimes.com/2012/09/19/the-redistribution-of-wealth, accessed 2013 August 5.

Fischer, J, Dyball, R, Fazey, I, Gross, C, Dovers, S, Ehrlich, PR, Brulle, RJ, Christensen, C, and Borden, RJ. 2012. Human behavior and sustainability. Frontiers in Ecology and the Environment 10(3):153–160.

Fisher, A. 2011. Growing Power Takes Massive Contribution from Wal-Mart: A Perspective on Money and the Movement. CivilEats. http://civileats.com/2011/09/16/growing-power-takes-massive-contribution-from-wal-mart-a-perspective-on-money-and-the-movement, accessed 2011 December 2.

Fisher, B, Turner, RK, and Morling, P. 2009. Defining and classifying ecosystem services for decision making. Ecological Economics 68(3):643–653.

Fitting, E. 2011. The Struggle for Maize: Campesinos, Workers, and Transgenic Corn in the Mexican Countryside. Durham, NC: Duke University Press.

Fitzgerald, D. 1990. The Business of Breeding: Hybrid Corn in Illinois, 1890–1940. Ithaca, NY: Cornell University Press.

FitzRoy, F, Franz-Vasdeki, J, and Papyrakis, E. 2012. Climate change policy and subjective well-being. Environmental Policy and Governance 22(3):205–216.

Flannery, KV, and Sabloff, JA, eds. 2009. The Early Mesoamerican Village. Updated edition. Walnut Creek, CA: Left Coast Press.

Fliessbach, A, Oberholzer, HR, Gunst, L, and Mader, P. 2007. Soil organic matter and biological soil quality indicators after 21 years of organic and conventional farming. Agriculture Ecosystems and Environment 118(1–4):273–284.

Foer, JS. 2009. Eating Animals. New York: Little Brown.

Foley, JA, et al. 2011. Solutions for a cultivated planet. Nature 478(7369):337–342.

Food and Water Watch. 2012. Why Walmart Can't Fix the Food System. Washington, DC: Food and Water Watch. http://documents.foodandwaterwatch.org/doc/FoodandWaterWatchReportWalmart022112.pdf, accessed 2013 July 20.

Food Circle (Food Circles Networking Project). 2012. A Vision of a Localized Food System. University of Missouri Extension. http://foodcircles.missouri.edu/vision.htm, accessed 2012 May 19.

Fortes, M. 1949. The Web of Kinship Among the Tallensi. London: Oxford University Press for the International African Institute.

Fox, J, and Haight, L. 2010. Subsidizing Inequality: Mexican Corn Policy since NAFTA. Woodrow Wilson International Center for Scholars and University of California, Santa Cruz. http://www.wilsoncenter.org/publication/subsidizing-inequality-mexican-corn-policy-nafta-0, accessed 2013 August 1.

FRAC (Food Research and Action Center). 2010. Working Document: Hunger and Obesity? Making the Connections. http://frac.org/pdf/proceedings05.pdf, accessed 2010 May 31.

Francis, CA, and Callaway, MB. 1993. Crop improvement for future farming systems. Pages 1–18 in Callaway, MB and Francis, CA, eds. Crop Improvement for Sustainable Agriculture. Lincoln: University of Nebraska Press.

Franck, S, von Bloh, W, Muller, C, Bondeau, A, and Sakschewski, B. 2011. Harvesting the sun: New estimations of the maximum population of planet Earth. Ecological Modelling 222(12):2019–2026.

Frankel, OH, Brown, AHD, and Burdon, JJ. 1995. The Conservation of Plant Biodiversity. Cambridge, UK: Cambridge University Press.

Friedmann, H. 2009. Feeding the empire: The pathologies of globalized agriculture. Socialist Register 41.

Fuglie, KO, Heisey, PW, King, JL, Pray, CE, Day-Rubenstein, K, Schimmelpfennig, D, Wang, SL, and Karmarkar-Deshmuk, R. 2011. Research Investments and Market Structure in the Food Processing, Agricultural Input, and Biofuel Industries Worldwide. Report no. ERR-130. Washington, DC: U.S. Dept. of Agriculture, Economic Research Service.

Fujiyoshi, PT, Gliessman, SR, and Langenheim, JH. 2007. Factors in the suppression of weeds by squash interplanted in corn. Weed Biology and Management 7(2):105–114.

Fukuoka, M. 1978. The One-Straw Revolution: An Introduction to Natural Farming. Translated by Chris Pierce, Tsune Kurosawa, and Larry Korn. Emmaus, PA: Rodale Press. First published 1975 in Japanese as Shizen Nōhō Wara Ippon No Kakumei by Hakujusha Co., Ltd.

Fuller, D, Asouti, E, and Purugganan, M. 2012. Cultivation as slow evolutionary entanglement: Comparative data on rate and sequence of domestication. Vegetation History and Archaeobotany 21(2):131–145.

G8. 2008a. Chair's Summary, Hokkaido Toyako Summit, G8 Information Center, University of Toronto. www.g8.utoronto.ca/summit/2008hokkaido/2008-summary.html, accessed 2009 December 12.

———. 2008b. G8 Hokkaido Toyako Summit: Chair's Summary of the G8 Development Ministers' Meeting. G8 Information Center, University of Toronto. www.mofa.go.jp/policy/economy/summit/2008/other/g8_develop_gs.html, accessed 2010 September 10.

———. 2010. Muskoka Accountability Report. G8 Information Center. http://g8.gc.ca/g8-summit/accountability/University of Toronto, accessed 2010 September 10.

———. 2011. Deauville Accountability Report. G8 Commitments on Health and Food Security: State of Delivery and Results. Information Center, University of Toronto. www.g8.utoronto.ca/summit/2011deauville/deauville-2011-deauville-accountability-report.pdf, accessed 2011 June 2.

G8 Summit. 2008. G8 Leaders Statement on Global Food Security, Hokkaido Toyako Summit, July 8, 2008. Information Center, University of Toronto. www.g8.utoronto.ca/summit/2008hokkaido/2008-food.html, accessed 2008 August 28.

Gadema, Z, and Oglethorpe, D. 2011. The use and usefulness of carbon labelling food: A policy perspective from a survey of UK supermarket shoppers. Food Policy 36(6):815–822.

Galloway, JN, Aber, JD, Erisman, JW, Seitzinger, SP, Howarth, RW, Cowling, EB, and Cosby, BJ. 2003. The nitrogen cascade. BioScience 53(4):341–356.

Galloway, JN, Townsend, AR, Erisman, JW, Bekunda, M, Cai, ZC, Freney, JR, Martinelli, LA, Seitzinger, SP, and Sutton, MA. 2008. Transformation of the nitrogen cycle: Recent trends, questions, and potential solutions. Science 320(5878):889–892.

Galluzzi, G, Eyzaguirre, P, and Negri, V. 2010. Home gardens: Neglected hotspots of agro-biodiversity and cultural diversity. Biodiversity and Conservation 19(13):3635–3654.

Gamson, WA. 1968. Power and Discontent. Homewood, IL: Dorsey Press.

Garrigan, D, and Hammer, MF. 2006. Reconstructing human origins in the genomic era. Nature Reviews Genetics 7(9):669–680.

Gates Foundation. 2011. Why the Foundation Funds Research in Crop Biotechnology. Bill and Melinda Gates Foundation. www.gatesfoundation.org/agriculturaldevelopment/Pages /why-we-fund-research-in-crop-biotechnology.aspx, accessed 2011 August 18.

Gaut, BS, Wright, SI, Rizzon, C, Dvorak, J, and Anderson, LK. 2007. Opinion—Recombination: An underappreciated factor in the evolution of plant genomes. Nature Reviews Genetics 8(1):77–84.

GCGH (Grand Challenges in Global Health). 2010. Challenge 9: Create a Full Range of Optimal, Bioavailable Nutrients in a Single Staple Plant Species. www.grandchallenges.org /ImproveNutrition/Challenges/NutrientRichPlants, accessed 2011 January 2.

Gearhardt, AN, Grilo, CM, DiLeone, RJ, Brownell, KD, and Potenza, MN. 2011. Can food be addictive? Public health and policy implications. Addiction 106(7):1208–1212.

Gearhardt, AN, Yokum, S, Orr, PT, Stice, E, Corbin, WR, and Brownell, KD. 2011. Neural correlates of food addiction. Archives of General Psychiatry 68(8):808–816.

Geisler, WS. 2008. Visual perception and the statistical properties of natural scenes. Annual Review of Psychology 59:167–192.

Gepts, P. 2002. A comparison between crop domestication, classical plant breeding, and genetic engineering. Crop Science 42:1780–1790.

———. 2004. Crop domestication as a long-term selection experiment. Plant Breeding Reviews 24 (Part 2):1–44.

Giampietro, M, Cerretelli, G, and Pimentel, D. 1992. Energy analysis of agricultural ecosystem management: Human return and sustainability. Agriculture, Ecosystems and Environment 38:219–244.

Gibson, RW, Byamukama, E, Mpembe, I, Kayongo, J, and Mwanga, ROM. 2008. Working with farmer groups in Uganda to develop new sweet potato cultivars: Decentralisation and building on traditional approaches. Euphytica 159(1–2):217–228.

Gibson, RW, Lyimo, NG, Temu, AEM, Stathers, TE, Page, WW, Nsemwa, LTH, Acola, G, and Lamboll, RI. 2005. Maize seed selection by East African smallholder farmers and resistance to Maize streak virus. Annals of Applied Biology 147(2):153–159.

Giere, RN. 1999. Science without Laws. Chicago: University of Chicago Press.

Gigerenzer, G, and Todd, PM. 1999. Fast and frugal heuristics: The adaptive toolbox. Pages 3–34 in Gigerenzer, G, Todd, PM, and Group, AR, eds. Simple Heuristics That Make Us Smart. Oxford, UK: Oxford University Press.

Gignoux, CR, Henn, BM, and Mountain, JL. 2011. Rapid, global demographic expansions after the origins of agriculture. Proceedings of the National Academy of Sciences of the United States of America 108(15):6044–6049.

Gilinsky, NL, Gould, SJ, and German, RZ. 1989. Asymmetries of clade shape and the direction of evolutionary time. Science 243(4898):1613–1614.

Gill, M, Smith, P, and Wilkinson, JM. 2010. Mitigating climate change: The role of domestic livestock. Animal 4(3):323–333.

Gilland, B. 2006. Population, nutrition and agriculture. Population and Environment 28(1):1–16.

Gleick, PH, and Palaniappan, M. 2010. Peak water limits to freshwater withdrawal and use. Proceedings of the National Academy of Sciences 107(25):11155–11162.

Gliessman, S. 2013. Agroecology: Growing the roots of resistance. Agroecology and Sustainable Food Systems 37(1):19–31.

Glover, D. 2007. Monsanto and smallholder farmers: A case study in CSR. Third World Quarterly 28(4):851–867.

———. 2010a. The corporate shaping of GM crops as a technology for the poor. Journal of Peasant Studies 37(1):67–90.

———. 2010b. Exploring the resilience of Bt cotton's "pro-poor success story." Development and Change 41(6):955–981.

———. 2010c. Is Bt cotton a pro-poor technology? A review and critique of the empirical record. Journal of Agrarian Change 10(4):482–509.

Goetz, SJ, and Swaminathan, H. 2006. Wal-Mart and county-wide poverty. Social Science Quarterly 87(2):211–226.

Golden Rice Project. 2009. Golden Rice Is Part of the Solution. www.goldenrice.org, accessed 2009 February 23.

Gomiero, T, Pimentel, D, and Paoletti, MG. 2011. Environmental impact of different agricultural management practices: Conventional vs. organic agriculture. Critical Reviews in Plant Sciences 30(1–2):95–124.

González, A. 2005. Territory, autonomy and defending maize. Seedling (January):14–17.

Gonzalez, AD, Frostell, B, and Carlsson-Kanyama, A. 2011. Protein efficiency per unit energy and per unit greenhouse gas emissions: Potential contribution of diet choices to climate change mitigation. Food Policy 36(5):562–570.

Good, AG, and Beatty, PH. 2011. Fertilizing nature: A tragedy of excess in the commons. Plos Biology 9(8).

Goodland, R, and Anhang, J. 2009. Livestock and climate change. WorldWatch Nov–Dec:10–19.

Goodland, R, and Anhang, J. 2012. Comments to the editor on Livestock and greenhouse gas emissions: The importance of getting the numbers right, by Herrero et al. Animal Feed Science and Technology 172(3–4):252–256.

Goodman, D. 1975. The theory of diversity-stability relationships in ecology. The Quarterly Review of Biology 50:237–266.

Goody, J. 1980. Rice-burning and the Green-Revolution in northern Ghana. Journal of Development Studies 16(2):136–155.

Gottfried, R, Wear, D, and Lee, R. 1996. Institutional solutions to market failure on the landscape scale. Ecological Economics 18(2):133–140.

Gould, SJ. 2002. The Structure of Evolutionary Theory. Cambridge, MA: Harvard University Press.

Gould, SJ, Gilinsky, NL, and German, RZ. 1987. Asymmetry of lineages and the direction of evolutionary time. Science 236(4807):1437–1441.

Gouse, M, Pray, C, Schimmelpfennig, D, and Kirsten, J. 2006. Three seasons of subsistence insect-resistant maize in South Africa: Have smallholders benefited? Agbioforum 9:15–22.

Govaerts, B, Verhulst, N, Castellanos-Navarrete, A, Sayre, KD, Dixon, J, and Dendooven, L. 2009. Conservation agriculture and soil carbon sequestration: Between myth and farmer reality. Critical Reviews in Plant Sciences 28(3):97–122.

Gowdy, J. 2005. Toward a new welfare economics for sustainability. Ecological Economics 53:211–222.

———. 2007. Avoiding self-organized extinction: Toward a co-evolutionary economics of sustainability. International Journal of Sustainable Development and World Ecology 14(1):27–36.

———. 2008. Behavioral economics and climate change policy. Journal of Economic Behavior and Organization 68(3–4):632–644.

Gowdy, J, and Erickson, JD. 2005. The approach of ecological economics. Cambridge Journal of Economics 29(2):207–222.

Gowdy, J, Hall, C, Klitgaard, K, and Krall, L. 2010. What every conservation biologist should know about economic theory. Conservation Biology 24(6):1440–1447.

Gowdy, J, and Krall, L. 2009. The fate of Nauru and the global financial meltdown. Conservation Biology 23(2):257–258.

GRAIN. 2008. Getting out of the food crisis. Seedling (July):2–6.

———. 2009. The other "pandemic." Seedling (July):2–4.

GreenpeaceUSA. 2011. Greenwashing. http://stopgreenwash.org, accessed 2012 July 14.

Gressel, J. 2008. Genetic Glass Ceilings: Transgenics for Crop Biodiversity. Baltimore, MD: John Hopkins University Press.

Guo, JH, Liu, XJ, Zhang, Y, Shen, JL, Han, WX, Zhang, WF, Christie, P, Goulding, KWT, Vitousek, PM, and Zhang, FS. 2010. Significant acidification in major Chinese croplands. Science 327(5968):1008–1010.

Gurian-Sherman, D. 2008. CAFOs Uncovered: The Untold Costs of Confined Animal Feeding Operations. Cambridge, MA: Union of Concerned Scientists. www.ucsusa.org/food_and_agriculture/our-failing-food-system/industrial-agriculture/cafos-uncovered.html, accessed 2010 January 8.

Gustavsson, J, Cederberg, C, Sonesson, U, van Otterdijk, R, and Meybeck, A. 2011. Global Food Losses and Food Waste: Extent, Causes and Prevention. Rome: FAO. www.fao.org/ag/ags/ags-division/publications/publication/en/?dyna_fef%5Buid%5D = 74045, accessed 2011 December 12.

Guthman, J. 2004. Agrarian Dreams: The Paradox of Organic Farming in California. Berkeley: University of California Press.

Haidt, J. 2007. The new synthesis in moral psychology. Science 316(5827):998–1002.

Hallauer, AR, and Miranda, JB. 1988. Quantitative Genetics in Maize Breeding. 2nd ed. Ames: Iowa State University Press.

Hamilton, H. 2010. Why sustainability needs big business, and why that's not enough: The story of the Sustainable Food Lab. Sustainable Food Laboratory Newsletter, Fall 2010, www.sustainablefoodlab.org/news-october-10, accessed 2013 June 30.

Hansen, J, Sato, M, Kharecha, P, Beerling, D, Berner, R, Masson-Delmotte, V, Pagani, M, Raymo, M, Royer, DL, and Zachos, JC. 2008. Target atmospheric CO_2: Where should humanity aim? The Open Atmospheric Science Journal 2:217–231.

Hansen, J, Sato, M, and Ruedy, R. 2012. Perception of climate change. Proceedings of the National Academy of Science 109: E2415-E2423.

Hardaker, JB, Huirne, RBM, and Anderson, JR. 1997. Coping with Risk in Agriculture. Wallingford, Oxon, UK: CABI.

Hardin, G. 1968. The tragedy of the commons. Science 162:1243–1248.

Harding, S. 1998. Is Science Multicultural? Postcolonialisms, Feminisms, and Epistemologies. Bloomington: Indiana University Press.

Harlan, JR. 1992. Crops and Man. 2nd ed. Madison, WI: American Society of Agronomy, Inc. and Crop Science Society of America, Inc.

Harris, S. 2010. The Moral Landscape: How Science Can Determine Human Values. New York: Free Press.

Harrison, GG, Sharp, M, Manalo-LeClair, G, Ramirez, A, and McGarvey, N. 2007. Food Security among California's Low-Income Adults Improves, But Most Severely Affected Do Not Share in Improvement. Los Angeles: UCLA Center for Health Policy Research.

www.healthpolicy.ucla.edu/pubs/files/Food_Security_PB_082207.pdf, accessed 2010 January 15.

Harrower, MJ. 2008. Hydrology, ideology, and the origins of irrigation in ancient Southwest Arabia. Current Anthropology 49(3):497–510.

Harwood, J. 2012. Europe's Green Revolution and Others Since: The Rise and Fall of Peasant-Friendly Plant Breeding. Milton Park, Abingdon, Oxon, UK: Routledge.

Haussmann, BIG, Rattunde, HF, Weltzien-Rattunde, E, Traore, PSC, vom Brocke, K, and Parzies, HK. 2012. Breeding strategies for adaptation of pearl millet and sorghum to climate variability and change in West Africa. Journal of Agronomy and Crop Science 198(5):327–339.

Hawkes, C. 2008. Dietary implications of supermarket development: A global perspective. Development Policy Review 26(6):657–692.

Hazell, P, Poulton, C, Wiggins, S, and Dorward, A. 2007. The Future of Small Farms for Poverty Reduction and Growth. 2020 Discussion Paper No. 42. Washington, DC: International Food Policy Research Institute. www.ifpri.org/2020/dp/vp42.pdf, accessed 2010 June 4.

Hedrick, PW. 2005. Genetics of Populations. 3rd ed. Boston, MA: Jones and Bartlett.

Helm, D. 2011. Peak oil and energy policy: A critique. Oxford Review of Economic Policy 27(1):68–91.

Herring, RJ. 2007. The Genomics revolution and development studies: Science, poverty and politics. Journal of Development Studies 43(1):1–30.

———. 2008. Science and society—Opposition to transgenic technologies: Ideology, interests and collective action frames. Nature Reviews Genetics 9(6):458–463.

Hildebrand, PE. 1990. Modified stability analysis and on-farm research to breed specific adaptability for ecological diversity. Pages 169–180 in Kang, MS, ed. Genotype-by-Environment Interaction and Plant Breeding. Baton Rouge: Department of Agronomy, Louisiana Agricultural Experiment Station, Louisiana State University Agricultural Center.

Hill, H. 2008. Food Miles: Background and Marketing. Washington, DC: ATTRA—National Sustainable Agriculture Information Service, National Center for Appropriate Technology (NCAT). http://attra.ncat.org/attra-pub/PDF/foodmiles.pdf, accessed 2009 November 5.

Hill, J, Becker, HC, and Tigerstedt, PMA. 1998. Quantitative and Ecological Aspects of Plant Breeding. London: Chapman and Hall.

Hill, KR, Walker, RS, Bozicevic, M, Eder, J, Headland, T, Hewlett, B, Hurtado, AM, Marlowe, FW, Wiessner, P, and Wood, B. 2011. Co-residence patterns in hunter-gatherer societies show unique human social structure. Science 331(6022):1286–1289.

Hinrichs, CC. 2003. The practice and politics of food system localization. Journal of Rural Studies 19(1):33–45.

Hirsh, JB, and Dolderman, D. 2007. Personality predictors of consumerism and environmentalism: A preliminary study. Personality and Individual Differences 43(6):1583–1593.

Hodgson, K, Campbell, MC, and Bailkey, M. 2011. Urban Agriculture: Growing Healthy, Sustainable Places. APA Planning Advisory Service. http://www.planning.org/apastore/Search/Default.aspx?p=4146, accessed 2012 January 31.

Hoekstra, AY, and Chapagain, AL. 2008. Globalization of Water: Sharing the Planet's Freshwater Resources. Oxford, UK: Blackwell Publishing.

Hofstadter, D. 2007. I Am a Strange Loop. New York: Basic Books.

Hohmann-Marriott, MF, and Blankenship, RE. 2011. Evolution of Photosynthesis. Pages 515–548 in Merchant, SS, Briggs, WR, and Ort, D, eds. Annual Review of Plant Biology, vol 62.

Hoppe, RA, and Banker, DE. 2010. Structure and Finances of U.S. Farms: Family Farm Report, 2010 Edition. Economic Information Bulletin No. (EIB-66). Washington, DC: U.S. Dept. of Agriculture, Economic Research Service. www.ers.usda.gov/Publications/EIB66/EIB66.pdf, accessed 2013 March 30.

Howard, P. 2009. Visualizing consolidation in the global seed industry, 1996–2008. Sustainability 1(4):1266–1287.

HSPH (Harvard School of Public Health). 2011. Healthy Eating Plate and Healthy Eating Pyramid. http://www.hsph.harvard.edu/nutritionsource/pyramid/, accessed 2013 August 7.

Hu, G, Wang, L, Arend, S, and Boeckenstedt, R. 2011. An optimization approach to assessing the self-sustainability potential of food demand in the Midwestern United States. Journal of Agriculture, Food Systems, and Community Development 2:195–207.

Huang, HT, and Yang, P. 1987. The ancient cultured citrus ant. BioScience 37(9):665–671.

Huang, J, Hu, R, Pray, C, Qiao, F, and Rozelle, S. 2003. Biotechnology as an alternative to chemical pesticides: A case study of Bt cotton in China. Agricultural Economics 29:55–67.

Huang, JK, Hu, RF, Rozelle, S, and Pray, C. 2005. Debate over a GM rice trial in China: Response. Science 310(5746):232–233.

Huber, B. 2011. Walmart's fresh food makeover. The Nation, October 3. www.thenation.com/article/163396/walmarts-fresh-food-makeover, accessed 2013 June 2.

Hufford, MB, et al. 2012. Comparative population genomics of maize domestication and improvement. Nature Genetics 44(7):808–811.

Hughes, CE, Govindarajulu, R, Robertson, A, Filer, DL, Harris, SA, and Bailey, CD. 2007. Serendipitous backyard hybridization and the origin of crops. Proceedings of the National Academy of Sciences of the United States of America 104(36):14389–14394.

Hull, DL. 1988. Science as a Process: An Evolutionary Account of the Social and Conceptual Development of Science. Chicago: University of Chicago Press.

Hunt, RC. 1989. Appropriate social organization? Water user associations in bureaucratic canal irrigation systems. Human Organization 48:79–90.

IAASTD (International Assessment of Agricultural Knowledge, Science and Technology for Development). 2009. Synthesis Report with Executive Summary: A Synthesis of the Global and Sub-global IAASTD Reports. Washington, DC: Island Press.

ICABR (International Consortium on Agricultural Biotechnology Research). 2008. Conference theme. 12th ICABR Conference on the Future of Agricultural Biotechnology: Creative Destruction, Adoption, or Irrelevance? Ravello, Italy, 12–14 June 2008. www.economia.uniroma2.it/icabr/index.php?p=2, accessed 2013 August 7.

IFAD (International Fund for Agricultural Development). 2011. Proceedings of the Conference: Conference on New Directions in Smallholder Agriculture. Rome, Italy, 24–25 January 2011. Rome: IFAD. www.ifad.org/events/agriculture/doc/proceedings.pdf, accessed 2013 March 20.

Ilbery, B, and Maye, D. 2006. Retailing local food in the Scottish-English borders: A supply chain perspective. Geoforum 37(3):352–367.

Ingram, K, Roncoli, M, and Kirshen, P. 2002. Opportunities and constraints for farmers of West Africa to use seasonal precipitation forecasts with Burkina Faso as a case study. Agricultural Systems 74:331–349.

IPCC (Intergovernmental Panel on Climate Change). 2007a. Climate Change 2007: The Physical Science Basis. Contribution of Working Group I to the Fourth Assessment Report of the Intergovernmental Panel on Climate Change. Edited by Solomon, S, Qin, D, Manning, M, Chen, Z, Marquis, M, Averyt, KB, Tignor, M, Miller, HL. Cambridge, UK: Cambridge University Press.

———. 2007b. Summary for Policymakers. In Solomon, S, Qin, D, Manning, M, Chen, Z, Marquis, M, Averyt, KB, Tignor, M, and Miller, HL, eds. Climate Change 2007: The Physical Science Basis. Contribution of Working Group I to the Fourth Assessment Report of the Intergovernmental Panel on Climate Change. Cambridge, UK: Cambridge University Press.

Jackson, T. 2009a. Beyond the growth economy. Journal of Industrial Ecology 13(4):487–490.

———. 2009b. Prosperity without Growth: Economics for a Finite Planet. London: Earthscan.

———. 2011. Societal transformations for a sustainable economy. Natural Resources Forum 35(3):155–164.

Jaffee, D, and Howard, PH. 2010. Corporate cooptation of organic and fair trade standards. Agriculture and Human Values 27(4):387–399.

James, C. 2009. Executive Summary: Global Status of Commercialized Biotech/GM Crops, 2009. ISAAA Brief No. 41. Ithaca, NY: ISAAA (International Service for the Acquisition of Agri-biotech Applications). www.isaaa.org/resources/publications/briefs/41/executivesummary/pdf/Brief 41—Executive Summary—English.pdf, accessed 2009 February 24.

———. 2010. Executive Summary: Global Status of Commercialized Biotech/GM Crops, 2010. ISAAA Brief No. 42. Ithaca, NY: ISAAA. www.isaaa.org/resources/publications/briefs/42/executivesummary/pdf/Brief 42—Executive Summary—English.pdf, accessed 2011 August 18.

———. 2011. Executive Summary: Global Status of Commercialized Biotech/GM Crops, 2011. ISAAA Brief No. 43. Ithaca, NY: ISAAA. www.isaaa.org/resources/publications/briefs/43/executivesummary/pdf/Brief 43—Executive Summary—English.pdf, accessed 2012 August 18.

Janick, J. 2005. The origins of fruit, fruit growing, and fruit breeding. Plant Breeding Reviews 25:255–321.

Jelliffe, DB. 1972. Commerciogenic malnutrition? Nutrition Reviews 30(9):199–205.

Johnson, R. 2012. The U.S. Trade Situation for Fruit and Vegetable Products. Washington, DC: Congressional Research Service. http://www.fas.org/sgp/crs/misc/RL34468.pdf, accessed 2012 March 02.

Johnston, DT, Wolfe-Simon, F, Pearson, A, and Knoll, AH. 2009. Anoxygenic photosynthesis modulated Proterozoic oxygen and sustained Earth's middle age. Proceedings of the National Academy of Sciences of the United States of America 106(40):16925–16929.

Ju, XT, et al. 2009. Reducing environmental risk by improving N management in intensive Chinese agricultural systems. Proceedings of the National Academy of Sciences of the United States of America 106(9):3041–3046.

Kahneman, D. 2011. Thinking Fast and Slow. New York: Farrar, Straus and Giroux.

Kambara, KM, and Shelley, CL. 2002. The California Agricultural Direct Marketing Study. Davis, CA: U.S. Department of Agriculture, Agricultural Marketing Service, Transportation and Marketing Programs, Marketing Services Branch, and California Institute of Rural Studies. http://purl.access.gpo.gov/GPO/LPS36189, accessed 2013 August 1.

Kaplan, JO, Krumhardt, KM, Ellis, EC, Ruddiman, WF, Lemmen, C, and Goldewijk, KK. 2011. Holocene carbon emissions as a result of anthropogenic land cover change. Holocene 21(5):775–791.

Karp, L. 2005. Global warming and hyperbolic discounting. Journal of Public Economics 89(2–3):261–282.

Karsai, I, and Kampis, G. 2010. The crossroads between biology and mathematics: The scientific method as the basics of scientific literacy. BioScience 60(8):632–638.

Katz, SH, Hediger, ML, and Valleroy, LA. 1974. Traditional maize processing techniques in New World. Science 184(4138):765–773.

Kell, DB. 2011. Breeding crop plants with deep roots: Their role in sustainable carbon, nutrient and water sequestration. Annals of Botany 108(3):407–418.

———. 2012. Large-scale sequestration of atmospheric carbon via plant roots in natural and agricultural ecosystems: Why and how. Philosophical Transactions of the Royal Society B-Biological Sciences 367(1595):1589–1597.

Kelleher, JP. 2012. Energy policy and the social discount rate. Ethics, Policy and Environment 15(1):45–50.

Keller, EF. 1983. A Feeling for the Organism: The Life and Work of Barbara McClintock. New York: W. H. Freeman.

———. 2000. The Century of the Gene. Cambridge, MA: Harvard University Press.

Kelley, TG, Rao, PP, Weltzien, E, and Purohit, ML. 1996. Adoption of improved cultivars of pearl millet in an arid environment: Straw yield and quality considerations in western Rajasthan. Experimental Agriculture 32:161–171.

Kepler. 2013. Kepler: A Search for Habitable Planets. NASA. http://kepler.nasa.gov, accessed 2013 January 15.

Khan, A, Martin, P, and Hardiman, P. 2004. Expanded production of labor-intensive crops increases agricultural employment. California Agriculture 58:35–39.

Khan, SA, Mulvaney, RL, Ellsworth, TR, and Boast, CW. 2007. The myth of nitrogen fertilization for soil carbon sequestration. Journal of Environmental Quality 36(6):1821–1832.

Kleidon, A. 2010. Life, hierarchy, and the thermodynamic machinery of planet Earth. Physics of Life Reviews 7(4):424–460.

Klein, N. 2007. The Shock Doctrine: The Rise of Disaster Capitalism. New York: Metropolitan Books (Henry Holt).

Kloppenburg, J. 1988. First the Seed: The Political Economy of Plant Biotechnology, 1492–2000. Cambridge, UK: Cambridge University Press.

———. 2010. Impeding dispossession, enabling repossession: Biological open source and the recovery of seed sovereignty. Journal of Agrarian Change 10(3):367–388.

Koinange, EMK, Singh, SP, and Gepts, P. 1996. Genetic control of the domestication syndrome in common bean. Crop Science 36(4):1037–1045.

Koster, EP. 2009. Diversity in the determinants of food choice: A psychological perspective. Food Quality and Preference 20(2):70–82.

Kristeller, JL, and Johnson, T. 2005. Science looks at spirituality—Cultivating loving kindness: A two-stage model of the effects of meditation on empathy, compassion, and altruism. Zygon 40(2):391–407.

Kuhn, TS. 1970 [1962]. The Structure of Scientific Revolutions. In International Encyclopedia of Unified Science, 2nd ed. Chicago, IL: University of Chicago Press.

Kumar, KA. 2010. Local Knowledge and Agricultural Sustainability: A Case Study of Pradhan Tribe in Adilabad District. Working Paper No. 81. Begumpet, Hyderabad, India: Centre for Economic and Social Studies. www.cess.ac.in/cesshome/wp/WP_81.pdf, accessed 2013 March 4.

Lacy, S, Cleveland, DA, and Soleri, D. 2006. Farmer choice of sorghum varieties in southern Mali. Human Ecology 34:331–353.

Lagasse, L, and Neff, R. 2010. Balanced Menus: A Pilot Evaluation of Implementation in Four San Francisco Bay Area Hospitals. Center for a Livable Future, Johns Hopkins School of Public Health. www.noharm.org/lib/downloads/food/BMC_Report_Final.pdf, accessed 2010 April 20.

Lal, R. 2010. Enhancing eco-efficiency in agro-ecosystems through soil carbon sequestration. Crop Science 50(2):S120-S131.

Lande, R, Engen, S, and Saether, BE. 2009. An evolutionary maximum principle for density-dependent population dynamics in a fluctuating environment. Philosophical Transactions of the Royal Society B-Biological Sciences 364(1523):1511–1518.

Larsen, CS. 1995. Biological changes in human populations with agriculture. Annual Review of Anthropology 24:185–213.

———. 2006. The agricultural revolution as environmental catastrophe: Implications for health and lifestyle in the Holocene. Quaternary International 150:12–20.

Leach, M, Fairhead, J, and Fraser, J. 2012. Green grabs and biochar: Revaluing African soils and farming in the new carbon economy. Journal of Peasant Studies 39(2):285–307.

Lee, RB. 2012. The Dobe Ju/'Hoansi. Case Studies in Cultural Anthropology. 4th ed. Belmont, CA: Wadsworth Publishing.

Lee, RB., and Daly, R, eds. 2000. The Cambridge Encyclopedia of Hunters and Gatherers. Cambridge, UK: Cambridge University Press.

Leigh, GJ. 2004. The World's Greatest Fix: A History of Nitrogen and Agriculture. New York: Oxford University Press.

Lele, SM. 1991. Sustainable development: A critical-review. World Development 19(6):607–621.

Leopold, A. 1970 [1949]. A Sand County Almanac. New York: Random House.

Letourneau, DK, et al. 2011. Does plant diversity benefit agroecosystems? A synthetic review. Ecological Applications 21(1):9–21.

Leung, H, Zhu, YY, Revilla-Molina, I, Fan, JX, Chen, HR, Pangga, I, Cruz, CV, and Mew, TW. 2003. Using genetic diversity to achieve sustainable rice disease management. Plant Disease 87(10):1156–1169.

Levidow, L. 2003. Precautionary risk assessment of Bt maiz: What uncertainties? Journal of Invertebrate Pathology 83:113–117.

Levitt, SD, and List, JA. 2008. Economics: Homo economicus evolves. Science 319(5865):909–910.

LFS (Local Food Systems, Inc.). 2013. Local Food Systems Connects High-volume Buyers to Reliable and Aggregated Supply. http://localfoodsystems.com, accessed 2013 January 15.

Li, CY, et al. 2009a. Crop diversity for yield increase. PLoS ONE 4(11) e8049.

Li, J, Li, D, Sun, Y, and Xu, M. 2012. Rice blast resistance gene Pi1 identified by MRG4766 marker in 173 Yunnan rice landraces. Rice Genomics and Genetics 3(3):13–18.

Li, J, Xin, Y, and Yuan, L. 2009b. Hybrid Rice Technology Development: Ensuring China's Food Security. IFPRI Discussion Paper 00918. Washington, DC: IFPRI (The International Food Policy Research Institute). www.ifpri.org/sites/default/files/publications/ifpridp00918.pdf, accessed 2010 April 10.

Lin, BB. 2011. Resilience in agriculture through crop diversification: Adaptive management for environmental change. BioScience 61(3):183–193.

Lin, BB., Perfecto, I, and Vandermeer, J. 2008. Synergies between agricultural intensification and climate change could create surprising vulnerabilities for crops. Bioscience 58(9):847–854.

Llorens, E, Comas, J, Marti, E, Riera, JL, Sabater, F, and Poch, M. 2009. Integrating empirical and heuristic knowledge in a KBS to approach stream eutrophication. Ecological Modelling 220(18):2162–2172.

Loevinsohn, ME, Bandong, JB, and Aviola II, AA. 1993. Asynchrony in cultivation among Philippine rice farmers: Causes and prospects for change. Agricultural Systems 41:419–439.

Long, SP, and Ort, DR. 2010. More than taking the heat: Crops and global change. Current Opinion in Plant Biology 13(3):240–247.

Louette, D, and Smale, M. 2000. Farmers' seed selection practices and maize variety characteristics in a traditional Mexican community. Euphytica 113:25–41.

Love Food Hate Waste. 2012. Help Save the Environment Simply by Wasting Less Food. WRAP (Waste and Resources Action Program). http://england.lovefoodhatewaste.com/content/help-save-environment-simply-wasting-less-food, accessed 2012 April 27.

Lovelock, J. 1986. Gaia: The world as living organism. New Scientist 112(1539):25–28.

Lu, YH, Wu, KM, Jiang, YY, Guo, YY, and Desneux, N. 2012. Widespread adoption of Bt cotton and insecticide decrease promotes biocontrol services. Nature 487(7407):362–365.

Lu, YH, Wu, KM, Jiang, YY, Xia, B, Li, P, Feng, HQ, Wyckhuys, KAG, and Guo, YY. 2010. Mirid bug outbreaks in multiple crops correlated with wide-scale adoption of Bt cotton in China. Science 328(5982):1151–1154.

Lundqvist, J, de Fraiture, C, and Molden, D. 2008. Saving Water: From Field to Fork: Curbing Losses and Wastage in the Food Chain. Stockholm International Water Institute, International Water Management Institute. www.siwi.org/documents/Resources/Papers/Paper_13_Field_to_Fork.pdf, accessed 2009 October 11.

Lusk, JL, and Norwood, FB. 2011. The Locavore's Dilemma: Why Pineapples Shouldn't Be Grown in North Dakota. www.econlib.org/library/Columns/y2011/LuskNorwoodlocavore.html, accessed 2012 May 19.

Lustig, RH. 2006. Childhood obesity: Behavioral aberration or biochemical drive? Reinterpreting the first law of thermodynamics. Nature Clinical Practice Endocrinology & Metabolism 2(8):447–458.

LVC (La Vía Campesina). 2008. La Vía Campesina Policy Documents, 5th Conference, Mozambique, 16th to 23rd October, 2008. Jakarta, Indonesia: LVC. http://viacampesina.org/downloads/pdf/policydocuments/POLICYDOCUMENTS-EN-FINAL.pdf, accessed 2012 October 12.

———. 2012. Land Grabbing for Agribusiness in Mozambique: UNAC Statement on the ProSavana Programme. http://viacampesina.org/en/index.php/main-issues-main-menu-27/agrarian-reform-mainmenu-36/1321-land-grabbing-for-agribusiness-on-mozambique-unac-statement-on-the-prosavana-programme, accessed 2012 December 26.

Lynam, J, Beintema, N, and Annor-Frempong, I. 2012. Agricultural R&D: Investing in Africa's Future—Analyzing Trends, Challenges, and Opportunities. Reflections on the Conference. International Food Policy Research Institute (IFPRI) / Forum for Agricultural Research in Africa (FARA) www.ifpri.org/sites/default/files/publications/astifaraconfsynthesis.pdf, accessed 2013 May 4.

Lynam, JK, and Herdt, RW. 1992. Sense and sustainability: Sustainability as an objective in international agricultural research. Pages 205–224 in Moock, JL and Rhoades, RE, eds. Diversity, Farmer Knowledge, and Sustainability. Ithaca, NY: Cornell University Press.

Lyson, TA, and Barham, E. 1998. Civil society and agricultural sustainability. Social Science Quarterly 79(3):554–567.

Lyson, TA, Torres, RJ, and Welsh, R. 2001. Scale of agricultural production, civic engagement, and community welfare. Social Forces 80(1):311–327.

Lyson, TA, and Welsh, R. 1993. The production function, crop diversity, and the debate between conventional and sustainable agriculture. Rural Sociology 58(3):424–439.

———. 2005. Agricultural industrialization, anticorporate farming laws, and rural community welfare. Environment and Planning A 37(8):1479–1491.

Mabry, JB, and Cleveland, DA. 1996. The relevance of indigenous irrigation: A comparative analysis of sustainability. Pages 227–260 in Mabry, JB, ed. Canals and Communities: Small-scale Irrigation Systems. Tucson: University of Arizona Press.

MacArthur, RH. 1972. Geographical Ecology: Patterns in the Distribution of Species. New York: Harper and Row.

MacArthur, RH, and Wilson, EO. 1967. The Theory of Island Biogeography. Princeton, NJ: Princeton University Press.

Mäder, P, Fließach, A, DuBois, D, Gunst, L, Fried, P, and Niggli, U. 2002a. Soil fertility and biodiversity in organic farming. Science 296:1694–1700.

———. 2002b. Organic farming and energy efficiency. Science 298(5600):1891–1891.

———. 2002c. The ins and outs of organic farming. Science 298(5600):1889–1890.

Malakoff, D. 2011. Are more people necessarily a problem? Science 333(6042):544–546.

Malley, ZJU, Mzimbiri, MK, and Mwakasendo, JA. 2009. Integrating local knowledge with science and technology in management of soil, water and nutrients: Implications for management of natural capital for sustainable rural livelihoods. International Journal of Sustainable Development and World Ecology 16(3):151–163.

Malthus, TR. 1992 [1803]. An Essay on the Principle of Population: Or a View of Its Past and Present Effects on Human Happiness; With an Inquiry into Our Prospects Respecting the Future Removal or Mitigation of the Evils which It Occasions. Selected and introduced by Donald Winch. 2nd ed. Cambridge, UK: Cambridge University Press.

Markowitz, EM. 2012. Is climate change an ethical issue? Examining young adults' beliefs about climate and morality. Climatic Change 114(3–4):479–495.

Marlow, HJ, Hayes, WK, Soret, S, Carter, RL, Schwab, ER, and Sabate, J. 2009. Diet and the environment: Does what you eat matter? American Journal of Clinical Nutrition 89(5):S1699–S1703.

Marshall, GR. 2004. Farmers cooperating in the commons? A study of collective action in salinity management. Ecological Economics 51(3–4):271–286.

Martinez, S, et al. 2010. Local Food Systems: Concepts, Impacts, and Issues. Economic Research Report Number 97. United States Department of Agriculture, Economic Research Service. www.ers.usda.gov/Publications/ERR97/ERR97.pdf, accessed 2011 January 14.

Matson, PA, Naylor, R, and Ortiz-Monasterio, I. 1998. Integration of environmental, agronomic, and economic aspects of fertilizer management. Science 280(5360):112–115.

Matsuoka, Y, Vigouroux, Y, Goodman, MM, Sanchez, GJ, Buckler, E, and Doebley, J. 2002. A single domestication for maize shown by multilocus microsatellite genotyping. Proceedings of the National Academy of Sciences 99:6080–6084.

May, RM. 1974. Stability and Complexity in Model Ecosystems. Monographs in Population Biology. 2nd ed. Princeton, NJ: Princeton University Press.

McIntyre, S. 2011. Hansen, WG3 and Green Kool-aid. Climate Audit. http://climateaudit.org/2011/08/13/hansen-and-ipccs-green-kool-aid, accessed 2013 January 15.

McKibben, B. 2010. Eaarth: Making a Life on a Tough New Planet. New York: Times Books.

McMichael, P. 2009. The world food crisis in historical perspective. Monthly Review: An Independent Socialist Magazine 61(3):32–47.

McNaughton, SJ. 1988. Diversity and stability. Nature 333:204–205.

McWethy, DB, Whitlock, C, Wilmshurst, JM, McGlone, MS, Fromont, M, Li, X, Dieffenbacher-Krall, A, Hobbs, WO, Fritz, SC, and Cook, ER. 2010. Rapid landscape transformation in South Island, New Zealand, following initial Polynesian settlement. Proceedings of the National Academy of Sciences of the United States of America 107(50):21343–21348.

McWilliams, C. 2000 [1939]. Factories in the Field: The Story of Migratory Farm Labor in California. Berkeley: University of California Press.

McWilliams, JE. 2009. Just Food: Where Locavores Get It Wrong and How We Can Truly Eat Responsibly. New York: Little, Brown and Company.

———. 2012. Agnostic Carnivores and Global Warming: Why Enviros Go After Coal and Not Cows. Freakonomics. www.freakonomics.com/2011/11/16/agnostic-carnivores-and-global-warming-why-enviros-go-after-coal-and-not-cows, accessed 2012 May 31.

Meadows, DH, Meadows, DL, and Randers, Jí. 1992. Beyond the Limits: Confronting Global Collapse, Envisioning a Sustainable Future. Post Mills, VT: Chelsea Green Publishing Company.

Medin, DL, and Atran, S. 1999. Introduction. Pages 1–15 in Medin, DL and Atran, S, eds. Folkbiology. Cambridge, MA: MIT Press.

Mehra, S. 1981. Instability in Indian Agriculture in the Context of the New Technology. Research Report. Washington, DC: International Food Policy Research Institute.

Merrey, DJ. 1987. The local impact of centralized irrigation control in Pakistan: A sociocentric perspective. Pages 352–372 in Little, PD, Horowitz, MM, and Nyerges, AE, eds. Lands at Risk in the Third World: Local Level Perspectives. Boulder, CO: Westview Press.

Meyer, RS, DuVal, AE, and Jensen, HR. 2012. Patterns and processes in crop domestication: An historical review and quantitative analysis of 203 global food crops. New Phytologist 196(1):29–48.

Miller, HI, Morandini, P, and Ammann, K. 2008. Is biotechnology a victim of anti-science bias in scientific journals? Trends in Biotechnology 26(3):122–125.

Minten, B, Reardon, T, and Sutradhar, R. 2010. Food prices and modern retail: The case of Delhi. World Development 38(12):1775–1787.

Mkumbira, J, Chiwona-Karltun, L, Lagercrantz, U, Mahungu, NM, Saka, J, Mhone, A, Bokanga, M, Brimer, L, Gullberg, U, and Rosling, H. 2003. Classification of cassava into "bitter" and "cool" in Malawi: From farmers' perception to characterisation by molecular markers. Euphytica 132(1):7–22.

Mohapatra, S, Rozelle, S, and Huang, JK. 2006. Climbing the development ladder: Economic development and the evolution of occupations in rural China. Journal of Development Studies 42(6):1023–1055.

Mondelaers, K, Aertsens, J, and Van Huylenbroeck, G. 2009. A meta-analysis of the differences in environmental impacts between organic and conventional farming. British Food Journal 111(10):1098–1119.

Monsanto. 2009. Growing Hope in Africa. www.monsanto.com/responsibility/our_pledge /stronger_society/growing_hope_africa.asp, accessed 2009 April 4.

Mooney, SD, et al. 2011. Late Quaternary fire regimes of Australasia. Quaternary Science Reviews 30(1–2):28–46.

Moran, DD, Wackernagel, MC, Kitzes, JA, Heumann, BW, Phan, D, and Goldfinger, SH. 2009. Trading spaces: Calculating embodied Ecological Footprints in international trade using a Product Land Use Matrix (PLUM). Ecological Economics 68(7):1938–1951.

Morse, S, Bennett, R, and Ismael, Y. 2006. Environmental impact of genetically modified cotton in South Africa. Agriculture Ecosystems and Environment 117(4):277–289.

Mortimore, M, and Tiffen, M. 1994. Population growth and a sustainable environment. Environment 36:10–32.

Mougi, A, and Kondoh, M. 2012. Diversity of interaction types and ecological community stability. Science 337(6092):349–351.

Mueller, UG, Gerardo, NM, Aanen, DK, Six, DL, and Schultz, TR. 2005. The evolution of agriculture in insects. Annual Review of Ecology Evolution and Systematics 36: 563–595.

Murphy, K, Lammer, D, Lyon, S, Carter, B, and Jones, SS. 2005. Breeding for organic and low-input farming systems: An evolutionary-participatory breeding method for inbred cereal grains. Renewable Agriculture and Food Systems 20(1):48–55.

Murphy, T. 2006. The Universe: Size, Shape, and Fate. San Diego: UCSD. http://physics.ucsd .edu/~tmurphy/phys10/universe.pdf, accessed 2011 June 20.

Naeem, S, Duffy, JE, and Zavaleta, E. 2012. The functions of biological diversity in an age of extinction. Science 336(6087):1401–1406.

Nestle, M. 2001. Food company sponsorship of nutrition research and professional activities: A conflict of interest? Public Health Nutrition 4(5):1015–1022.

———. 2002. Food Politics: How the Food Industry Influences Nutrition and Health. Berkeley: University of California Press.

———. 2006. Food marketing and childhood obesity: A matter of policy. New England Journal of Medicine 354(24):2527–2529.

Netting, RM. 1981. Balancing on an Alp: Ecological Change and Continuity in a Swiss Mountain Community. Cambridge: Cambridge University Press.

———. 1993. Smallholders, Householders: Farm Families and the Ecology of Intensive, Sustainable Agriculture. Stanford, CA: Stanford University Press.

Neumayer, E. 2004. Weak versus Strong Sustainability: Exploring the Limits of Two Opposing Paradigms. 2nd ed. Cheltenham, UK: Edward Elgar.

Neven, D, Odera, MM, Reardon, T, and Wang, HL. 2009. Kenyan supermarkets, emerging middle-class horticultural farmers, and employment impacts on the rural poor. World Development 37(11):1802–1811.

Niu, SW, Ding, YX, Niu, YZ, Li, YX, and Luo, GH. 2011. Economic growth, energy conservation and emissions reduction: A comparative analysis based on panel data for 8 Asian-Pacific countries. Energy Policy 39(4):2121–2131.

Nord, M, Andrews, M, and Carlson, S. 2009. Household Food Security in the United States, 2008. Economic Research Report No. (ERR-83). US Department of Agriculture, Economic Research Service. www.ers.usda.gov/publications/err83, accessed 2009 November 26.

Nordhaus, T, and Shellenberger, M. 2007. Break Through: From the Death of Environmentalism to the Politics of Possibility. Boston: Houghton Mifflin.

Norgaard, RB. 1994. Development Betrayed: The End of Progress and a Coevolutionary Revisioning of the Future. London: Routledge.

———. 2010. Ecosystem services: From eye-opening metaphor to complexity blinder. Ecological Economics 69(6):1219–1227.

Norton, BG. 1987. Why Preserve Natural Variety? Princeton, NJ: Princeton University Press.

———. 1991. Toward Unity among Environmentalists. New York: Oxford University Press.

———. 2005. Sustainability: A Philosophy of Adaptive Ecosystem Management. Chicago: University of Chicago Press.

Norton, JB, Sandor, JA, White, CS, and Laahty, V. 2007. Organic matter transformations through arroyos and alluvial fan soils within a native American agroecosystem. Soil Science Society of America Journal 71(3):829–835.

Norum, KR. 2008. PepsiCo recruitment strategy challenged. Public Health Nutrition 11(2):112–113.

NRC (National Research Council of the National Academies). 1972. Genetic Vulnerability of Major Crops. Washington, DC: National Academy of Sciences.

———. 1996. Understanding Risk: Informing Decisions in a Democratic Society. Washington, DC: National Academy Press.

———. 2002. Environmental Effects of Transgenic Plants: The Scope and Adequacy of Regulation. Washington, DC: National Academies Press.

———. 2010. Limiting the Magnitude of Future Climate Change. Washington, DC: National Academy Press.

Nygaard, S., et al. 2011. The genome of the leaf-cutting ant *Acromyrmex echinatior* suggests key adaptations to advanced social life and fungus farming. Genome Research 21(8):1339–1348.

Oakerson, RJ. 1992. Analyzing the commons: A framework. Pages 41–59 in Bromley, DW, ed. Making the Commons Work: Theory, Practice, and Policy. San Francisco, CA: Institute for Contemporary Studies Press.

Odling-Smee, FJ, Laland, KN, and Feldman, MW. 2003. Niche Construction: The Neglected Process in Evolution. Monographs in Population Biology. Princeton, NJ: Princeton University Press.

Olsen, KM, Caicedo, AL, Polato, N, McClung, A, McCouch, S, and Purugganan, MD. 2006. Selection under domestication: Evidence for a sweep in the rice *Waxy* genomic region. Genetics 173(2):975–983.

O'Neal, A, Pandian, A, Rhodes-Conway, S, and Bornbusch, A. 1995. Human economies, the land ethic, and sustainable conservation. Conservation Biology 9:217–220.

Oreskes, N. 2004. Science and public policy: What's proof got to do with it? Environmental Science and Policy 7(5):369–383.

Ortiz, R, and Golmirzaie, AM. 2004. Genotype by environment interaction and selection in true potato seed breeding. Experimental Agriculture 40(1):99–107.

Ortiz, R, Trethowan, R, Ferrara, GO, Iwanaga, M, Dodds, JH, Crouch, JH, Crossa, J, and Braun, HJ. 2007. High yield potential, shuttle breeding, genetic diversity, and a new international wheat improvement strategy. Euphytica 157(3):365–384.

Ory, MG, Jordan, PJ, and Bazzarre, T. 2002. The Behavior Change Consortium: Setting the stage for a new century of health behavior-change research. Health Education Research 17(5):500–511.

Ostrom, E. 1992. The rudiments of a theory of the origins, survival, and performance of common-property institutions. Pages 293–318 in Bromley, DW, ed. Making the Commons Work: Theory, Practice, and Policy. San Francisco, CA: Institute for Contemporary Studies Press.

———. 2010. Beyond markets and states: Polycentric governance of complex economic systems. American Economic Review 100(3):641–672.

Ostrom, E, Burger, J, Field, CB, Norgaard, RB, and Policansky, D. 1999. Revisiting the commons: Local lessons, global challenges. Science 284:278–282.

Paarlberg, R. 2008. Starved for Science: How Biotechnology Is Being Kept out of Africa. Cambridge, MA: Harvard University Press.

Pacala, SW, Bulte, E, List, JA, and Levin, SA. 2003. False alarm over environmental false alarms. Science 301:1187–1188.

Pachauri, RK. 2008. Global Warning! The Impact of Meat Production and Consumption on Climate Change. London: Compassion in World Farming. www.ciwf.org.uk/includes /documents/cm_docs/2008/l/1_london_08sept08.pps—564,1, accessed 2013 July 31.

Pandey, S. 1989. Irrigation and crop yield variability: A review. Pages 234–241 in Anderson, JR and Hazell, PBR, eds. Variability in Grain Yields: Implications for Agricultural Research and Policy in Developing Countries. Baltimore, MD: The Johns Hopkins University Press.

Pandey, S, Vasal, SK, and Deutsch, JA. 1991. Performance of open-pollinated maize cultivars selected from 10 tropical maize populations. Crop Science 31:285–290.

Parks, S, and Gowdy, J. 2013. What have economists learned about valuing nature? A review essay. Ecosystem Services 3:e1–e10.

Paterson, AH, Lin, Y-R, Li, Z, Schertz, KF, Doebley, JF, Pinson, SR, Liu, S-C, Stansel, JW, and Irvine, JE. 1995. Convergent domestication of cereal crops by independent mutations at corresponding genetic loci. Science 269:1714–1718.

PCRM (Physicians Committee for Responsible Medicine). 2011. Vegetarian and Vegan Diets. www.pcrm.org/health/diets, accessed 2013 July 31.

Pelletier, N, and Tyedmers, P. 2010. Forecasting potential global environmental costs of livestock production, 2000–2050. Proceedings of the National Academy of Sciences of the United States of America 107(43):18371–18374.

Pennisi, E. 1999. Did cooked tubers spur the evolution of big brains? Science 283:2004–2005.

Pereira, T. 2009. Sustainability: An integral engineering design approach. Renewable and Sustainable Energy Reviews 13(5):1133–1137.

Perfecto, I, and Vandermeer, J. 2010. The agroecological matrix as alternative to the land-sparing/agriculture intensification model. Proceedings of the National Academy of Sciences of the United States of America 107(13):5786–5791.

Perfecto, I, Vandermeer, JH, and Wright, AL. 2009. Nature's Matrix: Linking Agriculture, Conservation and Food Sovereignty. London: Earthscan.

Peterhansel, C, and Offermann, S. 2012. Re-engineering of carbon fixation in plants: Challenges for plant biotechnology to improve yields in a high-CO_2 world. Current Opinion in Biotechnology 23(2):204–208.

Peters, CJ, Bills, NL, Lembo, AJ, Wilkins, JL, and Fick, GW. 2009. Mapping potential foodsheds in New York State: A spatial model for evaluating the capacity to localize food production. Renewable Agriculture and Food Systems 24(1):72–84.

Peterson, GD, Cummings, GS, and Carpenter, SR. 2003. Scenario planning: A tool for conservation in an uncertain world. Conservation Biology 17:358–366.

Pimm, SL. 1991. The Balance of Nature: Ecological Issues in the Conservation of Species and Communities. Chicago, IL: University of Chicago Press.

Pingali, P, and Rajaram, S. 1999. Global wheat research in a changing world: Options for sustaining growth in wheat productivity. Pages 1–18 in Pingali, PL, ed. Global Wheat Research in a Changing World: Challenges and Achievements. Mexico, D.F.: CIMMYT.

Pirog, R, and Benjamin, A. 2003. Checking the Food Odometer: Comparing Food Miles for Local versus Conventional Produce Sales to Iowa Institutions. Ames: Leopold Center for Sustainable Agriculture, Iowa State University. www.leopold.iastate.edu/pubs/staff/files/food_travel072103.pdf, accessed 2009 November 4.

Poehlman, JM, and Sleper, DA. 1995. Breeding Field Crops. 4th ed. Ames: Iowa State University Press.

Pollan, M. 2006. The Omnivore's Dilemma: A Natural History of Four Meals. New York: Penguin Press.

Popkin, BM. 2006. Global nutrition dynamics: The world is shifting rapidly toward a diet linked with noncommunicable diseases. American Journal of Clinical Nutrition 84(2):289–298.

———. 2007. Science and society: Understanding global nutrition dynamics as a step towards controlling cancer incidence. Nature Reviews Cancer 7(1):61–67.

Popp, A, Lotze-Campen, H, and Bodirsky, B. 2010. Food consumption, diet shifts and associated non-CO_2 greenhouse gases from agricultural production. Global Environmental Change: Human and Policy Dimensions 20(3):451–462.

Potts, R. 2013. Hominin evolution in settings of strong environmental variability. Quaternary Science Reviews 73(0):1–13.

Power, AG. 2010. Ecosystem services and agriculture: Tradeoffs and synergies. Philosophical Transactions of the Royal Society B: Biological Sciences 365(1554):2959–2971.

PRB (Population Reference Bureau). 2004. 2004 World Population Data Sheet. Washington, DC: Population Reference Bureau. www.prb.org/Publications/Datasheets/2004/2004WorldPopulationDataSheet.aspx, accessed 2013 August 1.

———. 2010. 2010 World Population Data Sheet. Washington, DC: Population Reference Bureau. www.prb.org/Publications/Datasheets/2010/2010wpds.aspx, accessed 2013 August 1.

———. 2011. 2011 World Population Data Sheet. Washington, DC: Population Reference Bureau. www.prb.org/pdf11/2011population-data-sheet_eng.pdf, accessed 2013 August 1.

———. 2012. 2012 World Population Data Sheet. Washington, DC: Population Reference Bureau. www.prb.org/Publications/Datasheets/2012/world-population-data-sheet.aspx, accessed 2013 August 1.

Pretty, JN, Noble, AD, Bossio, D, Dixon, J, Hine, RE, de Vries, F, and Morison, JIL. 2006. Resource-conserving agriculture increases yields in developing countries. Environmental Science and Technology 40(4):1114–1119.

Provine, WB. 1971. The Origins of Theoretical Population Genetics. Chicago: University of Chicago Press.

Pujol, B, David, P, and McKey, D. 2005. Microevolution in agricultural environments: How a traditional Amerindian farming practice favours heterozygosity in cassava (*Manihot esculenta* Crantz, Euphorbiaceae). Ecology Letters 8(2):138–147.

Purugganan, MD, and Fuller, DQ. 2009. The nature of selection during plant domestication. Nature 457(7231):843–848.

Putnam, J, Allshouse, J, and Scott, K. 2002. U.S. per capita food supply trends: More calories, refined carbohydrates, and fats. Food Review 25(3):2–15.

Qaim, M, and Zilberman, D. 2003. Yield effects of genetically modified crops in developing countries. Science 299:900–902.

Qiu, SJ, Ju, XT, Lu, X, Li, L, Ingwersen, J, Streck, T, Christie, P, and Zhang, FS. 2012. Improved nitrogen management for an intensive winter wheat/summer maize double-cropping system. Soil Science Society of America Journal 76(1):286–297.

Ragland, E, and Tropp, D. 2009. USDA National Farmers Market Manager Survey, 2006. Washington, DC: USDA, AMS. http://www.ams.usda.gov/AMSv1.0/getfile?dDocName=S TELPRDC5077203, accessed 2013 August 7.

Rand, DG, Dreber, A, Ellingsen, T, Fudenberg, D, and Nowak, MA. 2009. Positive interactions promote public cooperation. Science 325(5945):1272–1275.

Rankin, DJ. 2011. The social side of *Homo economicus*. Trends in Ecology and Evolution 26(1):1–3.

Rankin, P, Morton, DP, Diehl, H, Gobble, J, Morey, P, and Chang, E. 2012. Effectiveness of a volunteer-delivered lifestyle modification program for reducing cardiovascular disease risk factors. American Journal of Cardiology 109(1):82–86.

Ratnadass, A, Fernandes, P, Avelino, J, and Habib, R. 2012. Plant species diversity for sustainable management of crop pests and diseases in agroecosystems: A review. Agronomy for Sustainable Development 32(1):273–303.

Raupach, MR, and Canadell, JG. 2010. Carbon and the Anthropocene. Current Opinion in Environmental Sustainability 2(4):210–218.

Raven, JA, Andrews, M, and Quigg, A. 2005. The evolution of oligotrophy: Implications for the breeding of crop plants for low input agricultural systems. Annals of Applied Biology 146(3):261–280.

Raven, JA, Handley, LL, and Andrews, M. 2004. Global aspects of C/N interactions determining plant-environment interactions. Journal of Experimental Botany 55(394):11–25.

Raven, PH. 2005. Transgenes in Mexican maize: Desirability or inevitability? Proceedings of the National Academy of Sciences 102:13003–13004.

Reardon, T, Barrett, CB, Berdegue, JA, and Swinnen, JFM. 2009. Agrifood industry transformation and small farmers in developing countries. World Development 37(11):1717–1727.

Reay, DS, Davidson, EA, Smith, KA, Smith, P, Melillo, JM, Dentener, F, and Crutzen, PJ. 2012. Global agriculture and nitrous oxide emissions. Nature Climate Change 2(6):410–416.

Reddy, AGS, Kumar, K, Rao, D, and Rao, S. 2009. Assessment of nitrate contamination due to groundwater pollution in north eastern part of Anantapur District, A. P. India. Environmental Monitoring and Assessment 148(1–4):463–476.

Redford, KH, and Sanderson, SE. 1992. The brief, barren marriage of biodiversity and sustainability? Bulletin of the Ecological Society of America 73:36–39.

Renting, H, Marsden, TK, and Banks, J. 2003. Understanding alternative food networks: Exploring the role of short food supply chains in rural development. Environment and Planning A 35(3):393–411.

Revilla-Molina, IM, Bastiaans, L, Van Keulen, H, Kropff, MJ, Hui, F, Castilla, NP, Mew, TW, Zhu, YY, and Leung, H. 2009. Does resource complementarity or prevention of lodging contribute to the increased productivity of rice varietal mixtures in Yunnan, China? Field Crops Research 111(3):303–307.

Richerson, PJ, and Boyd, R. 2005. Not by Genes Alone: How Culture Transformed Human Evolution. Chicago: University of Chicago Press.

Richerson, PJ, Boyd, R, and Bettinger, RL. 2001. Was agriculture impossible during the Pleistocene but mandatory during the Holocene? A climate change hypothesis. American Antiquity 66(3):387–411.

Riebeek, H. 2010. Global Warming. NASA. http://earthobservatory.nasa.gov/Features /GlobalWarming, accessed 2011 October 15.

Roberts, W. 2011. Citywatch: Food's a Trip, Actually a Baker's Dozen of Trips. http://blogs. worldwatch.org/citywatch-food%E2%80%99s-a-trip-actually-a-baker%E2%80%99s-dozen-of-trips, accessed 2011 December 16.

Robertson, B, and Pinstrup-Andersen, P. 2010. Global land acquisition: Neo-colonialism or development opportunity? Food Security 2(3):271–283.

Rockefeller Foundation. 2007. Biotechnology, Breeding and Seed Systems for African Crops: Research and Product Development that Reaches Farmers. Rockefeller Foundation and Instituto de Investigação Agrária de Moçambique (IIAM) conference. 26–29 March 2007, Maputo, Mozambique. Nairobi, Kenya: Rockefeller Foundation.

Rogelj, J, McCollum, DL, Reisinger, A, Meinshausen, M, and Riahi, K. 2013. Probabilistic cost estimates for climate change mitigation. Nature 493(7430):79–83.

Rolls, BJ, Roe, LS, and Meengs, JS. 2010. Portion size can be used strategically to increase vegetable consumption in adults. American Journal of Clinical Nutrition 91(4): 913–922.

Romagosa, I, and Fox, PN. 1993. Genotype × environment interaction and adaptation. Pages 373–390 in Hayward, MD, Bosemark, NO, and Romagosa, I, eds. Plant Breeding: Principles and Prospects. London: Chapman and Hall.

Romney, AK, Boyd, JP, Moore, CC, Batchelder, WH, and Brazill, TJ. 1996. Culture as shared cognitive representations. Proceedings of the National Academy of Sciences of the United States of America 93(10):4699–4705.

Ronald, P. 2011. Plant genetics, sustainable agriculture and global food security. Genetics 188(1):11–20.

Rothschild, LJ. 2008. The evolution of photosynthesis . . . again? Philosophical Transactions of the Royal Society B: Biological Sciences 363(1504):2787–2801.

Rouquette, JR, Posthumus, H, Gowing, DJG, Tucker, G, Dawson, QL, Hess, TM, and Morris, J. 2009. Valuing nature-conservation interests on agricultural floodplains. Journal of Applied Ecology 46(2):289–296.

Roy, P, Nei, D, Orikasa, T, Xu, QY, Okadome, H, Nakamura, N, and Shiina, T. 2009. A review of life cycle assessment (LCA) on some food products. Journal of Food Engineering 90(1):1–10.

Roy, P, Orikasa, T, Thammawong, M, Nakamura, N, Xu, QY, and Shiina, T. 2012. Life cycle of meats: An opportunity to abate the greenhouse gas emission from meat industry in Japan. Journal of Environmental Management 93(1):218–224.

Ruddiman, WF, and Ellis, EC. 2009. Effect of per-capita land use changes on Holocene forest clearance and CO_2 emissions. Quaternary Science Reviews 28(27–28):3011–3015.

Ruf, T, and Valony, MJ. 2007. The contradictions in the integrated management of water resources in irrigated Mediterranean agriculture. Cahiers Agricultures 16(4):294–300.

Runge, CF, and Defrancesco, E. 2006. Exclusion, inclusion, and enclosure: Historical commons and modern intellectual property. World Development 34(10):1713–1727.

Running, SW. 2012. A measurable planetary boundary for the biosphere. Science 337(6101):1458–1459.

Sachs, JD. 2005. The End of Poverty: How We Can Make It Happen in Our Lifetime. London: Penguin.

Sachs, JD, et al. 2010. Monitoring the world's agriculture. Nature 466(7306):558–560.

Sagoff, M. 1993. Biodiversity and the culture of ecology. Bulletin of the Ecological Society of America 74:374–381.

Sahlins, MD. 1972. Stone Age Economics. Hawthorne, NY: Aldine de Gruyter.

Saltmarsh, N, and Wakeman, T. 2004. Local Links in a Global Chain: Mapping Food Supply Chains and Identifying Local Links in the Broads and Rivers Area of Norfolk. East Anglia Food Link. www.eafl.org.uk/downloads/LocalLinksMainWeb.pdf, accessed 2011 April 1.

Sanders, B, and Shattuck, A. 2011. Cutting through the Red Tape: A Resource Guide for Local Food Policy Practitioners and Organizers. Policy Brief No. 19. Oakland, CA: Food First/Institute for Food and Development Policy and Oakland Food Policy Council. www.foodfirst.org/sites/www.foodfirst.org/files/pdf/PB_19_Cutting_Through_the_Red_Tape.pdf, accessed 2012 January 25.

Saunders, C, and Barber, A. 2008. Carbon footprints, life cycle analysis, food miles: Global trade trends and market issues. Political Science 60(1):73–88.

Savitz, AW. 2006. The Triple Bottom Line: How Today's Best-Run Companies Are Achieving Economic, Social and Environmental Success—and How You Can Too. Hoboken, NJ: Jossey-Bass/Wiley.

Sayre, NF. 2008. The genesis, history, and limits of carrying capacity. Annals of the Association of American Geographers 98(1):120–134.

SBC ACO (Santa Barbara County Agricultural Commissioner's Office). 2009. 2008 Crop Report, Santa Barbara, California. http://www.countyofsb.org/uploadedFiles/agcomm/crops/2008.pdf, accessed 2009 November 10.

———. 2010. 2009 Crop Report, Santa Barbara, California. http://www.countyofsb.org/uploadedFiles/agcomm/crops/2009CR2.pdf, accessed 2010 May 26.

Schneider, JR. 2002. Selecting with farmers: The formative years of cereal breeding and public seed in Switzerland (1889–1936). Pages 161–187 in Cleveland, DA and Soleri, D, eds. Farmers, Scientists and Plant Breeding: Integrating Knowledge and Practice. Wallingford, Oxon, UK: CABI.

Schoneveld, GC, German, LA, and Nutakor, E. 2011. Land-based investments for rural development? A grounded analysis of the local impacts of biofuel feedstock plantations in Ghana. Ecology and Society 16(4): 10.

http://dx.doi.org/10.5751/ES-04424-160410, accessed 2013 August 7.

Schoonover, H, and Muller, M. 2006. Food without Thought: How U.S. Farm Policy Contributes to Obesity. Institute for Agriculture and Trade Policy. http://www.iatp.org/files/421_2_80627.pdf, accessed 2010 April 7.

Schultz, TR, and Brady, SG. 2008. Major evolutionary transitions in ant agriculture. Proceedings of the National Academy of Sciences of the United States of America 105(14):5435–5440.

Schultz, TR, Mueller, U, Currie, C, and Rehner, S. 2005. Reciprocal illumination: A comparison of agriculture in humans and ants. Pages 149–190 in Vega, F and Blackwell, M, eds. Ecological and Evolutionary Advances in Insect-Fungal Associations. New York: Oxford University Press.

Schultz, TW. 1964. Transforming Traditional Agriculture. New Haven, CT: Yale University Press.

Schumpeter, JA. 1975 [1942]. Capitalism, Socialism and Democracy. New York: Harper.

Schwartz, GT. 2012. Growth, development, and life history throughout the evolution of *Homo*. Current Anthropology 53:S395-S408.

Scialabba, NEH, and Muller-Lindenlauf, M. 2010. Organic agriculture and climate change. Renewable Agriculture and Food Systems 25(2):158–169.

Scott, JC. 1998. Seeing Like a State: How Certain Schemes to Improve the Human Condition Have Failed. New Haven, CT: Yale University Press.

SELC (The Sustainable Economies Law Center). 2013. UrbanAgLaw. Oakland, CA: SELC. www.urbanaglaw.org/, accessed 2013 January 15.

Sen, A. 1999. The possibility of social choice. American Economic Review 89(3):349–378.

———. 2000. The discipline of cost-benefit analysis. Journal of Legal Studies 29(2):931–952.

Seneweera, S. 2011. Effects of elevated CO_2 on plant growth and nutrient partitioning of rice (*Oryza sativa* L.) at rapid tillering and physiological maturity. Journal of Plant Interactions 6(1):35–42.

Serratos-Hernández, J-A, Gómez-Olivares, J-L, Salinas-Arreortua, N, Buendía-Rodríguez, E, Islas-Gutiérrez, F, and de-Ita, A. 2007. Transgenic proteins in maize in the soil conservation area of Federal District, Mexico. Frontiers in Ecology and the Environment 5(5):247–252.

Seufert, V, Ramankutty, N, and Foley, JA. 2012. Comparing the yields of organic and conventional agriculture. Nature 485:229–232.

SFL (Sustainable Food Laboratory). 2011. About Us. http://sustainablefood.org/about-us, accessed 2011 January 2.

Shapiro, JM. 2005. Is there a daily discount rate? Evidence from the food stamp nutrition cycle. Journal of Public Economics 89(2–3):303–325.

Sheppard, E, and Leitner, H. 2010. Quo vadis neoliberalism? The remaking of global capitalist governance after the Washington Consensus. Geoforum 41(2):185–194.

Shostak, M. 1981. Nisa, the Life and Words of a !Kung Woman. Cambridge, MA: Harvard University Press.

Shreck, A, Getz, C, and Feenstra, G. 2006. Social sustainability, farm labor, and organic agriculture: Findings from an exploratory analysis. Agriculture and Human Values 23(4):439–449.

Siedenburg, J. 2006. The Machakos case study: Solid outcomes, unhelpful hyperbole. Development Policy Review 24(1):75–85.

Silberberg, E. 1995. Principles of Microeconomics. Englewood Cliffs, NJ: Prentice Hall.

Simmonds, NW. 1979. Principles of Crop Improvement. London: Longman Group Ltd.

Simmonds, NW, and Smartt, J. 1999. Principles of Crop Improvement. 2nd ed. Oxford, UK: Blackwell Science Ltd.

Simon, JL. 1983. The Ultimate Resource. Princeton, NJ: Princeton University Press.

Simon, M. 2007. Appetite for Profit: How the Food Industry Undermines Our Health and How to Fight Back. New York: Nation Books.

———. 2012. Food Stamps, Follow the Money: Are Corporations Profiting from Hungry Americans? Eat Drink Politics. www.eatdrinkpolitics.com/wp-content/uploads/FoodStampsFollowtheMoneySimon.pdf, accessed 2013 January 14.

Simpson, RD. 2003. Internalizing Ecological Externalities: Some Surprising Results from Partial Combinations. Washington, DC: Resources for the Future.

Sinha, R, Cross, AJ, Graubard, BI, Leitzmann, MF, and Schatzkin, A. 2009. Meat intake and mortality: A prospective study of over half a million people. Archives of Internal Medicine 169(6):562–571.

Skidelsky, R. 2009. Keynes: The Return of the Master. New York: PublicAffairs.

Smale, M. 2002. Economics perspectives on collaborative plant breeding for conservation of genetic diversity on farm. Pages 83–105 in Cleveland, DA and Soleri, D, eds. Farmers, Scientists and Plant Breeding: Integrating Knowledge and Practice. Oxon, UK: CABI.

Smale, M, Reynolds, MP, Warburton, M, Skovmand, B, Trethowan, R, Singh, RP, Ortiz-Monasterio, I, and Crossa, J. 2002. Dimensions of diversity in modern spring bread wheat in developing countries from 1965. Crop Science 42:1766–1799.

Smil, V. 2004. Enriching the Earth: Fritz Haber, Carl Bosch, and the Transformation of World Food Production. Cambridge, MA: MIT Press.

Smith, BD. 2001. Documenting plant domestication: The consilience of biological and archaeological approaches. Proceedings of the National Academy of Science 98:1324–1326.

Smith, HA. 2011. Alone in the universe despite the growing catalog of extrasolar planets, data so far do not alter estimates that we are effectively on our own. American Scientist 99(4):320–327.

Smith, P, et al. 2007. Agriculture. Pages 497–540 in Metz, B, Davidson, OR, Bosch, PR, Dave, R, and Meyer, LA, eds. Climate Change 2007: Mitigation. Contribution of Working Group III to the Fourth Assessment Report of the Intergovernmental Panel on Climate Change. Cambridge, UK: Cambridge University Press.

Smith, P, and Olesen, JE. 2010. Synergies between the mitigation of, and adaptation to, climate change in agriculture. Journal of Agricultural Science 148:543–552.

Soleri, D. n.d. Farmers' selected maize seed shows shorter germination time and higher vigor. Unpublished ms.

Soleri, D, et al. n.d.. Farmers' perceptions of crop genotype-environment interactions in five locations around the world. Unpublished ms.

Soleri, D, and Cleveland, DA. 1993. Hopi crop diversity and change. Journal of Ethnobiology 13:203–231.

———. 2001. Farmers' genetic perceptions regarding their crop populations: An example with maize in the Central Valleys of Oaxaca, Mexico. Economic Botany 55:106–128.

———. 2004. Farmer selection and conservation of crop varieties. Pages 433–438 in Goodman, RM, ed. Encyclopedia of Plant and Crop Science. New York: Marcel Dekker.

———. 2005. Scenarios as a tool for eliciting and understanding farmers' biological knowledge. Field Methods 17:283–301.

———. 2009. Breeding for quantitative variables. Part 1: Farmers' and scientists' knowledge and practice in variety choice and plant selection. Pages 323–366 in Ceccarelli, S, Weltzien, E, and Guimares, E, eds. Participatory Plant Breeding. Rome: FAO (United Nations Food and Agriculture Organization), in collaboration with ICARDA (International Center for Agricultural Research in the Dry Areas) and ICRISAT (International Crops Research Institute for the Semi-Arid Tropics).

Soleri, D, Cleveland, DA, and Aragón Cuevas, F. 2006. Transgenic crop varieties and varietal diversity in traditionally based agriculture: The case of maize in Mexico. BioScience 56:503–513.

———. 2008. Food globalization and local diversity: The case of tejate, a traditional maize and cacao beverage from Oaxaca, Mexico. Current Anthropology 49:281–290, + online supplement.

Soleri, D, Cleveland, DA, Aragón Cuevas, F, Ríos Labrada, H, Fuentes Lopez, MR, and Sweeney, SH. 2005. Understanding the potential impact of transgenic crops in traditional agriculture: Maize farmers' perspectives in Cuba, Guatemala and Mexico. Environmental Biosafety Research 4:141–166.

Soleri, D, Cleveland, DA, Castro García, FH, and Aragon, F. n.d. Farm characteristics associated with maize diversity conservation in Oaxaca, Mexico. Unpublished ms.

Soleri, D, Cleveland, DA, Eriacho, D, Bowannie, F, Jr., Laahty, A, and Zuni community members. 1994. Gifts from the Creator: Intellectual property rights and folk crop varieties. Pages 21–40 in Greaves, T, ed. IPR for Indigenous Peoples: A Sourcebook. Oklahoma City, OK: Society for Applied Anthropology.

Soleri, D, Cleveland, DA, Glasgow, GE, Sweeney, SH, Aragón Cuevas, F, Ríos Labrada, H, and Fuentes Lopez, MR. 2008. Testing economic assumptions underlying research on transgenic food crops for Third World farmers: Evidence from Cuba, Guatemala and Mexico. Ecological Economics 67:667–682.

Soleri, D, Cleveland, DA, Smith, SE, Ceccarelli, S, Grando, S, Rana, RB, Rijal, D, and Ríos Labrada, H. 2002. Understanding farmers' knowledge as the basis for collaboration with plant breeders: Methodological development and examples from ongoing research in Mexico, Syria, Cuba, and Nepal. Pages 19–60 in Cleveland, DA and Soleri, D, eds. Farmers, Scientists and Plant Breeding: Integrating Knowledge and Practice. Wallingford, Oxon, UK: CABI.

Soleri, D, and Smith, SE. 2002. Rapid estimation of broad sense heritability of farmer-managed maize populations in the Central Valleys of Oaxaca, Mexico, and implications for improvement. Euphytica 128:105–119.

Soleri, D, Smith, SE, and Cleveland, DA. 2000. Evaluating the potential for farmer and plant breeder collaboration: A case study of farmer maize selection in Oaxaca, Mexico. Euphytica 116:41–57.

SOLIBAM (Strategies for Organic and Low-input Integrated Breeding and Management Collaborative Project). 2011. Newsletter 1. http://www.solibam.eu/modules/wfdownloads/singlefile .php?cid=5&lid=11, accessed 2013 August 7.

Sorman, AH, and Giampietro, M. 2013. The energetic metabolism of societies and the degrowth paradigm: Analyzing biophysical constraints and realities. Journal of Cleaner Production 38:80–93.

Sotelo, A, Soleri, D, Wacher, C, Sanchez-Chinchillas, A, and Argote, RM. 2012. Chemical and nutritional composition of tejate, a traditional maize and cacao beverage from the Central Valleys of Oaxaca, Mexico. Plant Foods for Human Nutrition 67(2):148–155.

Souza, E, Myers, JR, and Scully, BT. 1993. Genotype by environment interaction in crop improvement. Pages 192–233 in Callaway, MB and Francis, CA, eds. Crop Improvement for Sustainable Agriculture. Lincoln: University of Nebraska Press.

Sperling, L, Loevinsohn, ME, and Ntabomvura, B. 1993. Rethinking the farmer's role in plant breeding: Local bean experts and on-station selection in Rwanda. Experimental Agriculture 29:509–519.

Spiertz, JHJ. 2010. Nitrogen, sustainable agriculture and food security: A review. Agronomy for Sustainable Development 30(1):43–55.

Steffen, W, et al. 2011. The Anthropocene: From global change to planetary stewardship. Ambio 40(7):739–761.

Stephens, PA, Buskirk, SW, and del Rio, CM. 2007. Inference in ecology and evolution. Trends in Ecology and Evolution 22(4):192–197.

Sterman, JD. 2008. Risk communication on climate: Mental models and mass balance. Science 322(5901):532–533.

Sterman, JD, and Booth Sweeney, L. 2007. Understanding public complacency about climate change: Adults' mental models of climate change violate conservation of matter. Climatic Change 80(3–4):213–238.

Stoessel, F, Juraske, R, Pfister, S, and Hellweg, S. 2012. Life cycle inventory and carbon and water food print of fruits and vegetables: Application to a Swiss retailer. Environmental Science and Technology 46(6):3253–3262.

Stone, B, and Zhong, T. 1989. Changing patterns of variability in Chinese cereal production. Pages 35–59 in Anderson, JR and Hazell, PBR, eds. Variability in Grain Yields: Implications for Agricultural Research and Policy in Developing Countries. Baltimore, MD: Johns Hopkins University Press.

Stone, GD. 2007. Agricultural deskilling and the spread of genetically modified cotton in Warangal. Current Anthropology 48(1):67–103.

———. 2012. Constructing facts: Bt cotton narratives in India. Economic and Political Weekly 47(38).

Stoskopf, NC, Tomes, DT, and Christie, BR. 1993. Plant Breeding Theory and Practice. Boulder, CO: Westview Press.

Strom, S. 2012. Has "organic" been oversized? New York Times, July 7. www.nytimes.com/2012/07/08/business/organic-food-purists-worry-about-big-companies-influence.html, accessed 2013 July 5.

Stuart, T. 2009. Waste: Uncovering the Global Food Scandal. New York: W.W. Norton.

Swift, MJ, and Anderson, JM. 1994. Biodiversity and ecosystem function in agricultural systems. Pages 15–41 in Schulze, E-D and Mooney, HA, eds. Biodiversity and Ecosystem Function. Berlin, Germany: Springer-Verlag.

Syngenta FSA (Syngenta Foundation for Sustainable Agriculture). 2009. Improving the Livelihood of Smallholder Farmers. www.syngentafoundation.org, accessed 2011 January 3.

Tabashnik, BE, Wu, KM, and Wu, YD. 2012. Early detection of field-evolved resistance to Bt cotton in China: Cotton bollworm and pink bollworm. Journal of Invertebrate Pathology 110(3):301–306.

Tandon, S, Landes, MR, and Woolverton, A. 2011. The Expansion of Modern Grocery Retailing and Trade in Developing Countries. ERS Research Report Number 122. Washington, DC: USDA ERS. www.ers.usda.gov/media/118890/err122.pdf, accessed 2013 August 2.

Tang, SY. 1992. Institutions and Collective Action: Self-governance in Irrigation. San Francisco, CA: Center for Self-Governance, Institute for Contemporary Studies Press.

Taylor, N. 2012. Reversing Meat-eating Culture to Combat Climate Change. www.worldpreservationfoundation.org/Downloads/ReversingMeatEatingCultureCC_NikTaylor_140612.pdf, accessed 2013 August 4.

Thompson, E, Jr, Harper, AM, and Kraus, S. 2008. Think Globally—Eat Locally: San Francisco Foodshed Assessment. www.farmland.org/programs/states/ca/Feature Stories/San-Francisco-Foodshed-Report.asp, accessed 2010 March 19.

Thompson, PB. 1995. The Spirit of the Soil: Agriculture and Environmental Ethics. London: Routledge.

Tienhaara, K. 2012. The potential perils of forest carbon contracts for developing countries: Cases from Africa. Journal of Peasant Studies 39(2):551–572.

Tiffen, M, and Mortimore, M. 1993. Population-growth and natural-resource use: Do we need to despair of Africa. Outlook on Agriculture 22(4):241–249.

Tiffen, M, Mortimore, M, and Gichuki, F. 1994. More People, Less Erosion: Environmental Recovery in Kenya. Chichester, UK: John Wiley and Sons.

Tilman, D. 1996. Biodiversity: Population versus ecosystem stability. Ecology 77:350–363.

Tilman, D, Wedin, D, and Knops, J. 1996. Productivity and sustainability influenced by biodiversity in grassland ecosystems. Nature 379:718–720.

Todaro, MP. 1994. Economic Development. 5th ed. New York: Longman.

Toenniessen, G, Adesina, A, and DeVries, J. 2008. Building an Alliance for a Green Revolution in Africa. Annals of the New York Academy of Sciences 1136:233–242.

Townsend, MS, Aaron, GJ, Monsivais, P, Keim, NL, and Drewnowski, A. 2009. Less-energy-dense diets of low-income women in California are associated with higher energy-adjusted diet costs. American Journal of Clinical Nutrition 89(4):1220–1226.

Trawick, PB. 2001. Successfully governing the commons: Principles of social organization in an Andean irrigation system. Human Ecology 29(1):1–25.

Trebbin, A, and Franz, M. 2010. Exclusivity of private governance structures in agrofood networks: Bayer and the food retailing and processing sector in India. Environment and Planning A 42(9):2043–2057.

Tsubo, M, Mukhala, E, Ogindo, HO, and Walker, S. 2003. Productivity of maize-bean intercropping in a semi-arid region of South Africa. Water SA 29(4):381–388.

UC (University of California). 2011. ucbiotech.org. Berkeley: University of California, Division of Agricultural and Natural Resources, Statewide Biotechnology Workgroup. http://ucbiotech.org/, accessed 2011 September 5.

UCS (Union of Concerned Scientists). 2010. Driving the Fox from the Henhouse: Improving Oversight of Food Safety at the FDA and USDA. www.ucsusa.org/assets/documents/scientific_integrity/driving-fox-from-henhouse-food-safety-report.pdf, accessed 2011 November 30.

———. 2012. Heads They Win, Tails We Lose: How Corporations Corrupt Science at the Public's Expense. UCS. www.ucsusa.org/assets/documents/scientific_integrity/how-corporations-corrupt-science.pdf, accessed 2013 August 2.

UN (United Nations). 2012. Demographic Yearbook. http://unstats.un.org/unsd/demographic/products/dyb/dyb2.htm, accessed 2013 January 15.

UNDP (United Nations Development Program). 2001. Human Development Report 2001: Making New Technologies Work for Human Development. New York: UNDP. http://hdr.undp.org/en/reports/global/hdr2001, accessed 2013 August 1.

Ungar, PS, and Sponheimer, M. 2011. The diets of early hominins. Science 334(6053):190–193.

UNHCHR (Office of the United Nations High Commissioner for Human Rights). 2000. The Right to Food. Commission on Human Rights Resolution 2000/10. Geneva, Switzerland: UNHCHR. www.unhchr.ch/huridocda/huridoca.nsf/%28Symbol%29/E.CN.4.RES.2000.10.En?Opendocument, accessed 2012 April 2.

Unwin, T. 2007. No end to poverty. Journal of Development Studies 43(5):929–953.

Ura, K. 2008. Gross National Happiness. Thimphu, Bhutan: Center for Bhutan Studies. www.grossnationalhappiness.com/gnhIndex/intruductionGNH.aspx, accessed 2011 June 3.

US Census Bureau. 2010. USA Counties. http://censtats.census.gov/usa/usa.shtml, accessed 2010 March 7.

———. 2012. USA Counties. http://censtats.census.gov/usa/usa.shtml, accessed 2010 March 7.

US FDA (US Food and Drug Administration). 2012. Food Labeling Guidance and Regulatory Information. www.fda.gov/Food/LabelingNutrition/FoodLabelingGuidanceRegulatoryInformation/default.htm, accessed 2012 December 12.

USDA (United States Department of Agriculture). 2010. Dietary Guidelines for Americans, 2010. www.cnpp.usda.gov/Publications/DietaryGuidelines/2010/PolicyDoc/PolicyDoc.pdf, accessed 2011 February 13.

USDA APHIS BRS (USDA, Animal and Plant Health Inspection Service, Biotechnology Regulatory Services). 2008. Biotechnology Regulatory Services Strategic Plan (FY 2009 to FY 2014). Riverdale, MD: USDA. www.aphis.usda.gov/biotechnology/downloads/Final BRS Strategic Plan_07.28.08.pdf, accessed 2009 April 4.

USDA ERS (US Department of Agriculture, Economic Research Service). 2009. Food Security in the United States: Definitions of Hunger and Food Security. www.ers.usda.gov/Briefing/FoodSecurity/labels.htm, accessed 2009 November 10.

———. 2010a. County-Level Unemployment and Median Household Income for California. www.ers.usda.gov/data-products/county-level-data-sets/unemployment.aspx#.Uf6CPVNiHjF, accessed 2013 August 4.

———. 2010b. Data Sets, Food Availability. www.ers.usda.gov/Data/FoodConsumption
/FoodAvailIndex.htm, accessed 2010 March 7.

———. 2010c. Food CPI and Expenditures. www.ers.usda.gov/briefing/cpifoodandexpenditures,
accessed 2011 January 27.

———. 2011a. Food Availability (Per Capita) Data System. www.ers.usda.gov/Data
/FoodConsumption, accessed 2012 January 30.

———. 2011b. Global Food Markets. www.ers.usda.gov/Briefing/GlobalFoodMarkets, accessed
2012 January 30.

USDA NASS (United States Department of Agriculture, National Agricultural Statistics Service).
2003. 2002 Census of Agriculture, vol. 1. Geographic area series, vol. 2006. Washington,
DC: USDA. www.nass.usda.gov/census/census02/volume1/us/index1.htm, accessed 2008
November 14.

———. 2009. Geographic Area Series, vol. 1. Part 5, California State and County Data. 2007
Census of Agriculture. Washington, DC: USDA. www.agcensus.usda.gov/Publica-
tions/2007/Full_Report/Volume_1,_Chapter_2_County_Level/California, accessed 2009
August 20.

———. 2012a. Adoption of Genetically Engineered Crops in the U.S. USDA. www.ers.usda.gov
/data-products/adoption-of-genetically-engineered-crops-in-the-us.aspx, accessed 2013
January 13.

———. 2012b. Historical Census Publications. Washington, DC: USDA. www.agcensus.usda.
gov/Publications/Historical_Publications, accessed 2012 April 26.

———. 2013. Adoption of Genetically Engineered Crops in the U.S. USDA. www.ers.usda.gov
/data-products/adoption-of-genetically-engineered-crops-in-the-us.aspx, accessed 2013
January 13.

van de Wouw, M, van Hintum, T, Kik, C, van Treuren, R, and Visser, B. 2010. Genetic diversity
trends in twentieth century crop cultivars: A meta analysis. Theoretical and Applied
Genetics 120(6):1241–1252.

Vandermeer, J. 1992. The Ecology of Intercropping. Cambridge, UK: Cambridge University Press.

———. 2009. The Ecology of Agroecosystems. Sudbury, MA: Jones and Bartlett Publishers.

Venuprasad, R, Lafitte, HR, and Atlin, GN. 2007. Response to direct selection for grain yield
under drought stress in rice. Crop Science 47(1):285–293.

Verheul, J. 2011. Methane as a greenhouse gas: Why the EPA should regulate emissions from
animal feeding pperations and concentrated animal feeding operations under the Clean
Air Act. Natural Resources Journal 51(1):163–187.

Victor, PA, and Jackson, T. 2012. A commentary on UNEP's Green Economy Scenarios. Ecological
Economics 77:11–15.

Vince, G. 2011. An epoch debate. Science 333(6052):32–37.

Vitousek, PM, Ehrlich, PR, Ehrlich, AH, and Matson, PA. 1986. Human appropriation of the
products of photosynthesis. BioScience 36:368–373.

Vohs, KD, Mead, NL, and Goode, MR. 2006. The psychological consequences of money. Science
314(5802):1154–1156.

vom Brocke, K, Presterl, T, Christinck, A, Weltzien, RE, and Geiger, HH. 2002. Farmers' seed
management practices open up new base populations for pearl millet breeding in a
semi-arid zone of India. Plant Breeding 121:36–42.

vom Brocke, K, Weltzien, E, Christinck, A, Presterl, T, and Geiger, HH. 2003. Effects of farmers'
seed management on performance and adaptation of pearl millet in Rajasthan, India.
Euphytica 130:267–280.

Vucetich, JA, and Nelson, MP. 2010. Sustainability: Virtuous or vulgar? BioScience
60(7):539–544.

Wade, R. 1988. Village Republics: Economic Conditions for Collective Action in South India. Cambridge, UK: Cambridge University Press.

Waley, A. 1958. The Way and Its Power: A Study of the Tao tê Ching and Its Place in Chinese Thought. New York: Grove Press.

Walker, D. 1992. Energy, Plants and Man. 2nd ed. Brighton, East Sussex, UK: Oxygraphics.

Walker, P, Rhubart-Berg, P, McKenzie, S, Kelling, K, and Lawrence, RS. 2005. Public health implications of meat production and consumption. Public Health Nutrition 8(4):348–356.

Walmart. 2012. Locally Grown Products. http://walmartstores.com/sustainability/7985.aspx, accessed 2012 February 2.

Walter, C, and Stützel, H. 2009. A new method for assessing the sustainability of land-use systems (I): Identifying the relevant issues. Ecological Economics 68(5):1275–1287.

Waltz, E. 2009. Battlefield. Nature 461:27–32.

Walz, E. 2004. Final Results of the Fourth National Organic Farmers' Survey: Sustaining Organic Farms in a Changing Organic Marketplace. Santa Cruz, CA: Organic Farming Research Foundation. http://ofrf.org/sites/ofrf.org/files/docs/pdf/4thsurvey_results.pdf, accessed 2013 August 2.

Wang, D, Heckathorn, SA, Wang, XZ, and Philpott, SM. 2012. A meta-analysis of plant physiological and growth responses to temperature and elevated CO_2. Oecologia 169(1):1–13.

Wang, R-L, Stec, A, Hey, J, Lukens, L, and Doebley, J. 1999. The limits of selection during maize domestication. Nature 398:236–239.

Wang, S, Just, D, and Pinstrup-Andersen, P. 2006. Tarnishing Silver Bullets: Bt Cotton Adoption and the Outbreak of Secondary Pest Infestation in China. American Agricultural Economics Association Annual Meeting. Long Beach, CA. www.grain.org/research_files/SWang_tarnished.pdf, accessed 2007 July 15.

Wanner, H, et al. 2008. Mid- to late Holocene climate change: An overview. Quaternary Science Reviews 27:1791–1828.

Wansink, B, Painter, JE, and North, J. 2005. Bottomless bowls: Why visual cues of portion size may influence intake. Obesity Research 13(1):93–100.

WCED (World Commission on Environment and Development). 1987. Our Common Future. New York: Oxford University Press. www.un-documents.net/our-common-future.pdf, accessed 2013 August 4.

Weber, CL, and Matthews, HS. 2008. Food-miles and the relative climate impacts of food choices in the United States. Environmental Science and Technology 42(10):3508–3513.

Weinberg, S. 1994. Dreams of a Final Theory: The Scientist's Search for the Ultimate Laws of Nature. New York: Vintage/Random House.

Weis, T. 2010. The accelerating biophysical contradictions of industrial capitalist agriculture. Journal of Agrarian Change 10(3):315–341.

Wellhausen, EJ, Roberts, LM, Hernández, XE, and Manglesdorf, PC. 1952. Races of Maize in Mexico. Cambridge, MA: The Bussey Institution of Harvard University. First published 1951 in Mexico as Razas de Maíz en México.

Wells, HF, and Buzby, JC. 2008. Dietary Assessment of Major Trends in U.S. Food Consumption, 1970–2005. Economic Information Bulletin No. 33. Washington, DC: USDA Economic Research Service. www.ers.usda.gov/publications/eib33, accessed 2010 April 23.

Welsh, R, and Ervin, DE. 2006. Precaution as an approach to technology development: The case of transgenic crops. Science Technology and Human Values 31(2):153–172.

Weltzien, E, Smith, ME, Meitzner, LS, and Sperling, L. 2003. Technical and Institutional Issues in Participatory Plant Breeding—From the Perspective of Formal Plant Breeding: A Global Analysis of Issues, Results, and Current Experience. PPB Monograph. Cali, Columbia: PRGA Program Coordination Office, Centro Internacional de Agricultura Tropical.

West, HG, and Domingos, N. 2012. Gourmandizing poverty food: The Serpa Cheese Slow Food Presidium. Journal of Agrarian Change 12(1):120–143.

White House. 2012. Fact Sheet: G-8 Action on Food Security and Nutrition. www.whitehouse.gov /the-press-office/2012/05/18/fact-sheet-g-8-action-food-security-and-nutrition, accessed 2012 July 24.

Whitt, SR, Wilson, LM, Tenaillon, MI, Gaut, BS, and Buckler, ES. 2002. Genetic diversity and selection in the maize starch pathway. Proceedings of the National Academy of Sciences of the United States of America 99(20):12959–12962.

WHO (World Health Organization). 2005. Modern Food Biotechnology, Human Health and Development: An Evidence-Based Study. Geneva, Switzerland: Food Safety Department, World Health Organization. www.who.int/foodsafety/publications/biotech/biotech_en .pdf, accessed 2007 January 11.

Wiggins, S. 2009. Can the Smallholder Model Deliver Poverty Reduction and Food Security? FAC Working Paper No. 08. Brighton, UK: Future Agricultures Consortium Secretariat, Institute of Development Studies. http://opendocs.ids.ac.uk/opendocs/bitstream/handle/123456789/2338 /FAC_Working%20_Paper_08.pdf, accessed 2011 February 15.

Wilder, M, and Lankao, PR. 2006. Paradoxes of decentralization: Water reform and social implications in Mexico. World Development 34(11):1977–1995.

Williams, H, and Wikstrom, F. 2011. Environmental impact of packaging and food losses in a life cycle perspective: A comparative analysis of five food items. Journal of Cleaner Production 19(1):43–48.

Williams, JH, DeBenedictis, A, Ghanadan, R, Mahone, A, Moore, J, Morrow, WR, Price, S, and Torn, MS. 2012. The technology path to deep greenhouse gas emissions cuts by 2050: The pivotal role of electricity. Science 335(6064):53–59.

Wilson, D, and Roberts, J. 2012. How Washington went soft on childhood obesity. Reuters, April 27. www.reuters.com/article/2012/04/27/us-usa-foodlobby-idUSBRE83Q0ED20120427, accessed 2012 June 4.

Wilson, DS, O'Brien, DT, and Sesma, A. 2009. Human prosociality from an evolutionary perspective: Variation and correlations at a city-wide scale. Evolution and Human Behavior 30(3):190–200.

Wilson, DS, and Wilson, EO. 2008. Evolution "for the good of the group." American Scientist 96(5):380–389.

Wilson, EO. 1984. Biophilia. Cambridge, MA: Harvard University Press.

Winch, D. 1992. Introduction. Pages vii-xxxiii in Winch, D, ed. An Essay on the Principle of Population: Or a View of Its Past and Present Effects on Human Happiness; With an Inquiry into Our Prospects Respecting the Future Removal or Mitigation of the Evils Which It Occasions. Selected and introduced by Donald Winch. 2nd ed. Cambridge, UK: Cambridge University Press.

Wise, TA. 2010. Agricultural Dumping under NAFTA: Estimating the Costs of U.S. Agricultural Policies to Mexican Producers. Mexican Rural Development Research Report No. 7. Woodrow Wilson International Center for Scholars. www.ase.tufts.edu/gdae/Pubs/rp /AgricDumpingWoodrowWilsonCenter.pdf, accessed 2012 October 23.

Witcombe, JR, Gyawali, S, Sunwar, S, Sthapit, BR, and Joshi, KD. 2006. Participatory plant breeding is better described as highly client-oriented plant breeding: II. Optional farmer collaboration in the segregating generations. Experimental Agriculture 42(1):79–90.

Witt, H, Patel, R, and Schnurr, M. 2006. Can the poor help GM crops? Technology, representation and cotton in the Makhathini flats, South Africa. Review of African Political Economy 33:497–513.

Wittfogel, KA. 1957. Oriental Despotism: A Comparative Study of Total Power. New Haven, CT: Yale University Press.

Woldeamlak, A, Grando, S, Maatougui, M, and Ceccarelli, S. 2008. Hanfets, a barley and wheat mixture in Eritrea: Yield, stability and farmer preferences. Field Crops Research 109(1–3):50–56.

Wolfe, MS. 2000. Crop strength diversity. Nature 406:681–682.

Wood, S. 2013. Revisiting the US food retail consolidation wave: Regulation, market power and spatial outcomes. Journal of Economic Geography 13(2):299–326.

Wooten, H, and Ackerman, A. 2011. Seeding the City: Land Use Policies to Promote Urban Agriculture. Oakland, CA: The National Policy and Legal Analysis Network to Prevent Childhood Obesity (NPLAN). http://changelabsolutions.org/sites/changelabsolutions.org/files/Urban_Ag_SeedingTheCity_FINAL_%28CLS_20120530%29_20111021_0.pdf, accessed 2012 March 4.

World Bank. 2007. World Development Report 2008: Agriculture for Development. Washington, DC: World Bank. http://web.worldbank.org/WBSITE/EXTERNAL/EXTDEC/EXTRESEARCH/EXTWDRS/0,,contentMDK:23092617~pagePK:478093~piPK:477627~theSitePK:477624,00.html, accessed 2013 August 2.

———. 2010. World Development Report 2010: Development and Climate Change. Washington, DC: World Bank. http://web.worldbank.org/WBSITE/EXTERNAL/EXTDEC/EXTRESEARCH/EXTWDRS/0,,contentMDK:23079906~pagePK:478093~piPK:477627~theSitePK:477624,00.html, accessed 2013 August 2.

———. 2012a. Inclusive Green Growth : The Pathway to Sustainable Development. Washington, DC: World Bank. https://openknowledge.worldbank.org/bitstream/handle/10986/6058/9780821395516.pdf, accessed 2013 May 28.

———. 2012b. Turn Down the Heat: Why a 4°C Warmer World Must Be Avoided. Washington, DC: World Bank.

Worthington, M, Soleri, D, Aragon-Cuevas, F, and Gepts, P. 2012. Genetic composition and spatial distribution of farmer-managed *Phaseolus* bean plantings: An example from a village in Oaxaca, Mexico. Crop Science 52(4):1721–1735.

WPF (World Preservation Foundation). 2011. Reducing Shorter Lived Climate Forcers through Dietary Change. www.worldpreservationfoundation.org/Downloads/Livestock-Production-World-Preservation-Foundation.pdf, accessed 2012 February 2.

Wu, KM, Lu, YH, Feng, HQ, Jiang, YY, and Zhao, JZ. 2008. Suppression of cotton bollworm in multiple crops in China in areas with Bt toxin-containing cotton. Science 321(5896):1676–1678.

WWAP. 2012. Managing Water under Uncertainty and Risk. The United Nations World Water Development Reports. Paris: UNESCO.

Zajfen, V. 2008. Fresh Food Distribution Models for the Greater Los Angeles Region: Barriers and Opportunities to Facilitate and Scale Up the Distribution of Fresh Fruits and Vegetables. Findings from an Action Research Project of the Center for Food and Justice, a division of the Urban and Environmental Policy Institute, Occidental College. December 2006–March 2008: CFJ. www.uepitestsite.dreamhosters.com/wp-content/uploads/2012/11/Pub-Fresh_Food_Distro_Models_LA.pdf, accessed 2013, August 2.

Zervas, G, and Tsiplakou, E. 2012. An assessment of GHG emissions from small ruminants in comparison with GHG emissions from large ruminants and monogastric livestock. Atmospheric Environment 49:13–23.

Zeven, AC. 1998. Landraces: A review of definitions and classifications. Euphytica 104:127–139.

Zhang, CC, Yuan, WY, and Zhang, QF. 2012. RPL1, a gene involved in epigenetic processes regulates phenotypic plasticity in rice. Molecular Plant 5(2):482–493.

Zhu, YY, et al. 2000. Genetic diversity and disease control in rice. Nature 406(6797):718–722.

Zhu, YY, Fang, H, Wang, YY, Fan, JX, Yang, SS, Mew, TW, and Mundt, CC. 2005. Panicle blast and canopy moisture in rice cultivar mixtures. Phytopathology 95(4):433–438.

Zhu, YY, Wang, Y, Chen, H, and Lu, B-R. 2003. Conserving traditional rice varieties through management for crop diversity. BioScience 53:158–162.

Zilberman, D, Ameden, H, and Qaim, M. 2007. The impact of agricultural biotechnology on yields, risks, and biodiversity in low-income countries. Journal of Development Studies 43(1):63–78.

Zimmerer, KS. 1996. Changing Fortunes: Biodiversity and Peasant Livelihood in the Peruvian Andes. Berkeley: University of California Press.

Zlolniski, C. 2011. Water flowing north of the border: Export agriculture and water politics in a rural community in Baja California. Cultural Anthropology 26(4):565–588.

Zohary, D, and Hopf, M. 2000. Domestication of Plants in the Old World: The Origin and Spread of Cultivated Plants in West Asia, Europe, and the Nile Valley. 3rd ed. Oxford, UK: Oxford University Press.

Zong, Y, Chen, Z, Innes, JB, Chen, C, Wang, Z, and Wang, H. 2007. Fire and flood management of coastal swamp enabled first rice paddy cultivation in east China. Nature 449:459-U4.

INDEX

350.org, and maximum CO_2 concentration, 206

Achebe, Chinua, 103
adaptation of crop varieties (wide), 137
agricultural revolutions, 47–70, 49*fig.*, 50*table*; demand-side, 48; inputs and outputs ratios, 49; as multilineal evolutions, 51; supply-side, 48; three key human relationships, 48–49
agriculture: before humans, 52–53; goal of, 19; modern, scientific, 6; small-scale, 5, 6, 51; Third World, 51
agrifood: biotechnology, 266n14; classification of, resources, 184, 185*table*; private, resources, 184; agroecology, 169
agrifood system: alternative, xxii–xxiii, 2; animal portion of, contribution to climate change, 215–16, 217, 229–31; anthropogenic climate change and, 205, 209–10; boundaries of, 214, 225; concentration of mainstream, 212; contribution to climate change, 213–14; contribution to mitigating climate change, 214–15; disconnect, 235–37, 235*fig.*; mainstream, 2; nitrogen cycle of, 218–20
agroecosystems: industrial, 167–68; sustainable, 168–69; traditional, 165–67
alley cropping, 67
Alliance for a Green Revolution in Africa (AGRA), xix, 79; and corporate globalization, 236; and TGVs, 155
allogamous, 136
alternative emphasis, 1, 2

alternative perspective, 98; in agricultural ecosystems management, 162
altruism, 109, 264n5; and common property management, 195
Ambalapuzha Temple, 22
American Association for the Advancement of Science (AAAS), and knowledge, 74
American Farmland Trust, and localwashing, 244
Anthropocene epoch, 28, 56, 208; and mass extinctions, 164
anticommons, 192
application efficiency (AE), 40; of irrigation water, 197
assumptions, xxiii, 261n3, 263n3; analyzing, 81; in calculating HCC, 39–40; empirically based, 4; taxonomy, 85–86, 85*table*; underlying sustainability definitions, 98; value-bases, 4
atmosphere, evolution of Earth's, 208
Australia, and altruism in farmer management of CPRs, 195
autogamous, 136
Azolla fern, in rice paddies, 218

barley, *hanfets* in Ethiopia, 165, 176
Bayer CropScience, and localwashing, 244
behavior, 69
benefit-cost analysis, and *ex ante precautionary* approach, 114; and *ex post* trial and error approach, 114
Berry, Wendell, 230

diversity: levels in agricultural systems, 133*table*; and stability in ecology, 163–64; and yield, stability, and scale in agroecosystems, 172, 182; and yield and stability in agricultural ecosystems management, 160–82, 165, 166*fig.*

DNA, 135

domestication, 57–65, 59*fig.*, 131–32; and autogamy, 65; and brittle rachis, 57; ; definition, 57–58, 262n4; and food preferences, 65; syndrome, 64; and vegetative propagation, 65

economic: emphasis and globalization, 235; emphasis in sustainability, 99, 101; mainstream, system, 74

economics: cowboy, 17; degrowth, 95; ecological and environmental, 264n4; spaceman, 17; steady state, 95

ecosystem management, 66–67

efficiencies: in calculating HCC, 38–39, 44; and economic emphasis, 99; of human activity, 99; in industrial agroecosystems, 167; and irrigation, 197–98; in mitigating climate change, 206–7, 214; and stocks and flows, 106

empirically based assumptions, and climate change mitigation, 222–23, 223*fig. See also under* assumptions

energy: clean, 18; sources of, 18

entropy, 14

environmental alarms, sensitivity of, 114–15, 115*fig.*

environmental emphasis, sustainability of, 101–2

Environmental Protection Agency (EPA), estimate of agrifood system contribution to climate change, 213

epigenetic processes, 135, 265n3

epistemology. *See* knowledge

Eshel, Gidon and Pamela Martin, 231, 231*fig.*

European colonialists, and global economy, xviii

evolution: biological, 15, 17, 19, 51; of chemical and biochemical pathways, 18; human, 53; phenotypic, 16; social, 51; sociocultural, 15, 17, 19, 28

exports, and localization, 268n4

externalities: internalizing, 111–13; internalizing in common property management, 187, 190, 194–95; negative, 111; positive, 112

extinction, in the Anthropocene, 208

extraterrestrial intelligent life, 13

facilitation, 175–76; and rice polyculture, 178, 180

fairwash, 94

FAO, 118–19, 122; Global Plan of Action, 166

farm workers, xx, xxii, 113, 233, 238, 241, 244, 252. *See also* migrant farm workers

farmer: crop selection and sustainability indicators, 90; crop varieties (FVs), 61, 126, 130*fig.*, 265n1; knowledge, 78–81; knowledge and scientist knowledge, 263n4; plant breeding, 147–54; scenarios and, knowledge, 148; small-scale, 5; Third World and TGVs, 156–59

farmer crop varieties (FVs). *See* farmer

Farmer Direct Produce (FDP), case study of success, 250–51

fertilization, fusion of haploid gametes, 136

fields, matrix of, 168

fitness: biological, 15, 16; paradox, 16

food, 265n10; crisis, 19; future demand for, 2, 3*fig.*; insecurity and low income, 229; labels for climate change impact, 207; miles and GHGE, 240–41; miles and migrant labor, 241; miles and nutrition, 241; miles as indicator of sustainable agrifood systems, 93, 239–45; prices, 118–19, 118*fig.*; security, 119, 120*fig.*, 120–21; sovereignty, 118, 120*fig.*, 121; supply linear growth, 26; waste in industrial and Third Worlds, 225; waste reducing to mitigate climate change, 207, 224–26

foragers, 47, 53–55, 131; energy efficiency, 57

fossil fuels, and CO_2 emissions of agrifood systems, 218

fruits and vegetables, and nutrition, 92

fully privileged game, and CPM of CPRs, 194–95, 195*fig.*

G8 (Group of Eight), 97, 117, 119, 120, 122, 263n1, 264n2

Gaia hypothesis, 15

gardens, project: 160–62, 161*fig.6.1*; traditional, 160–62, 161*fig.6.2*; tropical home and diversity, 173

gene, 135

genetic diversity, 60–61

CALIFORNIA STUDIES IN FOOD AND CULTURE

Darra Goldstein, Editor

DATE DUE